T0372439

Fluid Flow in Fractured Rocks

Fluid Flow in Fractured Rocks

Robert W. Zimmerman and Adriana Paluszny
Department of Earth Science and Engineering
Imperial College London
London, UK

This edition first published 2024

Registered Offices
John Wiley & Sons, Inc., 111 River Street, Hoboken, NJ 07030, USA
John Wiley & Sons Ltd, The Atrium, Southern Gate, Chichester, West Sussex, PO19 8SQ, UK

For details of our global editorial offices, customer services, and more information about Wiley products visit us at www.wiley.com.

Wiley also publishes its books in a variety of electronic formats and by print-on-demand. Some content that appears in standard print versions of this book may not be available in other formats.

Library of Congress Cataloging-in-Publication Data

Names: Zimmerman, Robert Wayne, author. | Paluszny, Adriana, author.
Title: Fluid flow in fractured rocks / Robert W. Zimmerman and Adriana Paluszny.
Description: Hoboken, NJ : Wiley, 2024. | Includes index.
Identifiers: LCCN 2023039657 (print) | LCCN 2023039658 (ebook) | ISBN 9781119248019 (cloth) | ISBN 9781119248033 (adobe pdf) | ISBN 9781119248026 (epub)
Subjects: LCSH: Rocks–Fracture. | Fluid dynamics.
Classification: LCC TA706 .Z56 2024 (print) | LCC TA706 (ebook) | DDC 620.1/064–dc23/eng/20230909
LC record available at https://lccn.loc.gov/2023039657
LC ebook record available at https://lccn.loc.gov/2023039658

Cover image: © ROMAN DZIUBALO/Shutterstock
Cover design: Wiley

Set in 9.5/12.5pt STIXTwoText by Straive, Chennai, India
Printed and bound by CPI Group (UK) Ltd, Croydon, CR0 4YY

C9781119248019_181024

Contents

Preface

Several well-regarded books have been written over the years on fluid flow in porous media and subsurface hydrology. However, it is now recognized that *fractured* rocks are ubiquitous in the subsurface, and that several issues that arise during fluid flow and transport through fractured rocks are distinct from their analogues in non-fractured porous media. Moreover, flow and transport through fractured rocks are of great importance in many technological areas, such as, for example, energy production from geothermal or hydrocarbon reservoirs, subsurface nuclear waste disposal, carbon sequestration, and contaminant remediation. Hence, there is a need for a monograph/textbook that provides a thorough, rigorous, and authoritative introduction to this topic. Our aim in writing this book has been to address this need.

This book was intended to be written in sufficient detail so as to provide a rigorous and broad introduction to the field of fluid flow through fractured rocks. It is intended for readers with interests in hydrogeology, hydrology, water resources, structural geology, reservoir engineering, underground waste disposal, or other fields that involve the flow of fluids through fractured rock masses. To the extent possible, the mathematical models developed and discussed in the book are compared to experimental or field data or validated/tested against numerical simulations. The book contains 157 individual figures, of which 39 are either images of real fractures or contain actual laboratory or field data.

Chapter 1 introduces the geomechanical background to the nucleation and growth of fractures in rock and the multi-scale characterization of their geometric traits, such as aperture, length, roughness, and density. Chapter 2 provides a rigorous treatment of the mathematics of fluid flow through a single rock fracture, starting with the Navier–Stokes equations and carefully explaining the conditions under which these equations can be replaced by the simpler Stokes equations or Reynolds lubrication equation. The effects of normal and shear stresses on the transmissivity of a rock fracture are discussed in Chapter 3. Chapter 4 discusses the effects of inertia on fluid flow through a fracture and the resulting deviations from a Darcy-like linear relation between pressure drop and flowrate. Some of the coupled thermal, hydraulic, mechanical, and chemical interactions that may alter the transmissivity of a single fracture are described in Chapter 5. Chapter 6 introduces the concept of solute transport through single fractures and presents some models for the advection and dispersion of solutes.

Whereas most of the first six chapters focus on flow and transport through a single fracture, the second half of the book focuses on the behavior of fracture networks and fractured rock masses. Chapter 7 presents some analytical models for the macroscopic-scale permeability of porous/fractured rocks. Chapter 8 discusses various approaches to the numerical modeling of fluid flow through geologically realistic fracture networks and fractured rock masses. Chapter 9 presents "dual-porosity" models for fractured-porous media, starting from the classical model developed by Barenblatt and his collaborators. Chapter 10 then gives a detailed treatment of the matrix block shape factors that are crucial ingredients in dual-porosity models for both flow and transport. Various models for solute transport through fracture networks and fractured rock masses, both "Fickian" and "non-Fickian," are discussed in Chapter 11. Finally, Chapter 12 briefly discusses two-phase flow in single fractures and fractured rocks.

Although this book might be classified as a monograph, it is hoped that it can also serve as a textbook for master's-level or advanced undergraduate courses. For this purpose, several problems have been given at the end of each chapter. Some of the problems ask for a mathematical derivation of an equation that may have been presented in the text without a detailed derivation. Other problems ask for analysis of data or further analysis and/or application of some of the mathematical models presented in the book.

This book is intended to be self-contained. Most of the mathematical derivations are presented in sufficient detail so as not to require the reader to refer to the original sources. Nevertheless, extensive reference is made to important papers, theses, and books that have made contributions to the field and/or contain relevant information. The references in each chapter have been collected at the end of that chapter rather than in a single reference list, for the convenience of the reader.

Equations are numbered consecutively within each chapter, so that, for example, "Eq. (5.7)" denotes the seventh equation in Chapter 5. Tables and figures are also numbered in this same manner, *i.e.*, numbering is not restarted within each section of a chapter. All symbols and variables are defined in the text, as soon as they are first used. When a variable is first defined, its "dimensions" are listed in brackets, in terms of the SI units that would typically be used for that variable, rather than in terms of the "Mass-Length-Time" convention. Efforts have been made to adhere to a consistent nomenclature, sometimes at the cost of not using the same notation as was used in the original sources of some of the equations. The large number of variables mentioned in the book made it unavoidable that some letters be used to denote different properties in different chapters. To avoid confusion, a single List of Symbols has been included at the end of the book, which in particular explains the meaning of those symbols that are used for different purposes in different chapters.

To aid in the use of this book as a textbook, all figures can be freely downloaded (see 'About the Companion Page' on page xiii). The authors welcome feedback from readers or users of this book, including notification of any errors that may be detected.

London, UK
October 2023

Robert W. Zimmerman
Adriana Paluszny

Author Biographies

Robert W. Zimmerman earned BS and MS degrees from Columbia University and a PhD from the University of California at Berkeley. He has been a staff scientist at the Lawrence Berkeley National Laboratory, the Head of the Division of Engineering Geology and Geophysics at the Royal Institute of Technology in Stockholm, and is currently a Professor of Rock Mechanics at Imperial College in London. He has been the Editor-in-Chief of the *International Journal of Rock Mechanics and Mining Sciences* since 2007 and is co-author, with J.C. Jaeger and N.G.W. Cook, of *Fundamentals of Rock Mechanics*, 4th ed. (Wiley-Blackwell, 2007).

Adriana Paluszny earned an Ing. degree *cum laude* in Computational Engineering from Universidad Simón Bolívar in Caracas, Venezuela, and a PhD in Computational Geomechanics from Imperial College. She is an Associate Editor of the *Journal of Geophysical Research: Solid Earth*, was the inaugural recipient of the Chin-Fu Tsang Coupled Processes Award, and is the main developer of the Imperial College Geomechanics Toolkit. She is currently a Royal Society University Research Fellow, and a Reader in Computational Geomechanics at Imperial College in London.

To my children

Adriana Paluszny
London, UK
28 November 2023

About the Companion Website

This book is accompanied by a companion website.

www.wiley.com/go/zimmerman/fluidflowinfracturedrocks

This website include: Figures from the book

1

Genesis and Morphology of Fractures in Rock

1.1 What Are Fractures, and Why Are They Important?

Fractures are discontinuities in the mechanical integrity of brittle Earth materials. They break the mechanical continuity of the medium and provide high-speed conduits for fluids to flow throughout the subsurface. Consequently, they are crucial to understanding how fluids migrate in the subsurface. Fractures play a key role in controlling how otherwise low-permeability media can allow the migration of fluids through the subsurface. Therefore, fractures can effectively provide access to natural resources such as water, gas, minerals, and geothermal energy. In some cases, fractures may provide unwanted migration paths for stored fluids by breaking geological seals and enabling fluid mixing, leading to the pollution of drinking water resources. In other situations, fractures can facilitate fluid migration, lead to undesirable consequences such as induced seismicity, or possibly compromise the long-term subsurface disposal of hazardous waste.

"Fracture" is a general term that can be used to describe discontinuities formed by extension or shear, including cracks, joints, and faults. Fractures can form veins as a result of long-term mineralization or can be filled by the intrusion of another material, such as magma, to form dykes or other structures (Pollard and Aydin, 1988). Fractures rarely appear as stand-alone features in the subsurface, as deformation of brittle bodies often leads to the creation of multiple simultaneous breaks in rocks, resulting in several superimposed fractures and fault patterns that form complex multi-scale systems. These discontinuities in the rock matrix influence mechanical properties and the conduction of fluids through most low-permeability media. Fractures are discontinuities with imperfect surfaces, which represent weakness in a mechanical sense. Terms related to the description of these discontinuities can be classified into geological, geometric, topological, and mechanical (Peacock *et al.*, 2016). Fractures in rocks have a geological interpretation, usually tied to their setting, geometry, and chemical composition. Connectivity dictates the topological relationship between fractures, and mechanical deformation further changes a fracture's properties and its form.

From the mechanical point of view, fractures limit the strength of a rock mass, mostly due to a lack of cohesion. From the point of view of fluid flow, they represent preferred conduits for flow since the aperture of a fracture is usually much larger than the pores of the host rock. Pre-existing fractures can be exploited, as they can channel flow through a reservoir, and their observation can serve to provide clues about how rocks conducted fluids

Fluid Flow in Fractured Rocks, First Edition. Robert W. Zimmerman and Adriana Paluszny.

Companion website: www.wiley.com/go/zimmerman/fluidflowinfracturedrocks

Figure 1.1 (a) Fractures at the centimeter and meter scales cross-cut each other to form complex multi-scale patterns. (b) Fractures in limestone exposed in Somerset, UK.

in the past. Fractures can also be induced or enhanced to increase local fluid flow in order to turn an otherwise impermeable rock into a permeable medium. Figure 1.1 shows some examples of fractures that were formed due to folding of layered limestones off the coast of Somerset, UK.

Isolating the behavior of fractures in a rock formation often represents a bounding scenario for safety cases in evaluating the integrity of underground waste repositories. In the field of geological carbon dioxide storage, for example, a project is commonly considered unsafe if injection-induced overpressure may cause slip along fractures. In geological nuclear waste disposal, the limiting case is often considered to be one in which fluid flow occurs only within the connected system of fractures and whether the ensuing radionuclide transport within this network exceeds some threshold. It follows that fractures are a logical point at which to start investigations on the impact that human interference has on an embedded geological setting, in terms of its mechanical and hydraulic properties.

1.2 Formation of Fractures in Rock

Materials that are quasi-brittle, such as most rocks in the upper crust of the Earth, are subjected to stresses resulting from gravity as well as a variety of local and regional stresses such as tectonic stresses, burial, uplifting, and folding, along with chemical, thermal, and fluid flow-related stresses. As a result, rocks deform in several stages. First, they develop micro-cracks that occur on a small scale, often along the boundaries of individual grains of a rock. These micro-cracks start to grow and form preferential paths which, when aligned, become material flaws. Flaws induce stress concentrations at their tips, the leading edge

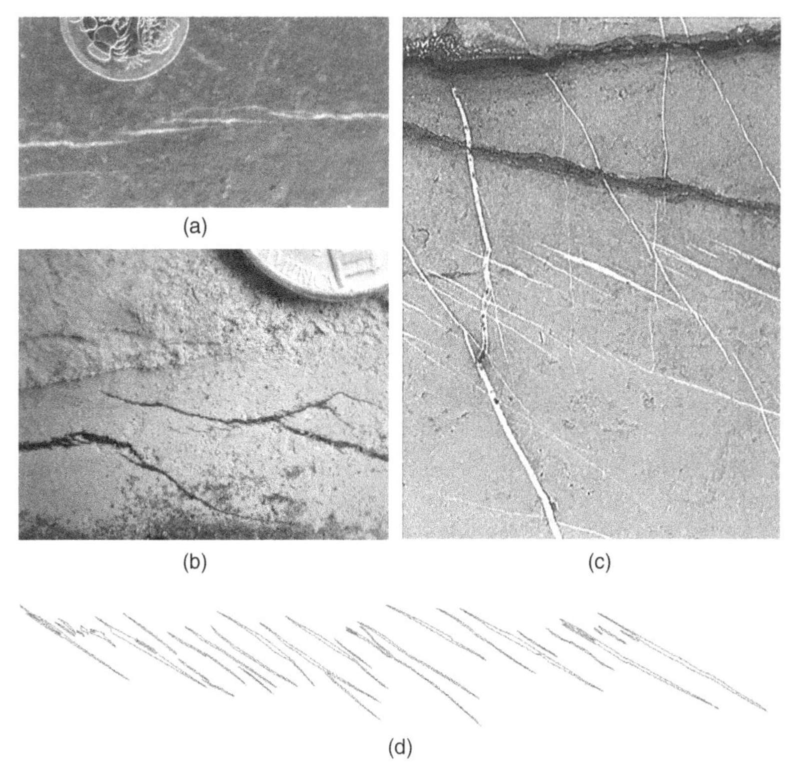

Figure 1.2 Fractures at the centimeter and millimeter scales in (a) and (c) limestone and (b) shale, respectively. The small fractures in (b) arise in one of the shale outcrops that cap the shallow Orcutt oil field in California, USA. In (a) and (c), fractures are filled with calcite, and in (b), fractures are tinted with naturally migrating hydrocarbon. The tracing of the fracture pattern in (c) for aperture quantification is shown in (d).

of the shape of the fracture, causing them to grow and extend into other areas of the rock. This self-organization occurs at larger and larger scales, up to the kilometer scale or larger. During this process, micro-fractures continue to form at different scales around fracture tips, as a result of stress field interactions, as well as other nonlocal chemical and thermal processes. This results in fracture growth across multiple scales to accommodate the ubiquitous deformation of the subsurface. Figure 1.2 shows several examples of interacting fractures. In all of these cases, the fractures are filled with material that delineates their shape, which is not always the case in the subsurface, but is convenient for visualization and interpretation purposes.

During or after fracture growth, the opposing fracture walls can be displaced in relation to one another. Under tension, fracture walls move directly apart (mode I), creating "thickness" or "aperture." Under shear (modes II and III), fracture walls slide against each other in a direction perpendicular to or parallel to the tip of the crack, respectively. A specific case of a displaced fracture is a "fault," which exhibits relative displacement of its walls and can appear under normal or shear deformation at scales that span from the centimeter up to the kilometer scale (Gudmundsson, 2000).

Fractures localize across large scales, usually along a preferential plane, and eventually link together to form larger features. From this moment onwards, stress is preferentially concentrated at the tips of the newly formed high-aspect ratio fracture, and growth drives the formation and linkage of larger discontinuities. As rocks are subjected to a variety of mechanical, hydraulic, thermal, and chemical changes over millions of years, many of these flaws form around pre-existing weaknesses in the matrix, pockets of low-integrity rock matrix that have resulted from localized processes. Heterogeneities leading to micro-fractures have been found to follow a Gaussian size distribution (Underwood, 1970), and in brittle rocks, they often appear as thin, penny-shaped microcavities distributed across the matrix (Herrmann, 1990).

Computerized tomography (CT) can reveal fractures with apertures up to five times smaller than the resolution provided by a scanner (Fig. 1.3), due to the strong density contrast between rock and gas/air. CT has been used to characterize micro-fracturing (Cnudde and Boone, 2013) and can yield three-dimensional images of micro-fractures embedded in porous rocks.

Many of the rocks in the upper crust, reaching a depth of around 50–70 km, are, in the most general, informal sense, "rigid," and are elasto-frictional, quasi-brittle materials. A brittle rock subjected to stresses will undergo elastic deformation, but if the stress surpasses the "strength" of the rock, the rock will undergo irreversible, nonlinear deformation, leading to the creation of fractures. Numerous failure criteria have been devised to describe the triggering of fracturing due to stress concentrations, including Mohr–Coulomb (Jaeger *et al.*, 2007), Hoek–Brown (Hoek and Brown, 1980), Drucker–Prager (Drucker and Prager, 1952), Mogi (Mogi, 1971), and their generalizations, derivations, and combinations (Bigoni and Piccolroaz, 2004). However, failure of a rock is rarely caused by the propagation of a single crack; instead, it is triggered by the coalescence of multiple aligned cracks that form during deformation (Hoek and Bieniawski, 1965). Furthermore, these types of failure criteria, based on the "continuum" stresses that are implicitly averaged over lengths much greater than those of individual pores or microcracks, cannot predict the complex crack paths that originate during crack propagation due to interaction with neighboring cracks (Brace and Bombolakis, 1963).

Failure of an initially intact, brittle material is usually a two-stage process that begins with diffuse, inelastic degradation of the material, also known as "damage." At a very small scale, damage can result from dislocations in the matrix of a crystal, localizing into micro-cracks within a grain of rock or between rock grains. Damage is followed by localization of the loss

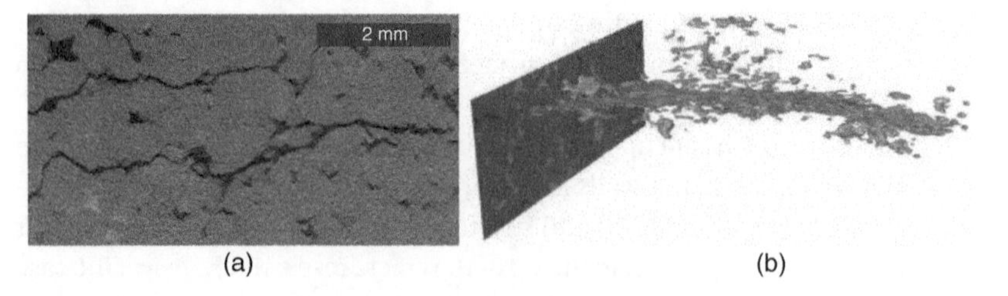

Figure 1.3 Micro-CT image of a fractured sandstone in (a), from which fracture surface can be extracted, as shown in (b), using standard segmentation techniques. Source: Iglauer *et al.* (2011) / Reproduced from John Wiley & Sons, Inc.

of integrity, leading to the growth of larger fractures. The length of "damage zones" ahead of fracture tips is a function of the grain size of the rock (Bažant and Kazemi, 1990). The size of this damage zone also depends on the heterogeneity distribution within the rock and on the influence of preexisting micro-fractures around fracture tips.

When studying fractures in the geological context, fractures are usually several orders of magnitude larger than the assumed near-tip fracture damage zone, allowing the assumption that the fracture process is a linear elastic process. The mechanical behavior of rocks can be described by idealizing the rock as a linear elastic, isotropic, and homogeneous medium, an approach that is referred to as *linear elastic fracture mechanics* (LEFM). Geological patterns such as pervasive extensional fracture patterns have been reproduced using LEFM, including fractures and faults in layered systems in two dimensions (Renshaw and Pollard, 1994; Schöpfer *et al.*, 2007) and three dimensions (Paluszny and Zimmerman, 2013), as well as single (Bremberg and Dhondt, 2009) and multiple (Paluszny and Zimmerman, 2011) fracture propagation and interaction in three dimensions.

1.3 Morphology of Single Fractures

At the millimeter scale, a rock will be composed of a few solid grains and pores. The behavior of these grains and pores can be approximated by idealized descriptions of spheres and ellipsoids, for example, using poroelasticity and effective medium theories (Zimmerman, 1991). Due to the range of grain sizes of sandstones (Boggs, 2012), at the meter scale, an intact piece of this type of rock can be expected to contain a large number of grains, pores, and additional heterogeneities. Importantly, this definition guarantees that, for a meter-scale sample, the sample size is several times larger than any individual grain of the rock. Thus, the scale at which the behavior of the constituting parts is described, and the scale at which these parts act together as a continuum, can be clearly separated. It follows that the behavior of such a rock at the meter scale will not be controlled by a single grain but by a representative number thereof. This separation of scales facilitates the derivation of governing equations that treat the ensemble of pores and solids as a continuum.

Rocks often contain a distribution of micro-heterogeneities that range from hard to soft inclusions, voids in the form of pores and vugs, micro-fractures, cemented veins, and fibers, among other features. Small-scale heterogeneities in rocks arise due to a combination of phenomena that take place during the formation of the rock and throughout its deformation and flow history. These heterogeneities can be due to the mechanical differences between the grains or crystals that initially formed the rock and their response to temperature, reactive flow, dissolution and precipitation, mineral replacement and deposition, and deformation of the rock over millions of years (Chen *et al.*, 2015). These differences lead to local stress concentrations which, when exceeding the local tensile strength of the rock, evolve into small discontinuities within and between grains, forming distributions of flaws that are present in most quasi-brittle rocks. This underlying variability has been characterized in the context of subsurface reservoir engineering, and it is now well established that it plays an important role in controlling the fluid flow and storage properties of the rock matrix.

At the meter and kilometer scales, heterogeneities can appear in the form of layers, fractures, or faults. Discrete breaks in quasi-brittle materials create discontinuities in the mechanical and fluid properties of the rock at many scales. Above the centimeter scale, discontinuities that form under tension are often referred to as "joints" or "cracks," whereas those that form under shear are often referred to as "faults" (Jaeger *et al.*, 2007). Joints are discontinuities that have not been subjected to shear, whereas fault walls have been subjected to shear displacement. These evolve to have a variety of structures and properties; in the broadest sense, both can be regarded as "fractures."

At smaller scales, geological materials exhibit much lower tensile strength than shear strength, and therefore, small tension fractures are ubiquitous. At larger scales, the pervasive presence of heterogeneities translates into lower shear strength, promoting the formation of large faults. Due to these two distinctions, smaller-scale tension discontinuities are often regarded as "fractures," whereas the term "fault" is often reserved for large-scale geological discontinuities that have experienced considerable relative displacement of their walls, either due to having been formed under shear or as the result of the transition of a fracture into a fault due to changes in the regional stresses. In particular, the walls of faults will have moved parallel to the plane of the discontinuity and, in the more general sense, faults are regarded as zones with related deformation structures surrounding the discontinuity, such as secondary fractures and crushing zones (Davatzes and Hickman, 2010). Fractures and faults can form under both extensional and compressional stress regimes, and both types can be found at a range of scales. Hence, small-scale faults are frequently observed in the field, as well as large-scale fracturing that can also be observed on the Earth's surface and on the surfaces of many other rocky and icy planets and satellites.

Multiple fractures can grow in the same direction in response to regional stresses, forming sets that, when superimposed, may form interconnected fracture networks. Growing fractures will coalesce against preexisting free boundaries or open fractures, establishing a geometric record of the relative age of the intersecting fracture sets. Thus, younger sets can be recognized, as they will be populated by shorter, more recent fractures that will have "abutted" against older, preexisting fractures. Fractures within a growing set will interact and intersect, modifying each other's growth orientation, length, and aperture. These interactions complicate the relationship between the orientations of these fractures and the regional stresses that originally led to their growth.

Fractures and faults in the subsurface have small aspect ratios, meaning that their thickness or aperture is many times smaller than their length. Faults frequently displace layers relative to one another, whereas fractures are often restricted by geological layers and display complex cross-cutting relationships. Fractures have varying permeabilities that can either promote or restrict flow, often depending on the current stress state and possible geochemical and mechanical processes that may have affected their internal structure. Typically, fractures have a variable aperture, intersect at small angles, and range in size over several orders of magnitude. Without loss of generality, rock discontinuities in general will be referred to as "fractures" in this book. The morphology of individual fractures and fracture networks changes as the rock mass deforms due to stress. These effects are discussed further in Chapters 3 and 8.

Fracture Shape

Stand-alone fractures can be considered as planar surface inclusions in a volumetric domain. In two-dimensional cut-planes of a fractured rock mass, such as outcrops and cliffs, fractures appear as lines and sets of lines that may be planar or curved and often align and organize to form larger, organized structures. In three dimensions, fractures can be approximated by planar disks or rectangular surfaces (Adler *et al.*, 2012). Their shapes are dictated by a combination of effects, due to the medium in which they grow, and other structures with which they interact. In layered media, fractures are quick to intersect the boundaries of the rock and form rectangular shapes. When intersecting each other in layered media, fractures often create a distribution of "blocks" that effectively subdivide the rock into smaller regions.

In monolithic rocks, such as granites, fractures tend to grow unimpeded, until reaching the boundaries of other fractures or discontinuities. In these media, fractures can be approximated by low-aspect ratio, flat spheroidal or ellipsoidal inclusions that are initially disk-shaped. As they grow, shapes may become sub-planar, curved, or even shaped like complex polyhedral surfaces. Once enough fractures intersect, they may become one larger fracture, or they may intersect to the point that they fragment the rock and subdivide the rock into smaller, disconnected sub-domains.

As fractures grow, their proximity locally overwrites regional stress conditions, effectively rotating stresses around the moving crack tips. This may lead to hooking, bending, intersection, or arrest of the fracture. When fractures interact during growth, their tips may hook against another fracture, as can be seen in Fig. 1.4a. Interactions can be systematically quantified using "interaction maps" that describe the effect of relative orientation as well as distance between fractures on interaction (Thomas *et al.*, 2017).

The results of multiple numerical simulations that quantify interaction between a static fracture and a secondary fracture that is located near the first fracture are summarized in Fig. 1.4. For each location, represented by a point on the graph, the intensity of the interaction is plotted between white (no interaction) and black (strong interaction). A one-meter-long static fracture is located at the origin. Each dot represents a simulation, with a second fracture that is centered at the dot. The graph summarizes the results of eighty numerical simulations. The gray-scale values of C_{I} and C_{II} capture the relative tensile and shear stress intensity concentration, respectively, of a system subjected to uniaxial extension, containing the static fracture and the secondary fracture (Thomas *et al.*, 2017). For tension, fractures placed to the right and above the static fracture yield the greatest interaction. For shear, the interaction is substantially lower than that in tension.

The parameter ε_{I} (not to be confused with strain) indicates whether the relative stress measure C_{I} is tensile (black) or compressive (white). The plots of ε_{I} show that in regions ahead of the tip, fractures promote each other's growth, whereas growth in the region above the fracture is inhibited. For secondary fractures placed close to the static fracture, the magnitude of stress concentration is much higher than if placed away from the fracture. Thus, fractures aligned with each other will tend to promote each other's growth, whereas fractures that are stacked parallel to each other will tend to inhibit each other's growth (Thomas *et al.*, 2017).

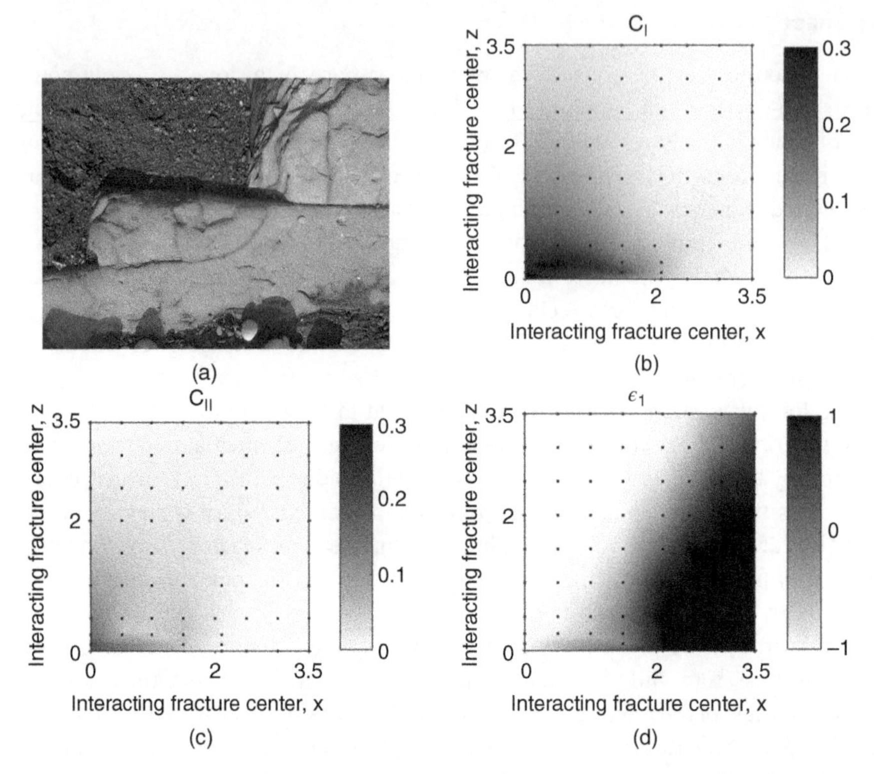

Figure 1.4 Interacting fractures. (a) A fracture hooks and coalesces with another fracture. Graphs (b) and (c) summarize the results of multiple numerical simulations that quantify interaction between a static fracture and a secondary fracture that is located near the first. The intensity of the interaction varies between no interaction (white) and strong interaction (black), quantified in terms of tensile (b) and shear (c) stresses. Dark areas in (d) show regions of tension, and light areas show regions of compression. See the text for further details.

Fracture Length

The length of a fracture is defined by the geometric extent of the curving plane that follows the fracture surface. In the two-dimensional case, fracture length is easily identified with the extent of the fracture and is also proportional to the fracture surface area. In three dimensions, the "length" must take into account the varying geometry that a fracture may present. If fractures are represented by circular disks, the length can be identified with the diameter. For fractures represented by rectangular shapes, length is defined as the measure of its longest side. For nonplanar, complex fracture shapes, length can be approximated using the definition of an equivalent radius or an equivalent extent measure, such as

$$L_f = 2\sqrt{A_f/\pi}, \tag{1.1}$$

where L_f [m] is the fracture "length" and A_f [m^2] is its surface area, when the fracture is thought of as a two-dimensional surface embedded in three-dimensional space.

Observing and characterizing fracture lengths presents a challenge, both in the subsurface and in surface outcrops. In layered media, the trace of the fracture on the rock face is representative of its length. In other cases, such as on the face of a tunnel or the edge of a cliff, intersecting fractures can be "traced." The trace length of a fracture is the intersection of the fracture with the wall at an arbitrary plane, and therefore its length may not be representative of the actual length of the fracture, in addition to being subject to sampling bias.

For a quasi-heterogeneous brittle material, such as most rocks, flaws and small cracks that initially approximately follow a Gaussian size distribution will concentrate stress at their tips during deformation, leading to growth. In particular, flaws with low aspect ratios (*e.g.*, less than 0.1) will tend to grow, align, and coalesce, forming small fractures and thereby leading to even more growth. As fractures grow, their lengths differentiate further, as some fractures are subjected to substantially more growth than others. This variability in lengths is a result not only of the *in situ* and deformation stresses but also of the interaction between the fractures, faults, and other heterogeneities in the rock. It follows that fracture length distributions are usually described using length–frequency relationships. These relationships are based primarily on geological outcrop observations, as fracture lengths cannot readily be measured in boreholes, nor can they be easily interpreted using geophysical imaging, using current techniques.

Fracture length distributions from the meter up to the kilometer scale follow a power-law distribution (Barton, 1995; Bonnet *et al.*, 2001). An example of this behavior can be observed at the Forsmark site (Munier, 2004) that has been chosen for the geological disposal of nuclear waste in Sweden. Detailed fracture trace mappings of the site illustrate the pervasive nature of faults and fractures. Fractures of tens of meters in length can be identified on the scale of tunnel excavations, whereas single-kilometer-long fault discontinuities span the extent of the entire site.

The size distribution across these scales can be approximated by a power-law distribution of the form

$$f(L) = bL^{-n}, \tag{1.2}$$

where L is the fracture length (written here without the subscript f, for notational simplicity), b is a proportionality coefficient for the relationship between amount of fractures and their length, and n is an exponent, sometimes referred to as the "fractal dimension" of the network. The latter term is often used to describe the exponent, despite the fact that fractures in a network can obey a power-law size distribution without forming a fractal geometry (Munier, 2004). It follows from Eq. (1.2) that $f(L)dL$ equals the number of fractures with lengths in the range of $[L, L + dL]$. The exponent n generally varies between 1 and 3.5, with a "typical value" around 2, with factors such as stress history, linkage and connectivity, scale, and sampling bias affecting the value. The value of b, also known as the density factor, varies over a much wider range: between 10^{-3} and 10^5 for faults and between 1 and 100 for fractures, as it depends on n and on the minimum and maximum values of the fracture radii within the network (Bonnet *et al.*, 2001).

The fact that fractures are consistently observed to follow a power-law size distribution suggests that, at any scale of investigation, a single fracture could be expected to control the observed domain. This hints at a lack of separation of scales between fractures and the domain of interest, which has led to the development of numerical models that represent fractures explicitly (Cundall, 1980; Long *et al.*, 1982) rather than as part of a homogenized continuum. This approach has been termed "discrete fracture modeling" (DFM), which indicates that fractures and intact rock are both explicitly represented and governed by separate sets of equations (*e.g.*, Geiger *et al.*, 2004).

Fracture Aperture

The aperture of a fracture, usually denoted by h [m], is a scalar quantity that reflects the physical separation between the walls of a fracture at any point in the nominal fracture plane. Apertures capture the variable distance between fracture walls along their geometry. Although many slightly different definitions of aperture have been proposed (*cf*., Oron and Berkowitz, 1998), the simplest and basic definition is that the aperture is the distance between the two opposing fracture faces, as measured perpendicularly to the nominal fracture plane. Apertures at depth can be observed and quantified when drilling boreholes and can be measured in outcrops. Fractures that are open will increase the permeability of a rock. Fractures that are "closed," with their walls in contact, are less permeable but still retain residual permeability, as explained in Chapter 3. Discontinuities can also seal if they are filled with mineral cement, resulting in a drastic reduction of their permeability (see Fig. 1.5a, b). The resulting veins have a thickness that reflects the aperture of the fracture at the time of precipitation (Vermilye and Scholz, 1995).

For a favorably oriented disk-shaped fracture under tension, apertures will be largest at the center and will reduce to zero at the fracture tips. Some analytical and numerical models assume that fractures have only a single aperture, which may be representative (for example, on average) of the aperture of the fracture. Some numerical models are able to capture apertures as a property that varies along the fracture surface and may change due to mechanical, thermal, or chemical changes in the rock, or due to fluid traveling through the rock (see Chapter 8). Due to the coalescence of smaller fractures into single larger fractures, as shown in Fig. 1.5a, the shape of the fracture may have variations in orientation that can translate into variations of the aperture.

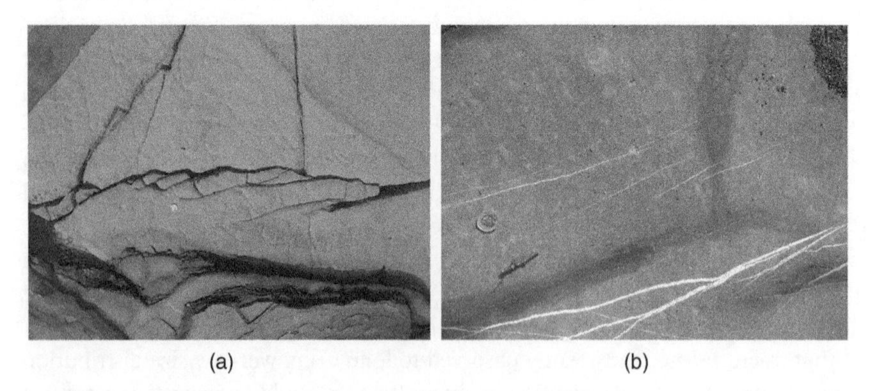

Figure 1.5 Fracture arrays. (a) Interacting fractures on a weathered rock form an array that has linked into a single larger fracture. (b) An array of fractures (top) has formed next to a larger fracture (bottom). These fractures are mineralized, and their apertures, in white, are captured by the precipitation of dissolved rock during regional deformation.

Fracture apertures, which control fracture permeability, are responsive to the history of deformation of the fracture surface and result in an uneven distribution of flow along the fracture surface in the form of channeling (Tsang and Tsang, 1989). This leads to a lower permeability of the fracture as compared to the length-dependent permeability assumption (Lang *et al.*, 2015). There are multiple factors , in addition to fracture network topology and connectivity, such as *in situ* stresses, which may strongly affect the enhancement or reduction of permeability during fracture growth and intersection (Paluszny and Matthäi, 2010).

The linear elastic deformation of the matrix predicts apertures that scale linearly with the length of the (isolated) fracture (Olson, 2003):

$$h_{\max} = \frac{(1-\nu)\sigma}{G} L_f, \tag{1.3}$$

where $h_{\max}$ is the maximum aperture of the fracture, σ is the effective driving stress, ν is Poisson's ratio, and G is the shear modulus of the rock matrix, which for an isotropic homogeneous material is related to the Young's modulus by $E = 2G(1+\nu)$. The *aspect ratio* of a fracture, α [–], is usually defined by

$$\alpha = h_{\max}/L_f. \tag{1.4}$$

Measurements and simulations over multiple scales yield a log-linear distribution of aperture-length distributions (Renshaw and Park, 1997) that follow a bi-linear distribution that shifts when the length of the fracture transitions from the small to the large scale. Based on field measurements of mineralized veins and igneous dykes in the field, aperture–length relationships can be expressed as

$$h_{\max} = C(L_f)^e, \tag{1.5}$$

where C is the pre-exponential constant, which ranges from 7×10^{-4} to 0.43, and e is the power-law scaling exponent, which ranges between 0.38 and 0.41 (Olson, 2003).

This shift is not attributed to a difference in the mechanical process of fracture growth, but rather to the complexity of the heterogeneities that emerge at larger scales that affect the manner in which stress perturbations induced by larger fractures affect smaller fractures. Specifically, it is observed (Renshaw and Park, 1997) that for the small scale, when $\log(L_f) \le \log(L_o)$, the behavior is super-linear, with $s_1 > 1$, and

$$\log(h_{\max}) = s_1[\log(L_f) - \log(L_o)] + \log(h_0), \tag{1.6}$$

where $s_1 > 1$ ranges from 1.76 to 2.54. For larger fractures, for which $\log(L_f) > \log(L_o)$, the behavior is approximately linear, with

$$\log(h_{\max}) = s_2[\log(L_f) - \log(L_o)] + \log(h_0) \approx [\log(L_f) - \log(L_o)] + \log(h_0), \tag{1.7}$$

where s_2 ranges between 0.70 and 1.28, with $\log(L_o) \in [-0.17, 1.26]$ $\log_{10}$-m. In addition, a sublinear relationship between length and aperture has also been reported, based on the measurement of mineralized fractures in the field. Figure 1.6 shows the length-to-aperture scaling laws that were proposed by Renshaw and Park (1997) based on various field data. The length-to-aperture ratio that corresponds to a uniform aspect ratio of 0.01 is shown, for comparison.

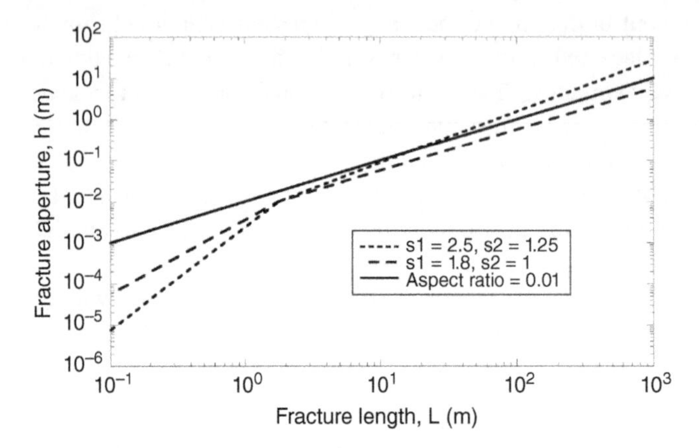

Figure 1.6 Length–aperture scaling law proposed by Renshaw and Park (1997), plotted from Eqs. (1.6) and (1.7), for the case $\{\log_{10}(L_o) = 0.25, \log_{10}(h_o) = -2\}$, for two different pairs of the parameters $\{s_1, s_2\}$. The case of a uniform aspect ratio of 0.01 is also plotted for comparison. Source: Adapted from Renshaw and Park (1997).

Fracture Surface Roughness

During their growth, fracture surfaces accrue small deviations from their original plane of nucleation. Fracture walls are not perfectly planar, and the small variations off the plane constitute the "roughness" of the fracture walls. Roughness exists at many length scales in rock fractures and can be approximated by a Gaussian height distribution and self-affine organization (Brown and Scholz, 1985). Self-affine surfaces form a fractal geometry local to the fracture surface, with statistical invariance under a scale transformation that has an anisotropic aspect ratio:

$$\Delta \boldsymbol{x} \to \zeta\, \Delta \boldsymbol{x}, \tag{1.8}$$

$$\Delta h \to \zeta^H \Delta h, \tag{1.9}$$

where $\Delta \boldsymbol{x} = (x, y)$ is the fracture in-plane coordinate vector, ζ is a constant that quantifies the magnitude of the transformation, h is the height of the fracture surface above some nominal plane, and H is the Hurst exponent, which lies between 0 and 1. The relation between the Hurst exponent, which describes the "jaggedness" of the surface, and the fractal dimension D_f of a two-dimensional surface, can be expressed as

$$H = 2 - D_f, \tag{1.10}$$

whereas for a three-dimensional surface, the relationship is

$$H = 3 - D_f. \tag{1.11}$$

Tensile fractures follow a nearly universal scaling exponent of $H \approx 0.8$, as measured for a range of different rock types and grain size distributions (*e.g.*, Poon *et al.*, 1992). The fact that the fractal geometry is statistically invariant implies that when portions of a surface profile are magnified, the same structure becomes apparent at the smaller scale again and again (see Figs. 1.7 and 1.8).

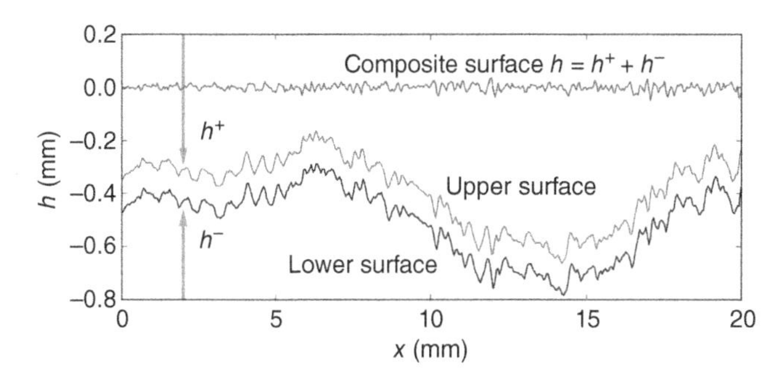

Figure 1.7 The roughness on the surfaces of fractures can be approximated by a Gaussian height distribution and self-affine surfaces. Self-affine surfaces form a fractal geometry that has statistical invariance under a scale transformation of anisotropic aspect ratio. Source: Lang *et al.* (2018)/ CCBY/Public domain.

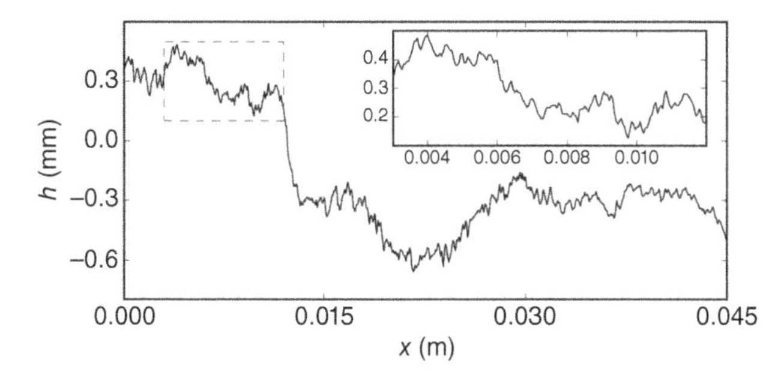

Figure 1.8 A trace of the roughness of a fracture surface. The inset is a zoomed image that illustrates the self-affine fractal nature of rock surfaces. This self-affine surface is a superposition of sine curves of decreasing wavelength and amplitude. Source: Adapted from Lang *et al.* (2016).

Surfaces of this kind can be represented as a series of superimposed sine curves of decreasing wavelength and amplitude. Based on Fourier Transform analysis, roughness can be defined according to wavelength, λ [m], or spatial frequency/wavenumber, q [m^{-1}], where

$$q = \frac{2\pi}{\lambda}, \tag{1.12}$$

and is associated with a measure of the roughness power, C. The slope of the resulting roughness power spectrum $C(q)$ on a log–log plot is related to the Hurst exponent, H. For an ideal self-affine surface, $C(q) \sim q^{-2(H+1)}$ (Persson *et al.*, 2005). The root-mean-square roughness of the surface, defined as

$$h_{rms} \equiv \langle h^2 \rangle^{1/2} = \left(2\pi \int qC(q)dq \right)^{1/2}, \tag{1.13}$$

controls the amplitude of the power spectrum.

When h_{rms} is measured on a fracture *segment* of length L, it is often found to follow a power-law distribution. Candela *et al.* (2012) found that roughness data, obtained from numerous fracture patches of sizes ranging from the mm to km scale, can be correlated with length as follows, in which h_{rms} and L are expressed in units of meters:

$$h_{rms} \approx 0.01 L^H, \tag{1.14}$$

where $H \approx 0.8$ in the direction normal to slip, and $H \approx 0.6$ in the direction parallel to slip. A subsequent study spanning eleven orders of magnitude, from millimeters to hundreds of kilometers, found a power-law exponent of $H = 0.77 \pm 0.23$ (Renard *et al.*, 2013).

When fractal surfaces are considered, their surface area increases, as smaller and smaller roughness wavelengths are included (see Fig. 1.8). Therefore, the geometrically determined area can be replaced by an "effective area" specific to the kind of process involved and the scale of investigation (Farin and Avnir, 1987).

Fracture Surface Contact

Fracture walls are usually at least partially in contact, due to the generally compressive nature of stresses in the subsurface. In cases where fracture walls are compressed, apertures may substantially be reduced, not only due to the mechanical reduction of the space between the fracture walls (see Chapter 3), but also due to the onset of chemical processes such as dissolution and the erosion of heterogeneities (see Chapter 5), which in some cases effectively seal fractures by drastically reducing their aperture and transmissivity. Although fracture wall contact is common in subsurface settings, the contact region between fracture walls is usually a small fraction of the "nominal area," whereas at a sufficiently small scale, the contact between two fracture walls under confining pressure can be considered to be planar (*e.g.*, Archard, 1957). Up to compressive stresses of about 30 MPa, as the load increases, the total contact area increases linearly (Greenwood and Williamson, 1966), resulting in the linear increase in frictional force as the normal load increases.

The size distribution of the discrete contact of a fracture follows a power-law distribution (Dieterich and Kilgore, 1996), a phenomenon that has been both experimentally observed and numerically reproduced (Borri-Brunetto *et al.*, 1999). Contact area of fractures exhibits a self-affine nature, and in the context of fluid flow through fractures, contact area ratio is an increasing function of compressive stress, and regions of contact may effectively render sectors of the fracture impermeable (see Chapter 3).

Due to the effect of scale on the measurement of fracture surface area, the contact area also depends on the chosen scale of investigation. Current imaging methods that are used to identify contact areas have an accuracy of up to 3.5 μm^2, whereas physical probing mechanisms usually have a submicron resolution (Pyrak-Nolte *et al.*, 1987; Yasuhara *et al.*, 2006). The fractal nature of surfaces in contact has been shown to apply down to the molecular scale (Luan and Robbins, 2005), indicating that the fractal pattern will affect the contact area distribution up to that scale. When considering numerical methods to capture contact between fractures, their surfaces are often represented discretely by a grid or lattice. The size of the elements of this grid will limit the size of the smallest representable roughness wavelength and will constrain the size of the smallest representable contact.

1.4 Morphology of Fracture Networks

There are three main sources of data for the geometric characterization of fractured rock masses (Berkowitz, 2002): geological mapping, geophysical imaging, and borehole data. Geological mapping typically relies on manual tracing of field fractures, including

Figure 1.9 Extensional fractures at the meter scale exposed at Kilve Beach, Somerset, UK. Fractures are layer-bound and they grow due to a combination of processes including folding, uplifting, and weathering. Each image is approximately 1 m wide at the lower boundary. Source: Peacock (2004) / Reproduced from Geological Society of London.

length, aperture, spacing, and shape measurements (Segall and Pollard, 1983). Geophysical techniques such as ground-penetrating seismic imaging are used to describe stratigraphic beds and faults at depth (Jouanna *et al.*, 1993). Borehole logging measurements and coring of rocks can be used to collect data on the density, aperture, and orientation at depth. These techniques focus on generating reservoir proxies and probability distributions that describe the geometry and topology of the otherwise occluded subsurface fractures. These data are integrated to generate geometric models of the subsurface, both as discrete and stochastic fracture networks. Figure 1.9 shows examples of fracture networks exposed at Kilve Beach, UK. Fractures appear at a variety of scales and in many places subdivide the rock into discrete blocks.

Discrete fracture models, or fracture analogues, aim to reproduce in detail the geometry of the fracture network, with the expectation that outcrops of rocks of interest exposed on the surface are representative of the fracture networks in the subsurface. Large meter-scale fracture networks have been traditionally mapped in the field (Belayneh *et al.*, 2007), and at larger scales, they can also be extracted from drone and satellite imaging (Weismüller *et al.*, 2020). Fracture networks mapped on the Earth's crust are, in certain circumstances, assumed to be analogous to subsurface patterns. The geometry of these fracture networks can be characterized by assuming that analogous buried fractured rock masses were formed under similar conditions, and, therefore, display similar topological and geometric characteristics.

In contrast, stochastically generated fracture networks aim to capture the statistics of the characterized network properties by generating sets of planar fractures of random orientation, size, and location (Dershowitz and Einstein, 1988). The main advantage of stochastic fracture networks is the speed at which two- and three-dimensional fracture datasets can be generated. However, purely stochastic approaches necessarily make various assumptions. For example, fractures are usually assumed to be planar, elliptically shaped, stand-alone features, as opposed to free-form intersecting surfaces that form in nature. Local boundary stresses, as determined by local fault and fracture interaction, are not taken into account, although they are key to fracture pattern formation (Pollard and Aydin, 1988). Fracture data measured in a specific borehole or in a specific sample may not be representative of the entire fractured rock mass.

Various alternatives coexist with the purely stochastic approach. For example, Srivastava *et al.* (2005) developed a sophisticated data-informed geostatistical model that honors geometric field data by pseudo-randomly growing cracks. Rives *et al.* (1994) devised a set of rules to generate fracture sets based on probability and geological constraints, including strain and curvature attributes. In contrast to statistic-based approaches, full geomechanical modeling reconstructs the geometry of a system based on the analysis of the events that formed it (Nelson, 2001). This approach seeks to reproduce topology and connectivity of fracture networks, as well as fracture sizes, apertures, and patterns that arise due to growth, interaction, and coalescence (Paluszny and Zimmerman, 2013).

Numerical codes attempt to reproduce the genesis of fracture patterns by modeling the process by which stresses affect the geometry of fractures in a rock. Olson and Pollard (1989) applied the boundary element method to numerically compute the energy release rates and estimate growth of a set of straight cracks and curving fractures. Ingraffea and Saouma (1985) introduced a finite element-based method for crack propagation. Belytschko and Black (1999) developed the extended finite element method, in which cracks are tracked independently of a discretizing mesh. Huang *et al.* (2003) demonstrated the applicability of this method to the simultaneous growth of multiple cracks.

Previous studies have given rise to two main approaches for fracture analysis: discrete and smeared, also known by the dichotomies of geometric/nongeometric, or grid/sub-grid. Both approaches have a domain of application. Discrete modeling is appropriate for simulating the growth of one or more dominant cracks, whereas the smeared approach is designed to model diffuse cracking patterns that arise due to the heterogeneity of rocks and other quasi-brittle materials (de Borst *et al.*, 2004).

Discrete fracture networks are characterized using various concepts. The next sub-section describes the techniques used to study spatial organization of the generated fractured rock analogs.

Observation and Quantification of Fractures

Geometric, topological, and geological models of fracture networks can be built by combining information sourced from field observations and measurements. Image logs capture the variability along a borehole, providing one-dimensional data used to define fracture orientation, density, and, in some cases, apertures. Apart from well data and photogrammetry, two-dimensional data can be collected from rocks exposed at the surface or accessible through pre-existing mines or tunnels. These geological analogs provide valuable information at the centimeter and meter scale – smaller than the current resolution of most *in situ* characterization methods – as well as relative positioning, orientations, and density of exposed fractures. Three-dimensional data can be obtained by directly inspecting extracted core using visual inspection, slabbing, and fragmentation measurement, or by more advanced techniques such as X-ray and medical tomography. Three-dimensional data can also be obtained by interpreting 3D seismic data from the subsurface. For growing fractures and faults, *in situ* acoustic imaging can record breakage events, delineating the structure of large subsurface discontinuities. Combined, these data can be used to construct a mechanically and geologically coherent geometric model that captures the effects of deformation on the site.

Persistence, Density, and Intensity

Due to the disparity between 1D, 2D, and 3D measurements of fractures, several measures have been defined to quantify the amount of fractures contained in a rock mass. In the most commonly used system, these measures are denoted by the "persistence" parameter P_{ij}, where the first subscript, i, indicates the dimension in which the space is measured, and the second subscript, j, indicates the dimension in which the fractures are measured (Rogers *et al.*, 2015; see also Dershowitz and Herda, 1992). The measures that correspond to $j = 0$ quantify the number of discrete fractures that would be observed in a given one-, two-, or three-dimensional region. Also called the linear, surface, and volume densities, these measures are defined as

$$P_{10} = n/L_D, \tag{1.15}$$

$$P_{20} = n/A_D, \tag{1.16}$$

$$P_{30} = n/V_D, \tag{1.17}$$

where n is the number of fractures, and $\{L_D, A_D, V_D\}$ are the length of a scanline, the area of a two-dimensional surface slice, or the volume of a three-dimensional domain. These measures have units of $[\mathrm{m}^{-1}]$, $[\mathrm{m}^{-2}]$, and $[\mathrm{m}^{-3}]$, respectively. In general, the SI units of P_{ij} will be $[\mathrm{m}^{j-i}]$.

The measure P_{21} quantifies the total length of discrete fracture traces that would be observed on a given two-dimensional surface, *i.e.*,

$$P_{21} = \frac{1}{A_D}\sum_{i=1}^{n} L_i. \tag{1.18}$$

Sometimes referred to as the areal fracture intensity, P_{21} has units of $[\mathrm{m/m^2}] = [\mathrm{m}^{-1}]$. Note that the two-dimensional slice may be planar, or may be curved, as in the case of a borehole or tunnel wall. If fractures are assumed to be two-dimensional surfaces, P_{31} has no meaning, since fractures will never appear as linear entities in a three-dimensional region.

The measure P_{32} quantifies the total surface area of fractures that would be observed in a given three-dimensional region:

$$P_{32} = \frac{1}{V_D}\sum_{i=1}^{n} A_i. \tag{1.19}$$

In this definition, the surface area of a fracture is only counted "once"; the upper and lower surfaces are not counted individually. Often referred to as the volumetric fracture density, P_{32} has units of $[\mathrm{m^2/m^3}] = [\mathrm{m}^{-1}]$.

If a fracture is thought of as a surface that has an aperture h associated with it, where the aperture is defined as the fracture's thickness in the direction locally normal to the nominal fracture surface, then two additional persistence measures can be defined. The measure P_{33}, defined by

$$P_{33} = \frac{1}{V_D} \sum_{i=1}^{n} A_i h_i, \tag{1.20}$$

represents the total volume of fracture void space contained in a three-dimensional region, and can be interpreted as the "fracture porosity of the rock mass." This parameter is obviously dimensionless.

Finally, another dimensionless measure, P_{22}, can be defined as the total area of fracture void space contained in a given two-dimensional slice:

$$P_{22} = \frac{1}{A_D} \sum_{i=1}^{n} L_i h_i. \tag{1.21}$$

However, this definition is problematic since, in general, the slicing plane will not be normal to the fracture plane, and so the apparent aperture, as observed on the slicing plane, will be greater than the actual aperture.

Estimates of P_{10} that are inferred from inspecting borehole and well data tend to sample a restricted region of the rock and often underestimate the actual fracture intensity. These estimates, therefore, must be appropriately scaled when mapped to two and three dimensions. Whereas measures such as P_{10}, P_{20}, and P_{30} capture the density of the network in the broadest sense, the literature also commonly refers to P_{32} as the "density" of a fracture network. Density is an important reference property when studying the effect of fractures on flow, as it is often used to populate stochastic fracture networks, it is measured in the field, and it is often monitored when numerically modeling the growth of fractures. The following sub-section reviews a further concept, the *dimensionless fracture density*, which is often used to characterize fracture networks when quantifying their flow properties.

Dimensionless Fracture Density

Particularly in the field of percolation theory (see Chapters 7 and 8), there is extensive use of the concept of dimensionless density. For two-dimensional and three-dimensional systems, the dimensionless fracture density can be defined as

$$\rho'_{2D} = \rho_{2D} \langle A_{ex} \rangle = P_{20} \langle A_{ex} \rangle, \tag{1.22}$$

$$\rho'_{3D} = \rho_{3D} \langle V_{ex} \rangle = P_{30} \langle V_{ex} \rangle, \tag{1.23}$$

where ρ' is the dimensionless density, ρ is the "absolute" density, as defined in Eqs. (1.16) and (1.17), and $\langle A_{ex} \rangle$, $\langle V_{ex} \rangle$ are the average *excluded area* and *excluded volume*, respectively (Balberg *et al.*, 1984). The excluded area or excluded volume defines the minimum area or volume that two fracture centers must not reside within, in order for the fracture objects not to overlap.

The average excluded area of a fracture network with a distribution of uniformly randomly oriented fracture lengths is defined as (Balberg *et al.*, 1984; Adler and Thovert, 1999)

$$\langle A_{ex} \rangle = \frac{2}{\pi} \langle L \rangle^2, \tag{1.24}$$

where $\langle L \rangle$ is the average length of the fractures. There is an alternate definition of excluded volume for high-density datasets where the system percolates with the assumption of constant apertures, which has been shown to achieve better agreement between 2D and 3D permeability measurements in the sampled systems. For percolating systems, excluded area can be defined as (Lang *et al.*, 2014)

$$\langle A_{ex} \rangle = \langle L \rangle^2, \tag{1.25}$$

This definition drops the geometric factor $2/\pi$, which serves to increase the 2D density of the trace maps, and conduces to an improved hydraulic equivalence between 3D and the associated 2D fracture networks.

In three dimensions, for disk-shaped fractures of varying radius, the average excluded volume can be defined as (Mourzenko *et al.*, 2005)

$$\langle V_{ex} \rangle = \frac{\pi^2}{2} \left\langle r_1 r_2^2 + r_2 r_1^2 \right\rangle, \tag{1.26}$$

where the average of the excluded volumes is computed for all combinations of fracture pairs, with radii $\{r_1, r_2\}$, lying within the volume of interest. For convex polygonal fractures, the excluded volume is defined as

$$\langle V_{ex} \rangle = \frac{1}{2} \langle a_f p_f \rangle, \tag{1.27}$$

where a_f is the fracture area, and p_f is the fracture perimeter (Bogdanov *et al.*, 2003).

Spacing

Spacing is a measure of how fractures are geometrically distributed within the rock volume. This parameter is instrumental in distinguishing clustered fracture networks from evenly-spaced networks. Spacing can be measured in analog outcrops and borehole images, and is the result of the mechanical interaction of fractures during growth. In a two-dimensional set of fractures, spacing is defined as the distance between parallel fractures. In a more general setting, spacing s can be defined as the mean distance between the center of a fracture and the nearest fracture neighbor. In three-dimensional domains, spacing s [m] can be defined as the minimum distance between two fractures i and j as

$$s = \min \{ \|\overrightarrow{P_i P_j}\| \}, \tag{1.28}$$

where P_i is a point on fracture i, P_j is a point on fracture j, $\|\overrightarrow{P_i P_j}\|$ is the length of the vector from P_i to P_j, and the minimum is computed over all possible pairs $\{P_i, P_j\}$. In the special case in which the two planes are parallel to each other, this definition reduces to the usual distance between two parallel planes.

In two dimensions, the *area method* is an average approximation of spacing, which defines spacing [m] as

$$s_{avg}^{2D} = \frac{A}{L_0 + \sum_{i=1}^{n} L_i}, \tag{1.29}$$

where L_0 is the height of the sampled area, and the summation is taken over all fractures in the sampling area. This is a compound method that estimates the saturation of poorly and

Table 1.1 Classification of discontinuity spacing.

Description	Spacing (mm)
Extremely close spacing	<20
Very close spacing	20–60
Close spacing	60–200
Moderate spacing	200–600
Wide spacing	600–2000
Very wide spacing	2000–6000
Extremely wide spacing	>6000

Source: ISRM (1978) / Elsevier.

well-developed fracture sets. Descriptors of spacing, according to the International Society of Rock Mechanics, range from "extremely close spacing" for spacings of less than 2 cm to "extremely wide spacing" for spacings of over 6 m (ISRM, 1978) (see Table 1.1).

An additional concept related to spacing is the "frequency" f [m^{-1}], defined as the number of fractures per meter of length, or $f = 1/s$, as measured by counting intersections of fractures with a linear scan line.

Connectivity

The connectivity χ [–] of a developing fracture set can be defined in relation to the initial fracture set as

$$\chi = 1 - (n_i/n_0), \tag{1.30}$$

where n_0 is the initial number of fractures, and n_i is the number of clusters (intersected fractures) at growth step i. Initially, $n_i = n_0$ and $\chi = 0$. When fractures are all connected to each other, they each form part of one large cluster, and $\chi = 1$. For growing fracture networks, connectivity is a measure of the fracture development stage. If a fracture is assumed to be planar, as they appear in most layered media, the length and relative orientation of two vicinal fractures will affect their connectivity. And so, length, spacing, and orientation correlate to density. For a poorly developed fracture network, $n_i \approx n_0$, and for a well-developed fracture network, $n_0 \gg n_i$. When all fractures are connected to each other, $\chi \to 1$. However, if the fractures are assumed to be nonplanar, connectivity can increase at a much more accelerated rate than density, as interacting fractures may hook and coalesce without substantially increasing in length.

Problems

1.1 What tensile stress must be applied to a fracture that has an effective radius of 1 m, in order to achieve an aperture of 3 mm, in a shale, a granite, a sandstone, and a carbonate? Use "typical" values of the elastic moduli for any calculations.

1.2 Micro-fractures capture the mechanical deformation state of the rock when formed. What is the orientation of micro-fractures ahead of a propagating fracture?

1.3 Under which conditions does a fracture grow under the influence of regional *in situ* stresses? Does this depend on the scale of the fracture?

1.4 Consider a 100 m^3 rock unit that contains ten circular fractures of 10 m radius, one hundred fractures of 0.1 m radius, and one thousand fractures of 0.01 m radius. Assume that each fracture has a uniform aperture of 100 μm. What are the P_{32} and P_{33} values of this fracture system? What would be the P_{32} and P_{33} values if each fracture had a uniform aperture and an aspect ratio of 0.001?

1.5 Consider one hundred random cut planes generated through the fracture network described in Problem 1.4. What are the corresponding P_{21} and P_{22} intensities of these 2D samples of the network? What happens to the 2D *vs.* 3D intensity contrast when the density of the 3D network is doubled?

1.6 Consider the fracture network described in Problem 1.4. What is the connectivity and spacing of the three-dimensional network? What is the corresponding two-dimensional connectivity of the two-dimensional fracture networks defined by the cut planes described in Problem 1.5?

References

Adler, P. M. and Thovert, J.-F. 1999. *Fractures and Fracture Networks*, Kluwer, Dordrecht.

Adler, P. M., Thovert, J.-F., and Mourzenko, V. V. 2012. *Fractured Porous Media,* Oxford University Press, Oxford.

Archard, J. F. 1957. Elastic deformation and the laws of friction. *Proceedings of the Royal Society A: Mathematical, Physical and Engineering Sciences*, 243(1233), 190–205.

Balberg, I., Anderson, C., Alexander, S., and Wagner, N. 1984. Excluded volume and its relation to the onset of percolation. *Physical Review B*, 30(7), 3933–43.

Barton, C. C. 1995. Fractal analysis of scaling and spatial clustering of fractures, in *Fractals in the Earth Sciences*, C. C. Barton and P. R. La Pointe, eds., Springer, Boston, pp. 141–78.

Bažant, Z. P. and Kazemi, M. T. 1990. Determination of fracture energy, process zone length and brittleness number from size effect, with application to rock and concrete. *International Journal of Fracture*, 44, 111–31.

Belayneh, M., Matthäi, S. K., and Cosgrove, J. W. 2007. The implication of fracture swarms in the Chalk of SE England on the tectonic history of the basin and their impact on fluid flow in high-porosity, low-permeability rocks, in *Geological Society Special Publication* 272, A. C. Ries, R. W. H. Butler, and R. H. Graham, eds., Geological Society, London, pp. 499–517.

Belytschko, T. and Black, T. 1999. Elastic crack growth in finite elements with minimal remeshing. *International Journal for Numerical Methods in Engineering*, 45(5), 601–20.

Berkowitz, B. 2002. Characterizing flow and transport in fractured geological media: a review. *Advances in Water Resources*, 25(8–12), 861–84.

Bigoni, D. and Piccolroaz, A. 2004. Yield criteria for quasi-brittle and frictional materials. *International Journal of Solids and Structures*, 41(11–12), 2855–78.

Bogdanov, I. I., Mourzenko, V. V. Thovert, J.-F., and Adler, P. M. 2003. Effective permeability of fractured porous media in steady state flow. *Water Resources Research,* 39(1), 1023.

Boggs, S. 2012. *Principles of Sedimentology and Stratigraphy*, 5th ed., Pearson Prentice-Hall, London.

Bonnet, E., Bour, O., Odling, N. E., Davy, P., *et al.* 2001. Scaling of fracture systems in geological media. *Reviews of Geophysics*, 39(3), 347–83.

Borri-Brunetto, M., Carpinteri, A., and Chiaia, B. 1999. Scaling phenomena due to fractal contact in concrete and rock fractures. *International Journal of Fracture*, 12(1–4), 221–38.

Brace, W. F. and Bombolakis, E. G. 1963. A note on brittle crack growth in compression. *Journal of Geophysical Research*, 68(12), 3709–13.

Bremberg D. and Dhondt G. 2009. Automatic 3-D crack propagation calculations: a pure hexahedral element approach versus a combined element approach. *International Journal of Fracture*, 157(1–2), 109–18.

Brown, S. R. and Scholz, C. H. 1985. Broad bandwidth study of the topography of natural rock surfaces. *Journal of Geophysical Research*, 90(B14), 12575.

Candela, T., Renard, F., Klinger, Y., Mair, K., *et al.* 2012. Roughness of fault surfaces over nine decades of length scales. *Journal of Geophysical Research*, 117(B8), B08409.

Chen L., Lu Y., Jiang S., Li J., *et al.* 2015. Heterogeneity of the Lower Silurian Longmaxi marine shale in the southeast Sichuan Basin of China. *Marine and Petroleum Geology*, 65, 232–46.

Cnudde, V. and Boone, M. N. 2013. High-resolution X-ray computed tomography in geosciences: a review of the current technology and applications. *Earth-Science Reviews*, 123, 1–17.

Cundall, P. A. 1980. *UDEC – A Generalized Distinct Element Program for Modelling Jointed Rock*, Report PCAR-1-80, Peter Cundall Associates, Virginia Water, UK.

Davatzes, N. C. and Hickman S. H. 2010. Stress fracture, and fluid-flow analysis using acoustic and electrical image logs in hot fractured granites of the Coso geothermal field, California, USA, in *Dipmeter and Borehole Image Log Technology, AAPG Memoir 92,* M. Poppelreiter, C. Garcia-Carballido, and M. Kraaijveld, eds., American Association of Petroleum Geologists, Tulsa, Oklahoma, pp. 259–93.

de Borst, R., Remmers, J. J. C., Needleman, A., and Abellan, M.-A. 2004. Discrete vs smeared crack models for concrete fracture: bridging the gap. *International Journal for Numerical and Analytical Methods in Geomechanics*, 28(7–8), 583–607.

Dershowitz, W. S. and Einstein, H. H. 1988. Characterizing rock joint geometry with joint system models. *Rock Mechanics and Rock Engineering*, 21, 21–51.

Dershowitz, W. S. and Herda, H. H. 1992. Interpretation of fracture spacing and intensity, in *Proceedings of the 33rd U.S. Rock Mechanics Symposium*, J. R. Tillerson and W. R. Wawersik, eds., A. A. Balkema, Rotterdam, pp. 757–66.

Dieterich, J. H. and Kilgore, B. D. 1996. Imaging surface contacts: power law contact distributions and contact stresses in quartz, calcite, glass and acrylic plastic. *Tectonophysics*, 256(1–4), 219–39.

Drucker, D. C. and Prager, W. 1952. Soil mechanics and plastic analysis or limit design. *Quarterly of Applied Mathematics*, 10(2), 157–65.

Farin, D. and Avnir, D. 1987. Reactive fractal surfaces. *Journal of Physical Chemistry*, 91(22), 5517–21.

Geiger, S., Roberts, S., Matthäi, S. K., Zoppou, C., *et al.* 2004. Combining finite element and finite volume methods for efficient multiphase flow simulations in highly heterogeneous and structurally complex geologic media. *Geofluids*, 4(4), 284–99.

Greenwood, J. A. and Williamson, J. B. P. 1966. Contact of nominally at surfaces. *Proceedings of the Royal Society Series A*, 295(1442), 300–19.

Gudmundsson, A. 2000. Fracture dimensions, displacements and fluid transport. *Journal of Structural Geology*, 22(9), 1221–31.

Herrmann, H. J. 1990. Introduction to basic notions and facts, in *Statistical Models for the Fracture of Disordered Media*, H. J. Herrmann and S. Roux, eds., Elsevier, Amsterdam, pp. 1–31.

Hoek, E. and Bieniawski, Z. T. 1965. Brittle rock fracture propagation in rock under compression. *International Journal of Fracture Mechanics*, 1, 137–55.

Hoek, E. and Brown, E. T. 1980. *Underground Excavation in Rock*, Institution of Mining and Metallurgy, London.

Huang, R., Sukumar, N., and Prevost, J. H. 2003. Modeling quasi-static crack growth with the extended finite element method. Part II: numerical applications. *International Journal of Solids and Structures*, 40(26), 7539–52.

Iglauer, S., Paluszny, A., Pentland, C. H., and Blunt, M. J. 2011. Residual CO_2 imaged with X-ray micro-tomography. *Geophysical Research Letters*, 38(21), L21403.

Ingraffea, A. R. and Saouma, V. 1985. Numerical modelling of discrete crack propagation in reinforced and plain concrete, in *Fracture Mechanics of Concrete: Structural Application and Numerical Calculation*, Sih, G. C. and Ditomasso, A., eds., Martinus Nijhoff, Leiden, pp. 171–225.

International Society for Rock Mechanics 1978. Suggested methods for the quantitative description of discontinuities in rock masses, *International Journal of Rock Mechanics and Mining Sciences*, 15(6), 319–68.

Jaeger, J. C., Cook, N. G. W., and Zimmerman, R. W. 2007. *Fundamentals of Rock Mechanics*, 4th ed., Wiley, Oxford.

Jouanna, P., Armangau, C., Batchelor, A. S., Bonazzi, D., *et al.* 1993. A summary of field test methods in fractured rocks, in *Flow and Contaminant Transport in Fractured Rock*, C.-F. Tsang, G. de Marsily, and J. Bear, eds., Academic Press, New York, pp. 437–543.

Lang, P. S., Paluszny, A., and Zimmerman, R. W. 2014. Permeability tensor of three-dimensional fractured porous rock and a comparison to trace map predictions. *Journal of Geophysical Research: Solid Earth*, 119, 6288–6307.

Lang, P. S., Paluszny, A., and Zimmerman, R. W. 2015. Hydraulic sealing due to pressure solution contact zone growth in siliciclastic rock fractures. *Journal of Geophysical Research: Solid Earth*, 120(6), 4080–4101.

Lang, P. S., Paluszny A., and Zimmerman R. W. 2016. Evolution of fracture normal stiffness due to pressure dissolution and precipitation, *International Journal of Rock Mechanics and Mining Sciences*, 88, 12–22.

Lang, P. S., Paluszny, A., Nejati, M., and Zimmerman, R. W. 2018. Relationship between the orientation of maximum permeability and intermediate principal stress in fractured rocks. *Water Resources Research*, 54, 8734–55.

Long, J. C. S., Remer, J. S., Wilson, C. R., and Witherspoon, P. A. 1982. Porous media equivalents for networks of discontinuous fractures. *Water Resources Research*, 18(3), 645–58.

Luan, B. and Robbins, M. O. 2005. The breakdown of continuum models for mechanical contacts. *Nature*, 435(7044), 929–32.

Mogi, K. 1971. Effect of the triaxial stress system on the failure of dolomite and limestone. *Tectonophysics*, 11(2), 111–27.

Mourzenko, V., Thovert, J.-F, and Adler, P. M. 2005. Percolation of three-dimensional fracture networks with power-law size distribution. *Physical Review E*, 72(3), 036103.

Munier, R. 2004. *Statistical Analysis of Fracture Data, Adapted for Modelling Discrete Fracture Networks* (*Version 2*). Technical Report R-04-66, Swedish Nuclear Fuel and Waste Management Company (SKB), Stockholm.

Nelson, R. A. 2001. *Geologic Analysis of Naturally Fractured Reservoirs*, 2nd ed., Gulf Professional Publishing, Houston.

Olson, J. and Pollard, D. D. 1989. Inferring paleostresses from natural fracture patterns: a new method. *Geology*, 17(4), 345–48.

Olson, J. E. 2003. Sublinear scaling of fracture aperture versus length: an exception or the rule? *Journal of Geophysical Research*, 108(B9), 2413.

Oron, A. P. and Berkowitz, B. 1998. Flow in rock fractures: the local cubic law assumption re-examined. *Water Resources Research*, 34(11), 2811–25.

Paluszny, A. and Matthäi, S. K. 2010. Impact of fracture development on the effective permeability of porous rocks as determined by 2-D discrete fracture growth modeling. *Journal of Geophysical Research*, 115(B2), B02203.

Paluszny, A. and Zimmerman, R. W. 2011. Numerical simulation of multiple 3D fracture propagation using arbitrary meshes. *Computer Methods in Applied Mechanics and Engineering*, 200(9–12), 953–66.

Paluszny, A. and Zimmerman, R. W. 2013. Numerical fracture growth modelling using smooth surface geometric deformation. *Engineering Fracture Mechanics*, 108(SI), 19–36.

Peacock D. C. P. 2004. Differences between veins and joints using the example of the Jurassic limestones of Somerset, in *Geological Society Special Publication 231*, J. W. Cosgrove and T. Engelder, eds., Geological Society, London, pp. 209–21.

Peacock D. C. P., Nixon, C. W., Rotevatn, A., Sanderson, D. J., *et al.* 2016. Glossary of fault and other fracture networks. *Journal of Structural Geology*, 92, 12–29.

Persson, B. N. J., Albohr, O., Tartaglino, U., Volokitin, A. I., *et al.* 2005. On the nature of surface roughness with application to contact mechanics, sealing, rubber friction and adhesion. *Journal of Physics: Condensed Matter*, 17(1), R1–R62.

Pollard, D. and Aydin, A. 1988. Progress in understanding jointing over the past century. *Geological Society of America Bulletin*, 100(8), 1181.

Poon, C. Y., Sayles, R. A., and Jones, T. A. 1992. Surface measurement and fractal characterization of naturally fractured rocks. *Journal of Physics D: Applied Physics*, 25(8), 1269–75.

Pyrak-Nolte, L. J., Myer, L. R., Cook, N. G. W., and Witherspoon, P. A. 1987. Hydraulic and mechanical properties of natural fractures in low permeability rock, in *Proceedings of the 6th International Congress of Rock Mechanics*, G. Herget and S. Vongpaisal, eds., Balkema, Rotterdam, pp. 225–31.

Renard, F., Candela, T., and Bouchaud, E. 2013. Constant dimensionality of fault roughness from the scale of micro-fractures to the scale of continents. *Geophysical Research Letters,* 40(1), 83–7.

Renshaw, C. E. and Pollard, D. D. 1994. Are large differential stresses required for straight fracture propagation paths? *Journal of Structural Geology*, 16(6), 817–22.

Renshaw, C. E. and Park, J. C. 1997. Effect of mechanical interactions on the scaling of fracture length and aperture. *Nature*, 386(6624), 482–84.

Rives, T., Rawnsley, K. D., and Pettit, J.-P. 1994. Analogue simulation of natural orthogonal joint set formation in brittle varnish. *Journal of Structural Geology*, 16(3), 419–29.

Rogers, S., Elmo, D., Webb G., and Catalan, A. 2015. Volumetric fracture intensity measurement for improved rock mass characterisation and fragmentation assessment in block caving operations. *Rock Mechanics and Rock Engineering*, 48, 633–49.

Schöpfer, M. P. J., Childs C., and Walsh J. J. 2007. Two-dimensional distinct element modelling of the structure and growth of normal faults in multi-layer sequences: 1. Model calibration, boundary conditions, and selected results. *Journal of Geophysical Research*, 112(B10), B10401.

Segall, P. and Pollard, D. D. 1983. Joint formation in granitic rock of the Sierra Nevada. *Geologic Society of America Bulletin*, 94(5), 563–75.

Srivastava, R. M., Frykman, P., and Jensen, M. 2005. Geostatistical simulation of fracture networks, in *Proceedings of Geostatistics Banff 2004*, O. Leuangthong and C. V. Deutsch, eds., Springer Netherlands, Heidelberg, pp. 295–304.

Thomas, R. N., Paluszny, A., and Zimmerman, R. W. 2017. Quantification of fracture interaction using stress intensity factor variation maps. *Journal of Geophysical Research: Solid Earth*, 122(10), 7698–7717.

Tsang, Y. W. and Tsang, C. F. 1989. Flow channeling in a single fracture as a two-dimensional strongly heterogeneous permeable medium. *Water Resources Research*, 25(9), 2076–80.

Underwood, E. E. 1970. *Quantitative Stereology*, Addison-Wesley, Boston.

Vermilye, J. M. and Scholz, C. H. 1995. Relation between vein length and aperture. *Journal of Structural Geology*, 17(3), 423–34.

Weismüller, C., Prabhakaran, R., Passchier, M., Urai, J. L., *et al.* 2020. Mapping the fracture network in the Lilstock pavement, Bristol Channel, UK: manual versus automatic. *Solid Earth*, 11, 1773–1802

Yasuhara, H., Polak, A., Mitani, Y., Grader, A., *et al.* 2006. Evolution of fracture permeability through fluid-rock reaction under hydrothermal conditions. *Earth and Planetary Science Letters*, 244(1–2), 186–200.

Zimmerman, R. W. 1991. *Compressibility of Sandstones*, Elsevier, Amsterdam.

2

Fluid Flow in a Single Fracture

2.1 Introduction

The basic problem in the field of fluid flow through fractured rocks is the flow of a viscous fluid through a single rock fracture. This problem is governed by the Navier–Stokes equations, which are a coupled set of four nonlinear partial differential equations. Although the form of these equations and their associated boundary conditions have been known for almost 200 years, exact analytical solutions can be obtained in very few situations. This difficulty is exacerbated by the fact that real rock fractures have very irregular geometries. Nevertheless, some understanding of flow through a single fracture can be obtained by examining idealized geometries – such as a fracture bounded by two smooth parallel walls, a fracture bounded by sinusoidally varying walls, or a fracture bounded by linearly converging-diverging walls – for which analytical solutions are obtainable. Further progress can be made by simplifying the Navier–Stokes equations to a more tractable form, such as the Stokes equations or the Reynolds lubrication equation.

The basic problem of fluid flow through a single rock fracture, driven by a pressure gradient, is discussed in this chapter. In Section 2.2, the exact solution of the Navier–Stokes equations is derived for pressure-driven flow between two parallel walls. This leads to the famous "cubic law," which states that the volumetric flux is proportional to the pressure gradient, inversely proportional to the fluid viscosity, and proportional to the cube of the fracture aperture. In Section 2.3, it is shown that, if the flowrate is in some sense "sufficiently low," the nonlinear Navier–Stokes equations for fluid flow through a fracture having a spatially varying aperture can be reduced to the mathematically linear Stokes equations. Additionally, it is shown in Section 2.4 that if the aperture does not vary "too abruptly," the Stokes equations can be further simplified into a single linear partial differential known as the Reynolds lubrication equation. This model is essentially equivalent to a conduction problem in a two-dimensional medium having a spatially varying conductivity, for which many theoretical bounds and approximate solutions are known.

Real fractures always have regions in which the two fracture walls are in contact. The effect of these contact regions on the hydraulic transmissivity of the fracture is discussed in Section 2.5. Available laboratory data on the transmissivity of fractures are reviewed in Section 2.6 and compared with some of the previously discussed analytical models. All of the models discussed in Sections 2.2–2.5 assume that the rock adjacent to the fracture is impermeable. In Section 2.7, the effect of a permeable matrix rock on the transmissivity

Fluid Flow in Fractured Rocks, First Edition. Robert W. Zimmerman and Adriana Paluszny.

Companion website: www.wiley.com/go/zimmerman/fluidflowinfracturedrocks

of a fracture is discussed and shown to usually be quite negligible. Finally, flow through a fracture that is filled with a permeable medium, such as the proppant material that is injected into hydraulically induced fractures, is discussed in Section 2.8.

2.2 The Navier–Stokes Equations and the Cubic Law

Fundamentally, the flow of a Newtonian viscous fluid through a rock fracture is governed by the Navier–Stokes equations. These are a set of three coupled nonlinear partial differential equations that can be written as (Batchelor 1967, p. 147)

$$\rho\left[\frac{\partial \mathbf{u}}{\partial t} + (\mathbf{u}\cdot\nabla)\mathbf{u}\right] = \mathbf{F} - \nabla p + \mu\nabla^2\mathbf{u}, \tag{2.1}$$

where $\mathbf{u} = (u_x, u_y, u_z)$ is the velocity vector, with units of [m/s]; $\mathbf{F}$ is the body-force vector acting on the fluid, per unit volume, with units of [N/m^3]; ρ is the fluid density, with units of [kg/m^3]; p is the pressure, with units of [Pa]; μ is the fluid viscosity, with units of [Pa s]; ∇ is the gradient operator; and ∇^2 is the Laplacian operator. The Navier–Stokes equations express the principle of conservation of linear momentum and incorporate a linear constitutive relation that relates the stress tensor to the rate of shear. The first term in the bracket on the left side of eq. (2.1) represents the acceleration of a fluid particle due to the fact that, at a fixed point in space, the velocity may change with time. The second term, the "advective acceleration," represents the acceleration that a particle would have, even in a steady-state flow field, by virtue of moving to a location at which the velocity is different. The forcing terms on the right side represent the applied body force, the pressure gradient, and the internal viscous forces. If the flow occurs under isothermal conditions, as will generally be assumed in this book, any mathematical solution that satisfies eq. (2.1) will also satisfy the equation of conservation of energy, and so no additional energy equation is needed (see Currie, 2003, p. 36).

In most cases, the only appreciable body force acting on the fluid is that due to gravity, in which case $\mathbf{F} = -\rho g\mathbf{e}_z$, where $\mathbf{e}_z$ is the unit vector pointing in the upward *vertical* direction, and g is the gravitational acceleration. If the density of the fluid is uniform, the gravity force can be eliminated from the Navier–Stokes equations by defining a *reduced pressure*, $\widehat{p} = p + \rho gz$ (Phillips, 1991, p. 26), in which case

$$\mathbf{F} - \nabla p = -\rho g\mathbf{e}_z - \nabla p = -(\nabla p + \rho g\mathbf{e}_z) = -\nabla(p + \rho gz) = -\nabla\widehat{p}. \tag{2.2}$$

The governing equations can therefore be written without the gravity term if the pressure is replaced by the reduced pressure. Henceforth, the Navier–Stokes equations will be written in terms of the reduced pressure, which, for simplicity of notation, will be denoted by p.

In steady-state conditions, $\partial\mathbf{u}/\partial t = 0$, and the Navier–Stokes equations (2.1) reduce to

$$\mu\nabla^2\mathbf{u} - \rho(\mathbf{u}\cdot\nabla\mathbf{u}) = \nabla p. \tag{2.3}$$

The physical interpretation of eq. (2.3) is that the pressure gradient serves as the driving force that causes advective acceleration, whereas the viscous forces tend to retard, or decelerate, the flow. The presence of the mathematically nonlinear advective acceleration term, $\rho(\mathbf{u}\cdot\nabla)\mathbf{u}$, generally makes the Navier–Stokes equations difficult to solve. Methods to circumvent these difficulties by reducing the Navier–Stokes equations to a simpler set of equations are discussed in Section 2.3.

When written out explicitly in component form, eq. (2.3) takes the form

$$\mu\left(\frac{\partial^2 u_x}{\partial x^2}+\frac{\partial^2 u_x}{\partial y^2}+\frac{\partial^2 u_x}{\partial z^2}\right)-\rho\left(u_x\frac{\partial u_x}{\partial x}+u_y\frac{\partial u_x}{\partial y}+u_z\frac{\partial u_x}{\partial z}\right)=\frac{\partial p}{\partial x}, \tag{2.4}$$

$$\mu\left(\frac{\partial^2 u_y}{\partial x^2}+\frac{\partial^2 u_y}{\partial y^2}+\frac{\partial^2 u_y}{\partial z^2}\right)-\rho\left(u_x\frac{\partial u_y}{\partial x}+u_y\frac{\partial u_y}{\partial y}+u_z\frac{\partial u_y}{\partial z}\right)=\frac{\partial p}{\partial y}, \tag{2.5}$$

$$\mu\left(\frac{\partial^2 u_z}{\partial x^2}+\frac{\partial^2 u_z}{\partial y^2}+\frac{\partial^2 u_z}{\partial z^2}\right)-\rho\left(u_x\frac{\partial u_z}{\partial x}+u_y\frac{\partial u_z}{\partial y}+u_z\frac{\partial u_z}{\partial z}\right)=\frac{\partial p}{\partial z}. \tag{2.6}$$

Equation (2.3), or eqs. (2.4)–(2.6), represent three equations for the four unknown variables: the three velocity components and the pressure. An additional equation that is required to mathematically close the system can be obtained by applying the principle of conservation of mass. For an incompressible fluid, this principle is equivalent to conservation of volume and can be expressed mathematically as

$$div\,\mathbf{u}\equiv\nabla\cdot\mathbf{u}\equiv\frac{\partial u_x}{\partial x}+\frac{\partial u_y}{\partial y}+\frac{\partial u_z}{\partial z}=0. \tag{2.7}$$

The compressibility of water is roughly 5×10^{-10}/Pa (Batchelor, 1967, p. 595), and so a pressure change as large as 10 MPa would alter the density by only 0.5%, which implies that the assumption of incompressibility is reasonable in practice. More rigorously, compressibility effects during fluid flow processes can be ignored if the velocity of the fluid is much less than the speed of sound in that fluid (Batchelor, 1967, p. 168). The speed of sound in water, over the pressure range of 1–100 MPa, and at all temperatures between the melting point and boiling point, is roughly 1500 m/s – much larger than the velocities that occur during the subsurface flow of water through fractured rocks. For practical purposes, therefore, compressibility effects can be ignored in subsurface flows involving liquids.

In order to solve the Navier–Stokes equations for the fluid contained in a specific region of space, the set of four coupled partial differential equations, eqs. (2.4)–(2.7), must be supplemented by boundary conditions. When a fluid is flowing through a fracture that is bounded by *impermeable* rock, the fluid must satisfy the "no-slip" boundary condition at the boundary between the fluid and the walls of the fracture. This boundary condition states that, in general, at an interface between a solid and a fluid, the velocity of the fluid must equal that of the solid. In the absence of seismic waves or large-scale rock deformation, this condition implies that both the normal component and the tangential component of the fluid velocity must be zero at the fracture walls if the adjacent rock is assumed to be impermeable. (Fluid flow between a fracture and adjacent *porous* rock is discussed in Section 2.7 and in Chapters 7–10). The condition that the *normal* component of the velocity is zero at the wall is a simple and obvious consequence of the assumption that the adjacent rock is impermeable. The condition of zero *tangential* velocity is not *a priori* obvious, but it has empirically been found to hold for all fluids, except for gases at very low pressures (Batchelor, 1967, p. 149). At the open faces of the fracture, either the pressure or the velocity profile must be specified. The boundary condition that is most commonly applied at the open faces of a fracture, for the purposes of estimating the fracture's transmissivity, is that of a specified uniform pressure.

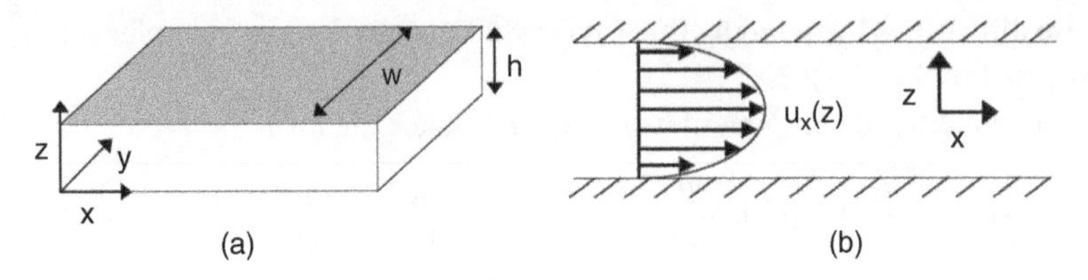

Figure 2.1 (a) A smooth-walled, "parallel plate" fracture having uniform aperture, h. The applied pressure gradient is aligned with the x-direction, and fluid enters and exits the fracture through the y–z plane. The walls of the fracture lie in the x–y plane (shaded area). (b) Parabolic velocity profile in the direction of the applied pressure gradient.

The simplest conceptual model of a fracture, for hydrological purposes, is that of two smooth, parallel walls separated by a uniform aperture, h (Fig. 2.1a). This geometry is often referred to as the "parallel plate" model. For this geometry, the Navier–Stokes equations can be solved *exactly* for the case in which there is a uniform pressure gradient within the fracture plane. To solve this problem, the x-axis is aligned with the pressure gradient, the y-axis taken to be perpendicular to the pressure gradient within the plane of the fracture, and the z-axis is taken to be normal to the fracture plane, with the fracture walls located at $z = \pm h/2$. Note that z is used here, and in most of Chapters 2–5, to denote the coordinate normal to the fracture plane, which is not necessarily the vertical direction. Since gravity has been eliminated from the governing equations by use of reduced pressure, the orientation of the fracture plane with respect to gravity is not relevant to the present discussion.

Although the advective acceleration term in general will render the Navier–Stokes equations nonlinear, this term *vanishes identically* in the parallel plate problem. This simplification arises from the fact that the velocity vector points in the x-direction, but the velocity varies only in the z-direction. Hence, the velocity gradient, which by definition points in the direction in which the velocity varies, points in the z-direction. The vectors $\mathbf{u}$ and $\nabla\mathbf{u}$ are therefore orthogonal to each other, and so the nonlinear term $\rho(\mathbf{u}\cdot\nabla)\mathbf{u}$ in eq. (2.3) is zero. Specifically, the velocity vector takes the form $\mathbf{u} = [u_x(z), 0, 0]$, and since the pressure gradient can be written as $\nabla p = [(\partial p/\partial x), 0, 0)]$, all terms in eqs. (2.5), (2.6) vanish, and eq. (2.4) reduces to

$$\mu\frac{\partial^2 u_x}{\partial z^2} = \frac{\partial p}{\partial x}. \tag{2.8}$$

Integration of this equation twice with respect to z yields

$$u_x = \frac{1}{2\mu}\left(\frac{\partial p}{\partial x}\right)z^2 + C_1 z + C_2, \tag{2.9}$$

where C_1 and C_2 are constants of integration. Imposition of the conditions that $u_x(z = \pm h/2) = 0$ leads to $C_1 = 0$ and $C_2 = -(\partial p/\partial x)h^2/8\mu$, and so the velocity profile (Fig. 2.1b) is given by Zimmerman and Bodvarsson (1996)

$$u_x = -\frac{1}{2\mu}\frac{\partial p}{\partial x}[(h/2)^2 - z^2], \tag{2.10}$$

with $u_y = u_z = 0$. Since $\partial p/\partial x$ was assumed to be constant, the velocity u_x as given by eq. (2.10) does not vary with x, and so it can readily be verified that this velocity profile satisfies the conservation of mass equation (2.7).

The total volumetric flux, with units of [m^3/s], is found by integrating the velocity across the width of the fracture:

$$Q_x = w \int_{-h/2}^{+h/2} u_x dz = -\frac{w}{2\mu}\frac{\partial p}{\partial x}\int_{-h/2}^{+h/2} \left[(h/2)^2 - z^2\right] dz = -\frac{wh^3}{12\mu}\frac{\partial p}{\partial x}, \tag{2.11}$$

where w is the depth of the fracture in the y-direction, normal to the pressure gradient. The term $wh^3/12$, with units of [m^4], is known as the fracture *transmissivity*, and is usually denoted by T. The transmissivity can be defined in general by the equation. $Q_x = -(T/\mu)(\partial p/\partial x)$, where T will take on a different form for fractures whose geometries deviate from that of the parallel plate model.

This expression for the hydraulic transmissivity of a channel bounded by two smooth parallel plates, $T = wh^3/12$, was first derived by the French mechanician Joseph Boussinesq (Boussinesq, 1868). As the transmissivity is proportional to the cube of the aperture, this result is known as the "cubic law." According to this law, the volumetric flux is proportional to the magnitude of the pressure gradient and inversely proportional to the fluid viscosity. The minus sign in eq. (2.11) indicates that the fluid flows in the direction in which the pressure decreases, *i.e.*, from higher to lower pressures. Although the name "cubic law" gives prominence to the prediction that T varies as the aperture to the third power, the cubic dependence of transmissivity on aperture is actually a necessary consequence of dimensional analysis. In this sense, the main nontrivial content of Boussinesq's result is the multiplicative factor of 1/12, which could not have been predicted by dimensional considerations alone.

As seen above, the ability of a fracture to conduct fluid is most naturally quantified in terms of its transmissivity. However, it is sometimes useful to think of the fracture as being a thin porous layer having thickness h, in which case the area of the fracture normal to the fluid flux vector is given by $A = hw$, and eq. (2.11) can be written as

$$Q_x = -\frac{h^2}{12}\frac{A}{\mu}\frac{\partial p}{\partial x}. \tag{2.12}$$

This expression is similar in form to Darcy's law for flow through a porous medium (Bear, 1988, Section 5.1; Zimmerman, 2018, Section 1.1), which states that

$$Q_x = -\frac{kA}{\mu}\frac{\partial p}{\partial x}, \tag{2.13}$$

where k is the *permeability* of the porous medium, with units of [m^2]. Comparison of these two expressions shows that $h^2/12$ can be identified as the "permeability of the parallel plate fracture."

It is also occasionally useful to make reference to the term $h^3/12$, which is essentially the transmissivity of the fracture per unit length in the direction within the fracture plane that is transverse to the pressure gradient. This term can be referred to as the *hydraulic conductivity*. Although the terminology is not standardized in the literature, in this book the "transmissivity" of the fracture will refer, for the case of a parallel plate fracture, to $wh^3/12$, the "hydraulic conductivity" of the fracture will refer to $h^3/12$, and the "permeability" of the fracture will refer to $h^2/12$.

2.3 The Stokes Equations

For realistic fracture geometries in which the walls are neither smooth nor parallel, the full Navier–Stokes equations are very difficult to solve, either analytically or numerically. Consequently, in order to analyze flow in "rough-walled" rock fractures, various approximations are usually made to reduce the Navier–Stokes equations to simpler equations that are more tractable.

As the main mathematical difficulties with the Navier–Stokes equations arise from the nonlinearity that is embodied in the advective acceleration term, $\rho(\mathbf{u}\cdot\nabla)\mathbf{u}$, the simplification process starts by investigating the conditions under which these terms can be ignored. For this analysis, it is convenient and sufficient to consider a "one-dimensional" fracture in which the aperture varies only in the x-direction, which is taken to be the same direction as the pressure gradient (Fig. 2.2). In this situation, the y-component of the velocity, u_y, will be zero. Although the out-of-plane velocity u_z must be zero *on average*, its local value may be nonzero. Hence, for steady-state conditions, the two non-trivial Navier–Stokes equations are

$$\rho\left(u_x\frac{\partial u_x}{\partial x}+u_z\frac{\partial u_x}{\partial z}\right)=-\frac{\partial p}{\partial x}+\mu\left(\frac{\partial^2 u_x}{\partial x^2}+\frac{\partial^2 u_x}{\partial z^2}\right), \tag{2.14}$$

$$\rho\left(u_x\frac{\partial u_z}{\partial x}+u_z\frac{\partial u_z}{\partial z}\right)=-\frac{\partial p}{\partial z}+\mu\left(\frac{\partial^2 u_z}{\partial x^2}+\frac{\partial^2 u_z}{\partial z^2}\right). \tag{2.15}$$

Without attempting to solve these equations, or even necessarily choosing a specific fracture geometry, it is possible to conduct an order-of-magnitude analysis to decide if the inertia terms can be neglected, following the approach outlined by Schlichting (1968, p. 109). If U_x and U_z are the "characteristic" velocities in the x- and z-directions, λ is the characteristic length scale in the x-direction, and h_m is the mean aperture, then the magnitude of the four velocity gradient terms appearing in eq. (2.14) can be estimated as follows:

$$mag\left[\rho u_x\frac{\partial u_x}{\partial x}\right]\approx\frac{\rho U_x^2}{\lambda},\quad mag\left[\rho u_z\frac{\partial u_x}{\partial z}\right]\approx\frac{\rho U_x U_z}{h_m}, \tag{2.16}$$

$$mag\left[\mu\frac{\partial^2 u_x}{\partial x^2}\right]\approx\frac{\mu U_x}{\lambda^2},\quad mag\left[\mu\frac{\partial^2 u_x}{\partial z^2}\right]\approx\frac{\mu U_x}{h_m^2}. \tag{2.17}$$

The relative magnitudes of the two viscous terms, as given by eq. (2.17), depend on the ratio of mean aperture to wavelength, h_m/λ. But small wavelengths always correspond to small values of roughness (Brown and Scholz, 1985), and small-scale roughness at small spatial wavelengths is known to be irrelevant for laminar flow (Schlichting, 1968, p. 580). So, the relevant case is $\lambda > h_m$, from which it follows that the second term in eq. (2.17) is the

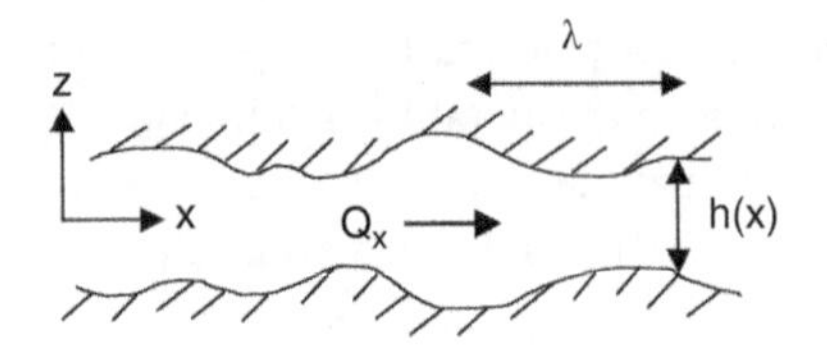

Figure 2.2 Generic rough-walled fracture, with a local aperture, $h(x)$, that varies only in the x-direction, and a "characteristic wavelength," λ.

larger of the two viscous terms. Therefore, for both inertia terms to be negligible compared to the dominant viscous term, it is necessary that

$$\frac{\rho U_x^2}{\lambda} \ll \frac{\mu U_x}{h_m^2}, \quad \frac{\rho U_x U_z}{h_m} \ll \frac{\mu U_x}{h_m^2}, \tag{2.18}$$

which can be written in dimensionless form as

$$\frac{\rho U_x h_m^2}{\mu \lambda} \ll 1, \quad \frac{\rho U_z h_m}{\mu} \ll 1. \tag{2.19}$$

To go further with this analysis, it is necessary to choose values for the characteristic velocities, U_x and U_z. An obvious choice for U_x is the mean velocity in the x-direction. But, as the *mean* value of u_z is necessarily zero, the mean value of u_z is *not* a sensible choice for the "characteristic velocity" in the z-direction. An estimate of the *absolute* value of u_z would be more useful in this context. Consider the simple case of a sawtooth fracture profile, as shown in Fig. 2.3a. This geometry is locally similar to the problem of flow between two smooth walls that converge (or diverge) to a point, which is the classical Jeffery–Hamel problem. According to the exact solution of this problem (White, 2006, pp. 183–87), at low velocities, the fluid flows in straight paths toward (or away from) the vertex, which yields the following (very) rough estimate:

$$\frac{U_z}{U_x} \approx \frac{h_{max} - h_{min}}{\lambda/2} \approx \frac{h_m}{\lambda}. \tag{2.20}$$

Making use of this estimate of $U_z = U_x h_m/\lambda$, both conditions in eq. (2.19) essentially reduce to

$$\frac{\rho U_x h_m}{\mu}\frac{h_m}{\lambda} = Re\frac{h_m}{\lambda} \ll 1, \tag{2.21}$$

where $Re = \rho U_x h_m/\mu$ is the *Reynolds number*, and $Re\, h_m/\lambda$ is a dimensionless parameter that is sometimes referred to as the "reduced Reynolds number" (Schlichting, 1968, p. 109).

This is about as far as it is possible to go with this type of rough order-of-magnitude analysis, which shows that the inertia terms will be negligible if the reduced Reynolds number is "small." Quantitative estimates of the magnitude of $Re\, h_m/\lambda$ that may be allowable before the inertial forces are non-negligible can be obtained by examining analytical solutions to the Navier–Stokes equations for simplified geometries that to some extent mimic the shape

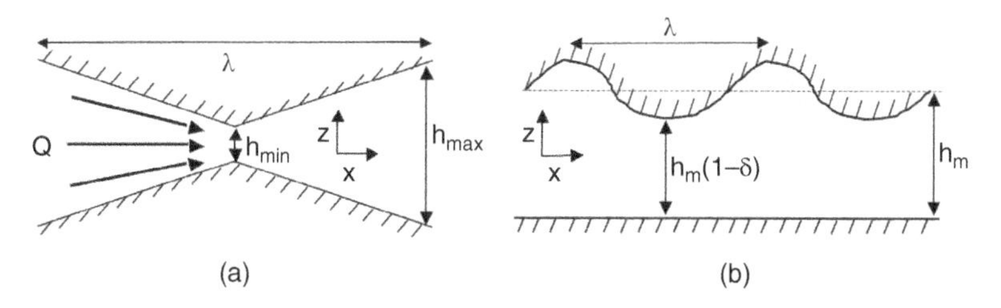

Figure 2.3 (a) Unit cell of a sawtooth-type fracture profile. According to the Jeffery–Hamel solution, the streamlines in a diverging or converging channel at low *Re* are radial lines converging on the vertex; this fact allows an estimation of the ratio U_z/U_x. (b) Fracture bounded by one smooth wall and one sinusoidal wall. The mean aperture is h_m, the wavelength is λ, and the relative amplitude of the aperture variation is δ.

of rough-walled fractures. One such solution that is useful in this context is the second-order perturbation solution derived by Hasegawa and Izuchi (1983) for flow through a channel that is bounded by one flat wall and one sinusoidal wall (Fig. 2.3b). The aperture of this fracture is described by

$$h(x) = h_m[1 + \delta \sin(2\pi x/\lambda)], \tag{2.22}$$

in which the dimensionless aperture variation δ may vary between 0 and 1. Hasegawa and Izuchi computed a few terms in the perturbation solution for the flow field, using Re and h_m/λ as their two "small" perturbation parameters. Written in the present notation, the transmissivity of this fracture was found to be given by

$$T = \frac{wh_m^3(1-\delta^2)^{5/2}}{12[1+(\delta^2/2)]}\left\{1 - \frac{3\pi^2(1-\delta^2)\delta^2}{5[1+(\delta^2/2)]}\left[1 + \frac{13(1-\delta^2)^5}{8085[1+(\delta^2/2)]^2}Re^2\right]\left(\frac{h_m}{\lambda}\right)^2\right\}. \tag{2.23}$$

The term outside of the curly brackets is the expression that results from integrating the "local" version of the cubic law along the length of the fracture; it is derived in Section 2.4. In order for the term that depends on the square of the reduced Reynolds number inside the {} brackets to be less than 10% of the dominant term, 1, it is necessary that

$$\frac{39\pi^2(1-\delta^2)^6\delta^2}{40{,}425[1+(\delta^2/2)]^3}Re^2\frac{h_m^2}{\lambda^2} < 0.1, \tag{2.24}$$

which is equivalent to the condition

$$\frac{(1-\delta^2)^3\delta}{[1+(\delta^2/2)]^{3/2}}Re\frac{h_m}{\lambda} < 3.24. \tag{2.25}$$

If $\delta = 0$, which reduces the sinusoidal geometry to the parallel plate geometry, this condition will be satisfied for *any* value of the Reynolds number. This result corresponds to the fact that the parabolic velocity profile, which leads to a constant transmissivity and a linear relation between the pressure gradient and the flow rate, will satisfy the Navier–Stokes equations for flow between smooth, parallel walls for *any* value of Re. At values of Re greater than about 2000, the parabolic velocity profile becomes unstable within a smooth-walled channel, and turbulent flow will occur. But such high flowrates are not of much relevance to fluid flow through subsurface rock fractures, except in certain exceptional circumstances.

The pre-factor on the left side of eq. (2.25) that depends on δ has a maximum value of 0.216. It follows that condition (2.25) will be satisfied for all values of roughness if

$$Re\frac{h_m}{\lambda} < 15. \tag{2.26}$$

According to this result, deviations in transmissivity due to inertial effects will remain less than 10% for reduced Reynolds numbers as high as 15. If the definition of "negligible" is taken to be a 1% deviation, criterion (2.26) would be replaced by $Re\,h_m/\lambda < 5$.

If the wavelength is long enough, eq. (2.26) shows that the nonlinear effects will be negligible at *any* value of Re, again ignoring the turbulence that would occur at very high values of Re. On the other hand, since appreciable roughness will not be expected to occur at wavelengths less than the mean aperture, the "worst case" in eq. (2.26) will occur when $h_m \approx \lambda$, in which case the allowable upper limit on Re is about 15. In summary, the foregoing analysis leads to the conclusion that, regardless of the details of the fracture geometry, Reynolds number effects should be negligible when $Re < 15$. This estimate of the critical value of Re is close to the value of 10 that was suggested by Oron and Berkowitz (1998), based on a similar type of order-of-magnitude analysis.

Skjetne *et al.* (1999) solved the Navier–Stokes equations for a simulated one-dimensional fracture at values of Re ranging from 0–52. They used a "self-affine" fracture with a Hurst roughness exponent of 0.8, as is suggested by numerous experimental measurements on fractures in various materials. It should be noted, however, that the mid-plane of their fracture exhibited large tortuosity at wavelengths of about $\lambda \approx 10h_m$, with amplitudes on the order of h_m. They found that the fracture transmissivity decreased by about 10% when $Re \approx 7$. This critical value of Re is of the same order of magnitude as the value of 15 derived above; the slight discrepancy may be due to the macroscopic tortuosity of their fracture.

The criterion given by eq. (2.26) is also in rough agreement with the experimental findings of Iwai (1976) for flow through a tension fracture in granite. Iwai's data are plotted in Fig. 2.4a in terms of the normalized transmissivity, $(h_H/h_m)^3$, where the "hydraulic aperture," h_H, is defined such that the transmissivity is given exactly by $T = wh_H^3/12$. The horizontal axis in Fig. 2.4a is the Reynolds number, Re, computed after dividing Iwai's Reynolds numbers by 2, so as to convert his values to the present definition, which is based on fracture aperture rather than half-aperture. The normalized transmissivity begins to decrease when Re is greater than about 10, as expected, although the trend is obscured by the large amount of scatter in the data at low values of Re, which occurs due to difficulty in accurately measuring the small flowrates and small pressure drops in this regime. In Fig. 2.4b, the data are re-plotted with all normalized transmissivities for $Re < 10$ replaced by their mean value, 0.817; this plot shows that the systematic deviations from linear behavior (*i.e.*, constant transmissivity) begin at about $Re \approx 10$.

This critical value of 10 is in rough agreement with the estimate derived above, as well as with the estimate derived by Oron and Berkowitz (1998) and the numerical simulation results of Skjetne *et al.* (1999). Hence, it seems that, for Reynolds numbers less than about

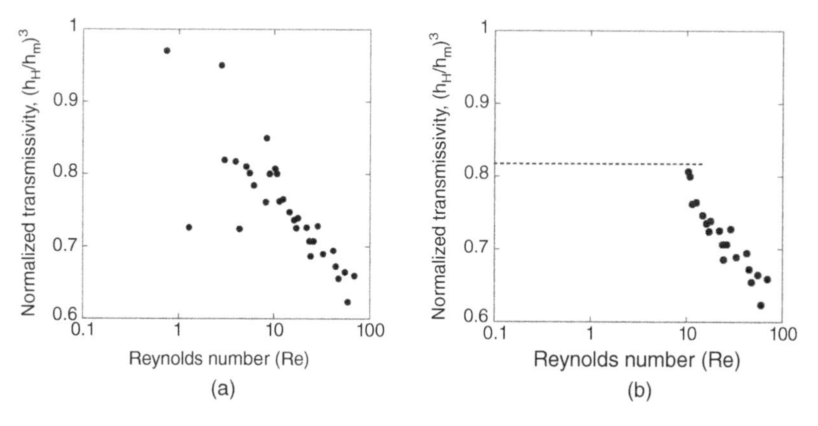

Figure 2.4 (a) Data from Iwai (1976), plotted as the ratio of the measured transmissivity to the cubic-law transmissivity calculated using the mean aperture. (b) Data re-plotted, with all normalized transmissivities for $Re < 10$ replaced by their mean value, 0.817.

10, the nonlinear effects inherent in the Navier–Stokes equations can be neglected, which leads to the so-called Stokes equations,

$$\nabla p = \mu \nabla^2 \mathbf{u}. \tag{2.27}$$

This set of three scalar equations must still be accompanied by the conservation of mass equation,

$$\frac{\partial u_x}{\partial x} + \frac{\partial u_y}{\partial y} + \frac{\partial u_z}{\partial z} = 0. \tag{2.28}$$

The Stokes equations are mathematically linear and are therefore easier to solve than the nonlinear Navier–Stokes equations. More fundamentally, because they are linear, the Stokes equations will always lead to a linear relation between pressure gradient and flowrate for any fracture geometry. This is equivalent to saying that the Stokes equations will lead to a fracture transmissivity that is independent of flowrate, *i.e.*, Darcy-like behavior.

2.4 The Reynolds Lubrication Equation

Although the Stokes equations are linear, they nevertheless comprise a set of four coupled partial differential equations and are therefore not trivial to solve. Consequently, these equations are usually reduced further to a single partial differential equation known as the *Reynolds lubrication equation*. This reduction can be made if the variations in aperture occur gradually within the plane of the fracture and can be motivated by an order-of-magnitude analysis that is similar to that which was used to reduce the Navier–Stokes equations to the Stokes equations.

The Stokes equations can be written in component form as

$$\frac{\partial^2 u_x}{\partial x^2} + \frac{\partial^2 u_x}{\partial y^2} + \frac{\partial^2 u_x}{\partial z^2} = \frac{1}{\mu}\frac{\partial p}{\partial x}, \tag{2.29}$$

$$\frac{\partial^2 u_y}{\partial x^2} + \frac{\partial^2 u_y}{\partial y^2} + \frac{\partial^2 u_y}{\partial z^2} = \frac{1}{\mu}\frac{\partial p}{\partial y}, \tag{2.30}$$

$$\frac{\partial^2 u_z}{\partial x^2} + \frac{\partial^2 u_z}{\partial y^2} + \frac{\partial^2 u_z}{\partial z^2} = \frac{1}{\mu}\frac{\partial p}{\partial z}. \tag{2.31}$$

Consider now a generic fracture that is bounded by two walls that are located at $z = h_1$ and $z = -h_2$, and whose aperture variation has characteristic wavelengths of λ_x and λ_y in the two directions. In the approximation process that leads from the Stokes equations to the Reynolds lubrication equation, the third Stokes equation, eq. (2.31), is ignored, based on the argument that since the mean value of u_z is zero, the local values will probably be negligibly small, and so eq. (2.31) will be satisfied by default.

The magnitudes of the second derivatives on the left side of eq. (2.29) can be estimated as

$$mag\left[\frac{\partial^2 u_x}{\partial x^2}\right] \approx \frac{U_x}{\lambda_x^2}, \quad mag\left[\frac{\partial^2 u_x}{\partial y^2}\right] = \frac{U_x}{\lambda_y^2}, \quad mag\left[\frac{\partial^2 u_x}{\partial z^2}\right] = \frac{U_x}{h_m^2}, \tag{2.32}$$

and similarly for eq. (2.30). If $\lambda_x^2 >> h_m^2$ and $\lambda_y^2 >> h_m^2$, the derivatives within the plane will be negligible compared to the derivative with respect to z, and eqs. (2.29) and (2.30) can be replaced by

$$\frac{\partial^2 u_x}{\partial z^2} = \frac{1}{\mu}\frac{\partial p}{\partial x}, \qquad \frac{\partial^2 u_y}{\partial z^2} = \frac{1}{\mu}\frac{\partial p}{\partial y}. \tag{2.33}$$

At any point (x, y) within the fracture plane, these two equations can be integrated twice with respect to z. Imposition of the no-slip boundary condition at the top and bottom walls, $z = h_1$ and $z = -h_2$, then leads to

$$u_x(x, y, z) = \frac{1}{2\mu}\frac{\partial p(x, y)}{\partial x}(z - h_1)(z + h_2), \tag{2.34}$$

$$u_y(x, y, z) = \frac{1}{2\mu}\frac{\partial p(x, y)}{\partial y}(z - h_1)(z + h_2). \tag{2.35}$$

This is essentially the same parabolic velocity profile that occurs for flow between parallel plates, except that the velocity vector is now aligned with the *local* pressure gradient, which is not necessarily collinear with the *global* pressure gradient.

The volumetric flux can be found by integrating the velocity across the fracture aperture. To do so, it is convenient to use a temporary variable, $\zeta = z + h_2$, that represents the vertical distance upwards from the bottom wall. The result is

$$h\bar{u}_x(x, y) = \int_{-h_2}^{h_1} u_x(x, y, z)dz = -\frac{h^3(x, y)}{12\mu}\frac{\partial p(x, y)}{\partial x}, \tag{2.36}$$

$$h\bar{u}_y(x, y) = \int_{-h_2}^{h_1} u_y(x, y, z)dz = -\frac{h^3(x, y)}{12\mu}\frac{\partial p(x, y)}{\partial y}, \tag{2.37}$$

where $h = h_1 + h_2$ is the total aperture, and the overbar indicates an average taken over the z direction.

Equations (2.36) and (2.37) represent an approximate solution to the Stokes equations but contain an unknown pressure field. A governing equation for the pressure field is found by appealing to the conservation of mass equation, $\nabla \cdot \mathbf{u} = 0$, which, however, applies to the local velocities, not the integrated values. But since $\nabla \cdot \mathbf{u} = 0$, the integral of $\nabla \cdot \mathbf{u}$ with respect to z must also be zero. Interchanging the order of the divergence and integration operations, which is valid as long as the velocity satisfies the no-slip boundary condition, then shows that the divergence of the z-integrated velocity, $\bar{\mathbf{u}}$, must also vanish (Zimmerman and Bodvarsson, 1996). Hence,

$$\frac{\partial(h\bar{u}_x)}{\partial x} + \frac{\partial(h\bar{u}_y)}{\partial y} = 0. \tag{2.38}$$

Insertion of eqs. (2.36) and (2.37) into eq. (2.38) yields, after canceling out the common factor 12μ,

$$\frac{\partial}{\partial x}\left(h^3\frac{\partial p}{\partial x}\right) + \frac{\partial}{\partial y}\left(h^3\frac{\partial p}{\partial y}\right) = 0, \tag{2.39}$$

which is known as the *Reynolds lubrication equation*. The Reynolds lubrication equation was first derived by the British engineer Osborne Reynolds (Reynolds, 1886) in the context of the flow of lubricating fluids through the small apertures between lubricated machinery parts. Equation (2.39) actually represents a special case of his equation for situations in which the two bounding surfaces are not moving relative to each other. A simpler but less rigorous derivation of eq. (2.39) can be made by first assuming *a priori* that the cubic law holds locally, at each location (x, y), and then invoking the principle of conservation of mass (Brown, 1987).

The key step in the reduction of the Stokes equations to the Reynolds equation was the assumption that $\lambda_x^2 >> h_m^2$ and $\lambda_y^2 >> h_m^2$, which would imply that the in-plane velocity derivatives are negligible compared to the velocity derivatives with respect to z. For the ratio of these terms to be roughly a factor of ten, this criterion requires that $h_m^2/\lambda_x^2 < 0.1$, *i.e.*, $h_m/\lambda_x < 0.3$, and likewise for λ_y. But this conclusion is based on an *a priori* estimate of the size of the various terms in the Stokes equations rather than on *solutions* to the Stokes equations. It is therefore instructive to again compare the criterion obtained by order-of-magnitude analysis with the predictions of the Hasegawa–Izuchi solution, eq. (2.23), for the one-dimensional fracture bounded by a smooth wall and a sinusoidal surface (Fig. 2.3b). Ignoring the terms that depend on *Re*, the transmissivity of this fracture is given to second-order in h_m^2/λ^2, as

$$T_{\text{Stokes}} = \frac{wh_m^3(1-\delta^2)^{5/2}}{12[1+(\delta^2/2)]}\left[1 - \frac{3\pi^2(1-\delta^2)\delta^2}{5(1+\delta^2/2)}\left(\frac{h_m}{\lambda}\right)^2\right]. \tag{2.40}$$

The term in front of the large square brackets is, in fact, the transmissivity that is predicted by the Reynolds lubrication equation. This is to be expected, since, as shown above, the Stokes equations reduce to the Reynolds equation in the limit as the wavelength of the aperture variations becomes infinite, in which case the h_m/λ term drops out of eq. (2.40).

It follows from eq. (2.40) that the criterion for the Reynolds and Stokes transmissivities to differ by no more than 10% is

$$\frac{3\pi^2(1-\delta^2)\delta^2}{5(1+\delta^2/2)}\left(\frac{h_m}{\lambda}\right)^2 < 0.1. \tag{2.41}$$

The pre-factor term that depends on the roughness parameter δ never exceeds about 1.1, and so the approximate condition for the Reynolds approximation to hold can be written as

$$\frac{h_m}{\lambda} < 0.3, \tag{2.42}$$

which agrees exactly with the results of the order-of-magnitude analysis presented above.

Insofar as flow through a fracture can be represented by the Reynolds equation, eq. (2.39) shows that the problem of finding the hydraulic conductivity of a variable-aperture fracture reduces to the problem of finding the effective conductivity of a heterogeneous two-dimensional conductivity field, with h^3 playing the role of the conductivity. The term $h^3/12$ can therefore be thought of as the "local hydraulic conductivity." For convenience in calculations, the factor 12 can be ignored and reinserted later if necessary. Using variational principles, it can be proven that the hydraulic aperture, defined so that $T = wh_H^3/12$, is bounded above and below as follows (Beran, 1968, p. 242):

$$\langle h^{-3}\rangle^{-1} \le h_H^3 \le \langle h^3\rangle, \tag{2.43}$$

where $\langle x \rangle \equiv x_m$ is the arithmetic mean value of the quantity x. The lower bound, $\langle h^{-3} \rangle^{-1}$, which is known as the *harmonic* mean of h^3, corresponds to a geometry in which the aperture varies only in the direction of flow, whereas the upper bound, $\langle h^3 \rangle$, which is the *arithmetic* mean of h^3, corresponds to aperture variation only in the direction transverse to the flow (Neuzil and Tracy, 1981; Silliman, 1989). These bounds are important because they are among the few results pertaining to flow in rough-walled fractures that are rigorously known, but they are usually too far apart to be quantitatively useful. For example, the hydraulic aperture of any real fracture is always less than the mean aperture, *i.e.*, $h_H^3 \leq \langle h \rangle^3$. But, since $\langle h \rangle^3 < \langle h^3 \rangle$ for any nonuniform aperture distribution, the Beran bounds alone are not sufficiently powerful to show that $h_H^3 \leq \langle h \rangle^3$.

Estimates of the hydraulic aperture can be obtained in terms of the mean $h_m \equiv \langle h \rangle$ and the standard deviation $\sigma_h \equiv \langle (h - h_m)^2 \rangle^{1/2}$ of the aperture distribution function. Elrod (1979) used Fourier transforms to solve the Reynolds equation for a fracture with an aperture having "sinusoidal ripples in two mutually perpendicular directions" and showed that, for the isotropic case,

$$h_H^3 = h_m^3 \left(1 - \frac{3}{2} \frac{\sigma_h^2}{h_m^2} + \ldots \right). \tag{2.44}$$

Zimmerman *et al.* (1991) considered the case of small regions of uni-directional ripples, which were then randomly assembled, and found results that agreed with eq. (2.44) up to second-order for both sinusoidal and sawtooth profiles. Furthermore, eq. (2.44) is consistent with the results derived by Landau and Lifshitz (1960, pp. 45–6), who required only that the aperture field (*i.e.*, conductivity field) be continuous and differentiable.

It is therefore very plausible that eq. (2.44) is accurate to second-order in the relative roughness parameter, σ_h/h_m. However, because it does not contain the higher-order terms in the series, it predicts physically unrealistic *negative* effective conductivities for sufficiently large values of the relative roughness. An alternative expression that agrees with eq. (2.44) up to second-order, but which does not yield unrealistic negative values for large values of the relative roughness, is (Renshaw, 1995)

$$h_H^3 = h_m^3 \left(1 + \frac{\sigma_h^2}{h_m^2} \right)^{-3/2}. \tag{2.45}$$

However, the hydraulic aperture predicted by eq. (2.44) will only become negative if $(\sigma_h/h_m)^2 > 2/3$, which is to say, if $(\sigma_h/h_m) > 0.816$. It is difficult to imagine a geologically plausible aperture profile with such a large value of relative roughness; see Problem 2.3. Note that, according to Brown (1987), "upper crustal conditions correspond roughly to $0.24 < (\sigma_h/h_m) < 0.5$."

It has often been suggested that the hydraulic aperture can be well approximated by the geometric mean value of the aperture distribution function, h_G, which is defined in general by $\ln h_G = \langle \ln h \rangle$, which is to say, $h_G = \exp\langle \ln h \rangle$. For example, it can be proven (Matheron, 1967) that the effective conductivity of a heterogeneous two-dimensional medium having a lognormal conductivity distribution is *exactly* equal to the geometric mean of the conductivity distribution. In the context of fracture flow, again ignoring the trivial factor of 1/12, this result can be expressed as $(h_H)^3 = \exp\langle \ln(h^3) \rangle$, which leads to

$$(h_H)^3 = \exp\langle \ln(h^3) \rangle = \exp\langle 3\ln(h) \rangle = \exp(3\ln h_G) = \exp\left[\ln(h_G)^3\right] = (h_G)^3, \qquad (2.46)$$

which is to say that the hydraulic aperture is equal to the geometric mean aperture. However, it is not clear that eq. (2.46) is more accurate than eq. (2.45) in the general case. For example, the numerical simulations of Piggott and Elsworth (1992) indicated that the geometric mean is a very poor predictor of the effective conductivity when the conductivity follows a *bimodal* distribution, *i.e.*, a distribution that has two local maxima.

For a fracture geometry in which the aperture varies only in the direction of the pressure gradient but not in the transverse direction, the Reynolds equation takes a simple form that lends itself to useful physical interpretations. After re-inserting the factors of $1/12\mu$ that had been canceled out to arrive at eq. (2.39), and setting the derivatives with respect to y to zero, eq. (2.39) takes the form

$$\frac{d}{dx}\left[\frac{h^3(x)}{12\mu}\frac{dp}{dx}\right] = 0. \qquad (2.47)$$

The term inside the parentheses is, according to eq. (2.36), equal to $-h\bar{u}_x(x)$, which can be denoted as $-q_x(x)$, the volumetric flux per unit length in the transverse direction. Equation (2.47) shows that this flux is constant along the length of the fracture, which merely reflects conservation of mass. So, eq. (2.47) can be written as

$$q_x = -\frac{h^3(x)}{12\mu}\frac{dp}{dx}. \qquad (2.48)$$

This equation can be rearranged and integrated from the inlet to the outlet of a fracture of finite length or over one wavelength of a fracture having a periodic profile, to find the overall pressure drop (Zimmerman *et al.*, 1991):

$$\int_0^L \frac{12\mu q_x}{h^3(x)}dx = -\int_0^L \frac{dp}{dx}dx,$$

$$12\mu q_x \int_0^L \frac{dx}{h^3(x)} = p(0) - p(L),$$

$$q_x = \frac{\Delta p}{12\mu L}\langle h^{-3} \rangle^{-1}. \qquad (2.49)$$

The hydraulic aperture of this "two-dimensional" fracture is therefore related to the harmonic mean of the cube of the aperture through the relation $h_H^3 = \langle h^{-3} \rangle^{-1}$. This is equivalent to saying that the global conductivity is equal to the harmonic mean of the local conductivities. This result lends itself to the interpretation that a fracture in which the aperture varies only in the direction of the pressure gradient acts like a set of infinitesimal hydraulic "resistors" arranged in series.

If the aperture varies only in the direction *perpendicular* to the applied pressure gradient, the hydraulic aperture can be shown (see Problem 2.4) to equal the cube root of the arithmetic mean of the cube of the aperture, *i.e.*, $h_H^3 = \langle h^3 \rangle$, and the fracture can be thought of as being composed of a set of resistors arranged in parallel (Neuzil and Tracy, 1981). However, it must be understood that these two heuristic results apply only if the aperture variations are sufficiently gradual such that the flow can be modeled by the Reynolds equation. In the more general case in which the full Stokes equations must be used, the concept of "local hydraulic conductivity" loses its meaning.

2.5 Effect of Contact Area

According to the results presented in Sections 2.3 and 2.4, regions of the fracture in which the aperture is small will cause the effective conductivity to decrease, as would be expected. Fluid flow through a fracture is also hindered by the presence of asperity regions at which the opposing fracture walls are in contact and the local aperture is zero. The geometric mean approximation will predict zero overall effective conductivity if there is a small but finite probability of having $h = 0$, and so this model breaks down if the fracture contains contact regions. Although the effect of contact regions on the overall effective conductivity can be approximated using eq. (2.44), with the aperture's mean and standard deviation calculated so as to include the regions of zero aperture within the fracture plane, more accurate models can be developed that treat contact regions differently than open regions (Walsh, 1981; Piggott and Elsworth, 1992; Zimmerman *et al.*, 1992).

If an effective hydraulic aperture, call it h_o, can be found for the "open" regions of the fracture, using, for example, the equations described in Section 2.4, the additional effect of contact regions can be modeled by assuming that the fracture consists of regions having aperture $h = h_o$, and regions of aperture $h = 0$ (Fig. 2.5). This conceptual model can also apply to a fracture that has been hydraulically fractured, where the proppant particles sometimes form discrete pillars that prop open the fracture (d'Huteau *et al.*, 2011; Wang and Elsworth, 2018). In this case, the "proppant pillars" themselves have a very low permeability compared to the open fracture and can be idealized as impermeable posts. If the flowrate is sufficiently low, *i.e.*, $Re\, h_m/\lambda < 1$, and the characteristic in-plane dimension a of the contact regions or proppant pillars is much larger than the aperture, *i.e.*, $h_o/a \ll 1$, then flow in the open regions is governed by eq. (2.39), with $h = h_o = \text{const}$, in which case the Reynolds equation reduces to Laplace's equation,

$$\nabla^2 p = \frac{\partial^2 p}{\partial x^2} + \frac{\partial^2 p}{\partial y^2} = 0. \tag{2.50}$$

Boundary conditions must be prescribed along the contours Γ_i in the (x,y) plane that form the boundaries of the contact regions. As no fluid can enter these regions, the component

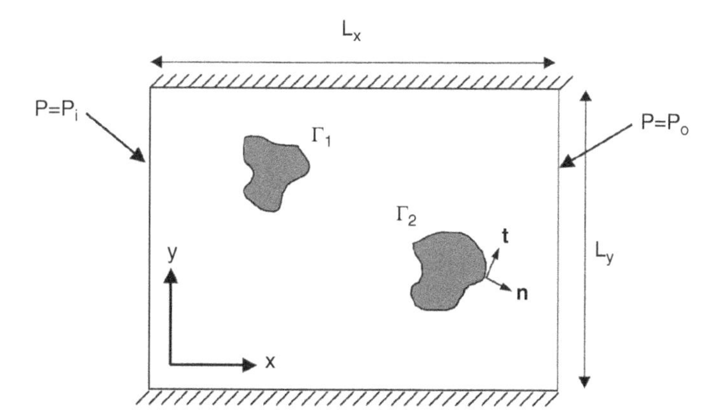

Figure 2.5 Schematic diagram of the Hele-Shaw model, with impermeable boundaries at $y = 0$ and $y = L_y$, constant-pressure boundaries at $x = 0$ and $x = L_x$, and impermeable boundaries along the contours of the asperity regions, Γ_i.

of the velocity vector that is normal to these contours must be zero. The velocity vector is parallel to the local pressure gradient, as shown by eqs. (2.36) and (2.37), and so the appropriate boundary condition for eq. (2.50), along each contour Γ_i, is that the component of the pressure gradient normal to the contour must be zero, *i.e.*,

$$\frac{\partial p}{\partial n} \equiv (\nabla p) \cdot \mathbf{n} = 0, \tag{2.51}$$

where $\mathbf{n}$ is the outward unit normal vector to Γ_i, and n is the coordinate in the direction of $\mathbf{n}$. This mathematical model of flow between two smooth parallel plates that are propped open by cylindrical posts, as described by eqs. (2.50) and (2.51), is known as the *Hele-Shaw model* (Bear, 1988, pp. 687–92).

Boundary condition (2.51) ensures that no flow enters the contact regions, but the *no-slip* boundary condition also requires that the *tangential* component of the velocity should vanish, which implies that the tangential component of the pressure gradient must vanish, *i.e.*, $(\nabla p)\cdot\mathbf{t} = 0$, where $\mathbf{t}$ is a unit vector perpendicular to $\mathbf{n}$, in the (x,y) plane. However, when solving Laplace's equation, it is not possible to simultaneously impose boundary conditions on both the normal and tangential components of the derivative (Bers *et al.*, 1964, pp. 152–4). So, solutions to the Hele-Shaw equations will generally not satisfy $(\nabla p)\cdot\mathbf{t} = 0$, and therefore do not account for viscous drag along the sides of the asperities. The relative error induced by this incorrect boundary condition is on the order of h_o/a (Thompson, 1968; Kumar *et al.*, 1991) and is negligible for real rock fractures. For example, Pyrak-Nolte *et al.* (1987) observed apertures in a fracture in crystalline rock that were on the order of 10^{-4}–10^{-4} m and contact dimensions (in the fracture plane) that were on the order of 10^{-1}–10^{-3} m, *i.e.*, at least ten times larger than the aperture. Gale *et al.* (1990) measured apertures and asperity dimensions on a natural fracture in a granite from Stripa, Sweden, under a normal stress of 8 MPa, and found $h \approx 0.1$ mm, and $a \approx 1.0$ mm. These observations imply that viscous drag along the sides of asperities will be negligible compared to the drag along the upper and lower fracture walls, which is consistent with the mathematical formulation of the Hele-Shaw model.

Walsh (1981) used the solution for potential flow (*i.e.*, flow governed by Laplace's equation) around a single circular obstruction of radius a (Carslaw and Jaeger, 1959, p. 426), along with the effective medium approximation proposed by Maxwell, to develop the following estimate of the influence of contact area on the fracture's hydraulic conductivity:

$$h_H^3 = h_o^3\left(\frac{1-c}{1+c}\right), \tag{2.52}$$

where c is the fraction of the fracture plane occupied by (circular) contact regions. This expression was shown numerically by Zimmerman *et al.* (1992) to be highly accurate for contact fractions up to at least 0.25. If the contact regions are in the form of randomly oriented ellipses of aspect ratio $\alpha \le 1$, an analysis similar to that conducted by Walsh leads to (Zimmerman *et al.*, 1992)

$$h_H^3 = h_o^3\left(\frac{1-\beta c}{1+\beta c}\right), \quad \text{where } \beta = \frac{(1+\alpha)^2}{4\alpha}. \tag{2.53}$$

As the ellipses become more elongated, the factor β increases and the hydraulic aperture decreases. Hence, elliptical contact regions lead to a greater decrease in hydraulic conductivity than do circular contact regions having the same area, since the factor β takes on its minimum value when $\alpha = 1$. In fact, the bounds derived by Hashin and Shtrikman (1962) imply that, for a given contact fraction c, circular contact regions will cause minimum possible decrease in hydraulic conductivity, compared to *all* possible shapes, not only ellipses. Thorpe (1992) analyzed the case in which the regions of zero conductivity were shaped like regular n-sided polygons (*i.e.*, triangles, squares, *etc.*) and found β values of 1.291 for triangles, 1.094 for squares, and 1.044 for pentagons, rapidly approaching 1 as $n \rightarrow \infty$ and the polygons became circles.

Although contact areas in fractures are not perfectly elliptical, Zimmerman *et al.* (1992) showed that eq. (2.53) may be used if the shapes of the actual contact areas are replaced by "equivalent" ellipses that have the same perimeter/area ratio. In the context of hydraulic fractures that are propped open, a more general model for proppant pillars that have *finite* hydraulic conductivity can be obtained using the results of Zimmerman (1996), who investigated the effective conductivity of a two-dimensional medium containing elliptical inclusions having finite conductivity.

2.6 Accuracy of the Lubrication Model

As explained in Sections 2.3 and 2.4, if the fracture surface and aperture profile are not "too rough" and the flowrate is not "too high," the Navier–Stokes equations can be reduced to the Reynolds lubrication equation, which can be solved easily using numerical methods and is also amenable to analytical treatment. The extent to which actual rock fractures satisfy the geometric constraints that allow the Navier–Stokes simulations to be reduced to the Reynolds lubrication equation, at low Reynolds numbers, $Re \ll 1$, will be discussed in this section. Deviations from the predictions of the various simplified models (Stokes equations, Reynolds lubrication equation) that occur at higher Reynolds numbers will be discussed in Chapter 4.

Consider a two-dimensional fracture in which the aperture varies sinusoidally in the direction of flow (Fig. 2.6a), according to

$$h(x) = h_m[1 + \delta \sin(2\pi x/\lambda)], \tag{2.54}$$

where h_m the mean aperture, λ is the wavelength, and δ is the relative amplitude of the aperture variation. The transmissivity that is predicted by direct integration of the Reynolds lubrication equation is (Zimmerman *et al.*, 1991)

$$T_{\text{Reynolds}} = \frac{wh_m^3(1-\delta^2)^{5/2}}{12[1+(\delta^2/2)]}. \tag{2.55}$$

Sisavath *et al.* (2003), following the approach of Kitanidis and Dykaar (1997), used a second-order perturbation solution to the Stokes equations, with h_m/λ as the small perturbation parameter, to find the following approximate expression for the transmissivity of this sinusoidal fracture:

$$T_{\text{Stokes}} = \frac{wh_m^3(1-\delta^2)^{5/2}}{12[1+(\delta^2/2)]}\left\{1 + \frac{3\pi^2(1-\delta^2)\delta^2}{5[1+(\delta^2/2)]}\left(\frac{h_m}{\lambda}\right)^2\right\}^{-1}. \tag{2.56}$$

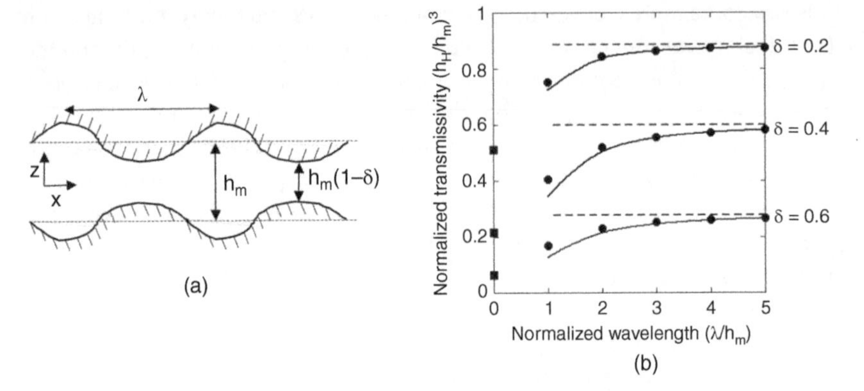

Figure 2.6 (a) Fracture bounded by two in-phase sinusoidal walls. The mean aperture is h_m, the wavelength is λ, and the relative amplitude of the aperture variation is δ. (b) Normalized transmissivity, as a function of the normalized wavelength λ/h_m, as predicted by the Reynolds lubrication equation (dashed lines), the second-order perturbation solution (solid lines), the stagnant limit lower bound (■), and finite element solutions of the Stokes equations (•) (Data taken from Sisavath *et al.*, 2003).

The "Stokes" transmissivity, eq. (2.56), lies below the Reynolds transmissivity, eq. (2.55), and converges to the value given by the Reynolds lubrication model as the wavelength increases.

Sisavath *et al.* (2003) also solved the Stokes equations for flow through this type of fracture using a finite element code, FLUIDITY. They imposed a parabolic velocity profile at the entrance of the fracture, which comprised three identical sinusoidal wavelengths in series. After verifying that the flow had become fully developed by the third wavelength, and that the pressure drop per wavelength had stabilized between the second and third wavelengths, the mean pressure drop and volumetric flowrate were computed by integrating the pressures and velocities over the inlet and outlet of the third wavelength. In order to achieve a mesh-independent result, a grid of $40 \times 500 = 20{,}000$ quadratic elements per wavelength was found to be sufficient.

Their results showed that, as expected from the discussion in Section 2.3, the lubrication model becomes increasingly accurate as the wavelength increases (Fig. 2.6b). The numerically computed transmissivities are always lower than those predicted by the lubrication model, with the discrepancy increasing as the wavelength decreases. The actual trend of the decrease in transmissivity with decreasing wavelength is reasonably well captured by the perturbation solution, eq. (2.56). For a fracture having $\delta = 0.2$, the Reynolds lubrication prediction is accurate to within 10% for all wavelengths that exceed about 1.5 times the mean aperture, *i.e.*, for $\lambda > 1.5h_m$. For a much rougher fracture having $\delta = 0.6$, the Reynolds lubrication model is accurate to within 10% only for $\lambda > 3h_m$. Since a value of δ as large as 0.6 would seem to be an upper bound on the level of aperture variation that might reasonably be expected in a rock fracture, these results imply that the Reynolds equation will be accurate to within 10% for $\lambda > 3h_m$. This conclusion agrees with the estimate given by eq. (2.42), which was obtained in Section 2.4 based on an order-of-magnitude analysis of the terms in the Navier–Stokes equations.

Also shown in Fig. 2.6b, for comparison, is the transmissivity computed from the "stagnant limit," in which fluid is assumed to flow only through the straight central channel defined by $|z| < h_m(1-\delta)/2$, and is stagnant (*i.e.*, has zero velocity) in the side pockets, where $h_m(1-\delta)/2 < |z| < h_m(1+\delta)/2$. For the sinusoidal geometry, the stagnant limit corresponds to a hydraulic aperture of $h_m(1-\delta)$. Brown *et al.* (1995) suggested that the stagnant limit would provide a lower bound on the actual transmissivity, and this indeed seems to be the case.

The sinusoidal fractures discussed above are not entirely realistic due to their lack of aperture variation in the transverse direction and their lack of smaller-scale roughness. It is therefore more instructive to compare the predictions of the lubrication model to either experimentally measured transmissivities of real fractures or to transmissivities computed by solving the Navier–Stokes equations for aperture profiles obtained from actual rock fractures.

Yeo *et al.* (1998) tested the accuracy of the lubrication model by conducting flow tests through a 20 cm × 20 cm fracture cast made from a natural fracture in a red Permian sandstone and then simulating fluid flow through this fracture using the Reynolds equation. A cast was made of the fracture aperture, and the apertures were measured on the cast, using the microscope of a "Miniload hardness tester" made by Leitz Wetzlar, to within an accuracy of ±1 μm. Measurements were made every 5 mm in both the x and y directions, resulting in 1600 aperture values. For the case of zero shear displacement, the mean aperture was 607 μm, the standard deviation was 160 μm and the aperture closely followed a normal distribution. For the experiments in which fluid was injected into a central injection hole in the fracture plane, and collected at the outer boundaries of the 20 cm × 20 cm fracture region, the effective hydraulic conductivity that was computed by solving the Reynolds lubrication equation using the finite element method, when normalized against $h_m^3/12$, was 0.66, whereas the experimentally measured normalized hydraulic conductivity was 0.48. It is worth noting that the hydraulic apertures computed by numerically solving the lubrication equation generally agreed reasonably well with the predictions of eq. (2.44). Hence, although the lubrication equation was much more accurate than the naïve use of the cubic law based on the mean aperture, *i.e.*, $h_m^3/12$, it nevertheless overestimated the hydraulic conductivity by 38%.

Similar investigations have been carried out by other researchers, with roughly similar results. For example, Konzuk and Kueper (2004) created a tensile fracture in a slab of dolomitic limestone, with a resulting nominal fracture plane that had a length of 203 mm in the direction of flow and a width of 240 mm in the transverse direction. Detailed aperture measurements across the entire fracture plane yielded an aperture distribution that could be fit reasonably well with a normal or a lognormal distribution, with a mean aperture of 417 μm and a standard deviation of 198 μm. Flow experiments were conducted at Reynolds numbers below about 10, and it was verified that nonlinear effects were small and could be corrected for by extrapolating the measured transmissivities down to $Re = 0$. They then solved the lubrication equation using a finite difference method and found a transmissivity that was about 1.9 times *larger* than the measured value.

The Reynolds lubrication equation is elegant and convenient to use. As this model is equivalent to a two-dimensional conductivity problem with a heterogeneous local conductivity field, it naturally lends itself to various bounds and approximate "upscaling" formulas that are known to be quite accurate. However, comparison with flow measurements and detailed Navier–Stokes simulations for flow through actual rock fractures tends to show that although the lubrication model offers a substantial improvement over the naïve cubic law based on the mean aperture, it will generally overestimate the transmissivity, in some cases by as much as a factor of 2.

2.7 Fracture in a Permeable Matrix

The various analyses presented thus far in this chapter have assumed that the matrix rock adjacent to the fracture is hydraulically impermeable, in which case the fracture walls serve as both no-flow and no-slip boundaries. Some matrix rocks, such as granites, have such low permeabilities, in the range of 10^{-20} m^2, that neglecting flow in the matrix would clearly seem to be an acceptable approximation. Other rocks, however, have matrix permeabilities that are sufficiently large such that flow in the matrix cannot be neglected *a priori*. For these rocks, a pressure gradient aligned along the fracture plane would give rise to a flow component in the same direction, within the matrix, leading to an "apparent" fracture transmissivity that is greater than that which would exist if the matrix were impermeable. This apparent fracture transmissivity will depend on the aperture of the fracture and on the permeability of the adjacent rock. The following treatment of this problem follows the analysis given by Berkowitz (1989), with minor notational changes and one simplification, as explained below. The more general situation in which fluid may flow from the fracture into the adjacent permeable matrix, due to a pressure gradient *normal* to the fracture plane, is discussed to various degrees in Chapters 7–10.

Imagine a parallel plate fracture of aperture h, embedded in a porous rock that has permeability k. The pressure gradient is assumed to be oriented in some direction parallel to the fracture plane. The x-axis is taken to lie parallel to the pressure gradient, and the z-axis is oriented perpendicular to the fracture plane (Fig. 2.7). In this problem, there will obviously be no flow in the y-direction, which is oriented within the fracture plane perpendicular to the pressure gradient. Flow within the fracture, which is to say within the

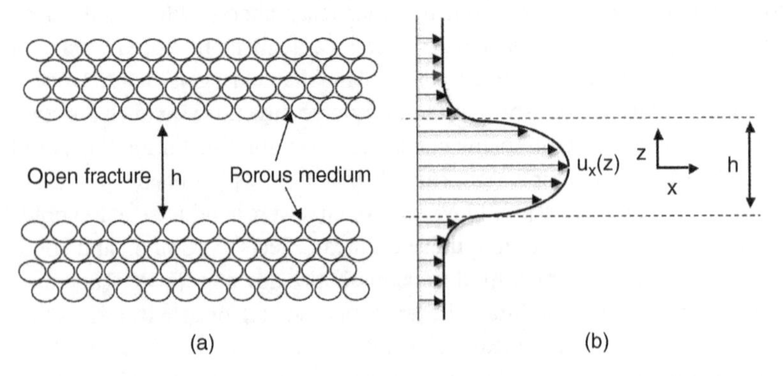

Figure 2.7 (a) Parallel plate fracture having aperture h, located in a permeable rock having permeability k. (b) Velocity profile in the fracture and the adjacent porous rock, according to eqs. (2.60) and (2.61).

region $-h/2 < z < h/2$, is governed by the Navier–Stokes equations, which take the form given by eq. (2.8):

$$\mu \frac{d^2 u_x}{dz^2} = \frac{dp}{dx}. \tag{2.57}$$

Note that the partial derivatives can be replaced by total derivatives in this problem, since u_x depends only on z, and p varies only with x.

Flow within the adjacent porous rock would normally be assumed to be governed by Darcy's law, $u_x = -(k/\mu)(dp/dx)$, where k is the permeability of the porous rock (Bear, 1988, Section 5.1). In the present problem, Darcy's law would predict a uniform velocity field of the form (u_x = constant, $u_y = u_z = 0$) within the porous rock, with no velocity gradient in the z-direction. However, it has been established experimentally that at the interface between a porous medium and a free fluid (*i.e.,* the fluid in the fracture), there exists a thin boundary layer within the porous medium in which the velocity varies with z (Beavers and Joseph, 1967). Such a gradient is inconsistent with Darcy's law, and so an extension of Darcy's law, known as the Brinkman equation, is often used to model the flow within a porous medium where it is in contact with a free fluid. Brinkman (1947) developed a generalized flow law for this type of problem, by assuming that both the Darcy pressure drop, $-\mu u_x/k$, and the Navier–Stokes pressure drop, $\mu(d^2u_x/dz^2)$, act simultaneously, thus arriving at the following equation:

$$-\frac{\mu}{k} u_x + \mu \frac{d^2 u_x}{dz^2} = \frac{dp}{dx}. \tag{2.58}$$

An "effective viscosity," $\overline{\mu} \neq \mu$, is sometimes used in the Navier–Stokes-like second-derivative term, to account for the fact that the rock is porous. The precise relationship between $\overline{\mu}$ and μ has long been a matter of debate (Kim and Russel, 1985; Martys, 2001), but for the purposes of developing a qualitative understanding of the effect of the permeable rock adjacent to the fracture, it will be assumed that $\overline{\mu} = \mu$.

The fluid velocity within the porous rock must approach the Darcy velocity at distances far from the fracture plane, and so the far-field boundary condition is

$$\lim_{z \to \pm\infty} u_x = -\frac{k}{\mu}\frac{dp}{dx} \equiv u_\infty. \tag{2.59}$$

It is assumed that both the fluid velocity and the shear stress are continuous at the interface between the fracture and the adjacent rock. The viscous shear stress is given by $\tau = \mu(du_x/dz)$, and so the two interface conditions are that both u_x and du_x/dz must be continuous at $z = \pm h/2$. Equations (2.57) and (2.58), along with the far-field boundary conditions (2.59), and the two sets of continuity conditions, can easily be solved (Berkowitz, 1989; see also Problem 2.5) to yield the following velocity profiles within the fracture, $-h/2 < z < h/2$, and within the porous rock, $|z| \geq h/2$ (Fig. 2.7b):

$$u_x(z) = u_\infty \left[1 + \frac{h}{2\sqrt{k}} + \frac{h^2}{8k} - \frac{z^2}{2k}\right], \text{for } |z| \leq h/2; \tag{2.60}$$

$$u_x(z) = u_\infty \left[1 + \frac{h}{2\sqrt{k}} \exp\left\{-\left[|z| - (h/2)\right] / \sqrt{k}\right\}\right], \text{ for } |z| > h/2. \tag{2.61}$$

The total flow along the fracture is found by integrating the velocity profile (2.60) across the fracture area that is normal to the flow:

$$Q_{fracture} = w\int_{-h/2}^{h/2} u_x(z)dz = w\,u_\infty \int_{-h/2}^{h/2}\left(1+\frac{h}{2\sqrt{k}}+\frac{h^2}{8k}-\frac{z^2}{2k}\right)dz$$
$$= w\,u_\infty\left(h+\frac{h^2}{2\sqrt{k}}+\frac{h^3}{8k}-\frac{h^3}{24k}\right), \tag{2.62}$$

where w is the width in the y-direction, in the fracture plane, normal to the flow. But from eq. (2.59), $u_\infty = -(k/\mu)(dp/dx)$, and so the volumetric flowrate through the fracture can be written as

$$Q_{fracture} = -w\frac{k}{\mu}\frac{dp}{dx}\left(h+\frac{h^2}{2\sqrt{k}}+\frac{h^3}{12k}\right) = -\frac{wh^3}{12\mu}\frac{dp}{dx}\left(1+\frac{6\sqrt{k}}{h}+\frac{12k}{h^2}\right). \tag{2.63}$$

The term outside the brackets on the right is the volumetric flowrate through a parallel plate fracture embedded in an *impermeable* rock, as given by eq. (2.11). The second and third terms inside the brackets represent an additional fractional flux that occurs in the fracture, due to the fluid in the adjacent matrix being "dragged along" by the fluid flowing through the fracture. Comparison of eq. (2.63) with eq. (2.11) shows that the apparent transmissivity of the fracture is given by (Timothy and Meschke, 2017; Singh and Cai, 2019; Hosseini and Khoei, 2021)

$$\frac{T(\text{fracture in a permeable rock})}{T(\text{fracture in an impermeable rock})} = 1+\frac{6\sqrt{k}}{h}+\frac{12k}{h^2}. \tag{2.64}$$

The second and third terms on the right side of eq. (2.64) are generally small. A rough estimate of their sizes can be obtained by appealing to the capillary tube model for the permeability of porous rock. In its simplest form, which assumes that the pore space consists of a random collection of tubular pores of diameter d, this model predicts that the permeability is given by $k = \phi d^2/96$, where ϕ is the porosity (Bear, 1988, p. 163). For rocks having porosities less than 0.2, for example, the term $6\sqrt{k}/h$ will be less than $0.1d/h$. Since the aperture of a fracture will typically be greater than the typical pore diameter of the host rock, the term $6\sqrt{k}/h$ will always be small, and the term $12k/h^2$ will be a smaller term of second order in d/h. Hence, the last term in eq. (2.64) can be neglected, yielding

$$\frac{T(\text{fracture in a permeable rock})}{T(\text{fracture in an impermeable rock})} = 1+\frac{6\sqrt{k}}{h}. \tag{2.65}$$

The Brinkman model also predicts the existence of an additional excess flux within the matrix, represented by the exponential term in eq. (2.61), due to the fluid in the rock matrix being "dragged along" by the fluid in the fracture. The additional total flux that occurs in the matrix is given by

$$\Delta Q_{matrix} = 2w\int_{h/2}^{\infty}\frac{u_\infty h}{2\sqrt{k}}e^{-[|z|-(h/2)]/\sqrt{k}}dz = u_\infty hw. \tag{2.66}$$

This is precisely equal to the flux that would have occurred within the region $-h/2<z<h/2$, if that region had been filled with matrix rock, rather than being occupied by the fracture. Hence, the total flux through the matrix, in the presence of the fracture, is *exactly the same* as it would have been if the fracture were not present.

2.8 Fracture Filled with Porous or Granular Material

The models presented thus far in this chapter have pertained to fractures that are "open," or which are obstructed at discrete locations by asperities at which the two rock faces are in contact. Some rock fractures, however, are filled with materials that can be considered, on length scales smaller than the fracture aperture, as being a porous medium having some permeability, k. For example, some fractures (see Fig. 1.5b) are filled with precipitated minerals, in which case the fractures are often called *veins* (Ramsay, 1980; Tokan-Lawal *et al.*, 2015). Some fractures are filled with fine-grained clay-like material known as fault gouge (Morrow *et al.*, 1984; Ikari *et al.*, 2009). When fractures are induced in a rock mass by hydraulic fracturing, for the purposes of increasing the flow of fluid to or from a well, granular material referred to as "proppant" is injected into the fracture to prevent it from closing due to the compressive *in situ* stresses (Sanematsu *et al.*, 2015). Each of these types of fractures can be idealized as a parallel plate fracture filled with a porous material (Fig. 2.8a).

Flow through such a "filled" fracture can be modeled using the same Brinkman equation that was used in Section 2.7 to model flow through an unfilled fracture adjacent to a porous, permeable rock. The governing differential equation of the Brinkman model is (Bird *et al.*, 1960, p. 150)

$$-\frac{\mu}{k}u_x + \mu\frac{d^2u_x}{dz^2} = \frac{dp}{dx}, \tag{2.67}$$

where k represents the permeability of the porous infilling material, as would be measured if this material were present in bulk form, rather than being confined within a narrow channel. The adjacent rock is considered to be impermeable, in which case no-slip boundary conditions are assumed to hold at the fracture walls, *i.e.*,

$$u_x = 0 \quad \text{at} \quad h = \pm h/2. \tag{2.68}$$

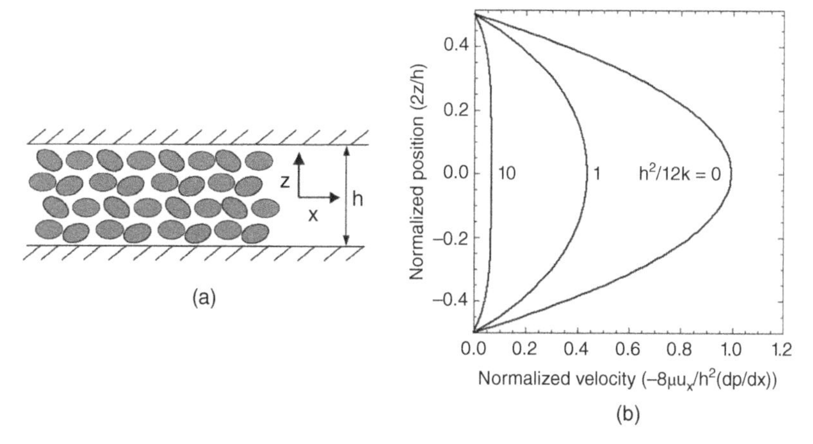

Figure 2.8 (a) Smooth-walled fracture of aperture h, filled with a permeable medium having permeability k. (b) Velocity profile, for different values of the ratio of "fracture permeability" to infill material permeability.

If there is a uniform pressure gradient along the x-axis of the fracture, the resulting velocity profile (Fig. 2.8b) is (Kumar *et al.*, 1991)

$$u_x(z) = -\frac{k}{\mu}\frac{dp}{dx}\left[1 - \frac{\cosh\left(z/\sqrt{k}\right)}{\cosh\left(h/2\sqrt{k}\right)}\right]. \tag{2.69}$$

If the permeability of the infill material is very large, the arguments of the two cosh terms become very small. Using the Taylor series expansion $\cosh \zeta \approx 1 + \zeta^2/2 + \ldots$ on both cosh terms then shows that the velocity profile reduces to

$$u_x(z) = -\frac{h^2}{8\mu}\frac{dp}{dx}[1 - (2z/h)^2], \tag{2.70}$$

which agrees with profile (2.10) for flow in the open space between two parallel plates. If the permeability of the infill material is very small, on the other hand, the arguments of both of the cosh terms become very large, in which case use of the asymptotic approximation $\cosh \zeta \approx e^{\zeta}/2$ in eq. (2.69) eventually leads to

$$u_x(z) = -\frac{k}{\mu}\frac{dp}{dx}, \tag{2.71}$$

except in an infinitesimally small boundary layer near the two fracture walls at $z = \pm h/2$, where the velocity must vanish. The profile given by eq. (2.71) corresponds to the "plug-flow" velocity profile that occurs for Darcy flow through a porous medium of permeability k. The velocity profile predicted by the Brinkman equation, eq. (2.69), therefore provides a smooth transition between the plug-like profile of Darcy flow in a porous medium, and the parabolic profile of Stokes flow in an open channel, as the permeability of the infill material varies (Fig. 2.8b).

The flux through this filled fracture is found by integrating the velocity across the area normal to the velocity, *i.e.*, across the y–z plane:

$$\begin{aligned} Q_x = 2w\int_0^{h/2} u_x(z)dz &= -\frac{2kw}{\mu}\frac{dp}{dx}\int_0^{h/2}\left[1 - \frac{\cosh\left(z/\sqrt{k}\right)}{\cosh\left(h/2\sqrt{k}\right)}\right]dz \\ &= -\frac{2kw}{\mu}\frac{dp}{dx}\left[z - \frac{\sqrt{k}\sinh\left(z/\sqrt{k}\right)}{\cosh\left(h/2\sqrt{k}\right)}\right]_0^{h/2} \\ &= -\frac{khw}{\mu}\frac{dp}{dx}\left[1 - \frac{2\sqrt{k}}{h}\tanh\left(h/2\sqrt{k}\right)\right]. \end{aligned} \tag{2.72}$$

Recalling from Section 2.2 that the "permeability" of an open parallel plate fracture is $h^2/12$, the term in brackets is seen to be a function of the ratio of permeability of the open fracture to the permeability of the infill material. If the permeability of the infill material is very large, the argument of the tanh term becomes small, and the use of the Taylor series $\tanh \zeta = \zeta - \zeta^3/3 + \ldots$ shows that the flux reduces to the value expected for flow through an open fracture,

$$Q_x = -\frac{h^3 w}{12\mu}\frac{dp}{dx}. \tag{2.73}$$

If the permeability of the infill material is very small, the argument of the tanh term becomes large, in which case tanh $\zeta \to 1$, and so the second term inside the brackets in eq. (2.72) goes to zero, and the flux reduces to the value expected for flow through a porous medium having permeability k and cross-sectional area $A = wh$ normal to the flow:

$$Q_x = -\frac{khw}{\mu}\frac{dp}{dx} = \frac{kA}{\mu}\frac{dp}{dx}. \tag{2.74}$$

The effective permeability of the fracture is readily found from eq. (2.72), when normalized against the permeability of the infilling material, to be given in the general case by (Fig. 2.9a)

$$\frac{k(\text{filled fracture})}{k(\text{infill material})} = \left[1 - \frac{2\sqrt{k}}{h}\tanh(h/2\sqrt{k})\right], \tag{2.75}$$

and when normalized against the permeability of the unfilled fracture, to be given by (Fig. 2.9b)

$$\frac{k(\text{filled fracture})}{k(\text{unfilled fracture})} = \frac{12k}{h^2}\left[1 - \frac{2\sqrt{k}}{h}\tanh(h/2\sqrt{k})\right]. \tag{2.76}$$

In either format, the normalized permeability depends only on the ratio of the permeability of the unfilled fracture, $h^2/12$, to the permeability of the infilling material, k.

The proppant particles that are injected into hydraulic fracture to prevent the fracture from closing up often are found to occupy discrete, isolated regions of the fracture plane (d'Huteau *et al.*, 2011; Wang and Elsworth, 2018), as in Fig. 2.5. If these "pillars" are thought of as forming a porous medium, then the Brinkman model might be appropriate for computing the effective permeability of the fracture, with k denoting the transverse permeability of a fiber-like medium that consists of parallel long pillars oriented in the z-direction (Kumar *et al.*, 1991). Detailed numerical solutions of the Stokes equations for flow through a fracture propped open by cylindrical pillars (Jasinski and Dabrowski, 2018) have shown that this Brinkman-based model is highly accurate, over a wide range

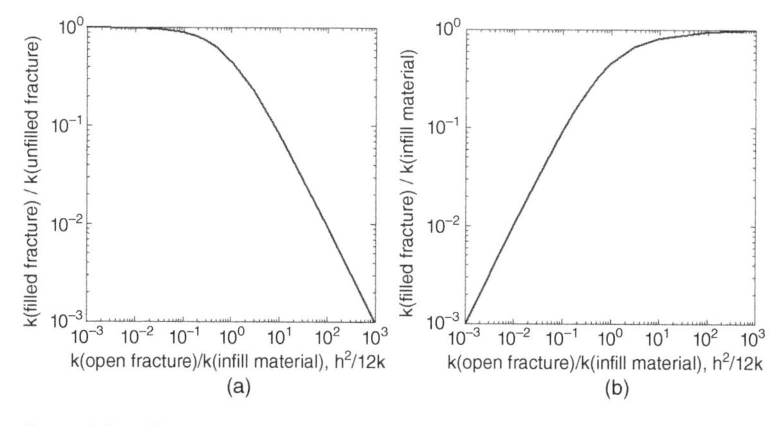

Figure 2.9 Effective permeability of a smooth-walled fracture of aperture h, filled with a material having permeability k, (a) normalized against the permeability of the unfilled fracture, and (b) normalized against the permeability of the infill material.

of the area fraction occupied by the proppant, and a wide range of ratios of pillar radius to aperture, which is equivalent to a wide range of the parameter $h^2/12k$.

Problems

2.1 Starting with eq. (2.8), and using a simple order-of-magnitude analysis such as that used in Section 2.3, derive an approximate expression for the transmissivity of a parallel-plate fracture without actually solving the differential equation. Hint: note that the velocity will vary from 0 to U_{max} over the region between $z = 0$ and $z = h/2$.

2.2 Consider a saw-tooth type fracture whose aperture varies linearly in the x-direction from h_{min} to h_{max}, over a distance $\lambda/2$, as in Fig. 2.3a. According to eq. (2.49), the hydraulic aperture, for the case of flow in the x-direction, is given by $\langle h^{-3}\rangle^{-1/3}$. Compute the hydraulic aperture, and plot it, normalized against the mean aperture, as a function of h_{min}/h_{max}, for $0.1 < (h_{min}/h_{max}) < 1$.

2.3 Consider a two-dimensional fracture having a sinusoidal aperture described by eq. (2.22). Calculate the relative roughness of this fracture, σ_h/h_m, as a function of δ, and then plot the normalized transmissivity, as given by eqs. (2.55) and (2.56), over the entire possible range of δ values.

2.4 Consider a two-dimensional fracture whose aperture varies only in the direction normal to the pressure gradient. Starting from the Reynolds equation, eq. (2.39), show that the hydraulic aperture is equal to the cube root of the arithmetic mean of the cube of the aperture, *i.e.*, $h_H^3 = \langle h^3\rangle$.

2.5 Fill in all steps in the derivation of the velocity profiles within the fracture and the adjacent permeable rock, as given by eqs. (2.60) and (2.61).

2.6 Consider a fracture having a hydraulic aperture of 100 μm, in a matrix rock that has permeability k, How large would k need to be in order for the effective transmissivity of the fracture to be 10% greater than it would be if the matrix rock were impermeable?

References

Batchelor, G. K. 1967. *An Introduction to Fluid Dynamics*, Cambridge University Press, Cambridge and New York.

Beran, M. J. 1968. *Statistical Continuum Theories*, Wiley Interscience, New York.

Bear, J. 1988. *Dynamics of Fluids in Permeable Media*, Dover Publications, Mineola, N.Y.

Beavers, G. S., and Joseph, D. D. 1967. Boundary conditions at a naturally permeable wall. *Journal of Fluid Mechanics*, 30, 197–207.

Berkowitz, B. 1989. Boundary conditions along permeable fracture walls: influence on flow and conductivity. *Water Resources Research*, 25(8), 1919–22.

Bers, L., John, F., and Schechter, M. 1964. *Partial Differential Equations*, Wiley Interscience, New York.

Bird, R. B., Stewart, W. E., and Lightfoot, E. N. 1960. *Transport Phenomena*, Wiley, New York.

Boussinesq, J. 1868. Mémoire sur l'influence des frottements dans les mouvements réguliers des fluids [Study of the influence of friction on the laminar motion of fluids]. *Journal de Mathématiques Pures et Appliqués*, 2(13), 377–424.

Brinkman, H. C. 1947. A calculation of the viscous force exerted by a flowing fluid on a dense swarm of particles. *Applied Scientific Research*, A1, 27–34.

Brown, S. R. 1987. Fluid flow through rock joints: the effect of surface roughness, *Journal of Geophysical Research*, 92(B2), 1337–47.

Brown, S. R., and Scholz, C. H. 1985. Broad bandwidth study of the topography of natural surfaces. *Journal of Geophysical Research*, 90(B12), 575–82.

Brown, S. R., Stockman, H. W., and Reeves, S. J. 1995. Applicability of the Reynolds equation for modeling fluid flow between rough surfaces. *Geophysical Research Letters*, 22(18), 2537–40.

Carslaw, H. S., and Jaeger, J. C. 1959. *Conduction of Heat in Solids*, 2nd ed., Clarendon Press, Oxford.

Currie, I. G. 2003. *Fundamental Mechanics of Fluids*, 3rd ed., Marcel Dekker, New York.

d'Huteau, E., Gillard, M., Miller, M., Peña, A., *et al.* 2011. Open-channel fracturing – a fast track to production. *Oilfield Review*, 23(3), 4–17.

Elrod, H. G. 1979. A general theory for laminar lubrication with Reynolds roughness. *Journal of Lubrication Technology*, 101(1), 8–14.

Gale, J., MacLeod, R., and Le Messurier, P. 1990. *Measurement of flowrate, solute velocities and aperture variation in natural fractures as a function of normal and shear stress, Stripa Project Report 90–11*, Swedish Nuclear Fuel and Waste Management Company, Stockholm.

Hasegawa, E. and Izuchi, H. 1983. On the steady flow through a channel consisting of an uneven wall and a plane wall. *Bulletin of the Japanese Society of Mechanical Engineers*, 26, 514–20.

Hashin, Z., and Shtrikman, S. 1962. A variational approach to the theory of the effective magnetic permeability of multiphase materials. *Journal of Applied Physics*, 33(10), 3125–31.

Hosseini, N., and Khoei, A. R. 2021. Modeling fluid flow in fractured porous media with the interfacial conditions between porous medium and fracture. *Transport in Porous Media*, 139, 109–29.

Ikari, M. J. Saffer, D. M., and Marone, C. 2009. Frictional and hydrologic properties of clay-rich fault gouge. *Journal of Geophysical Research*, 114, B05409.

Iwai, K. 1976. *Fundamental Studies in Fluid Flow through a Single Fracture*, Ph.D. dissertation, University of California, Berkeley.

Jasinski, L., and Dabrowski, M. 2018. The effective transmissivity of a plane-walled fracture with circular cylindrical obstacles, *Journal of Geophysical Research*, 123, 242–63.

Kim, S., and Russel, W. B. 1985. Modeling of porous media by renormalization of the Stokes equations. *Journal of Fluid Mechanics*, 154, 269–86.

Kitanidis, P. K., and Dykaar, B. B. 1997. Stokes flow in a slowly-varying two-dimensional pore. *Transport in Porous Media*, 26(1), 89–98.

Konzuk, J. S., and Kueper, B. H. 2004. Evaluation of cubic law based models describing single-phase flow through a rough-walled fracture. *Water Resources Research*, 40, W02402.

Kumar, S., Zimmerman, R. W., and Bodvarsson, G. S. 1991. Permeability of a fracture with cylindrical asperities. *Fluid Dynamics Research*, 7(3, 4), 131–37.

Landau, L. D. and Lifshitz, E. M. 1960. *Electrodynamics of Continuous Media*, Pergamon, New York.

Martys, N. S. 2001. Improved approximation to the Brinkman equation using a lattice Boltzmann method. *Physics of Fluids*, 13(6), 1807–10.

Matheron, G. 1967. *Elements pour une Théorie des Milieux Poreux [Elements of a Theory of Porous Media]*, Masson, Paris.

Morrow, C. A., Shi, L. Q., and Byerlee, J. D. 1984. Permeability of fault gouge under confining pressure and shear stress. *Journal of Geophysical Research*, 89(B5), 3193–3200.

Neuzil, C. E., and Tracy, J. V. 1981. Flow through fractures. *Water Resources Research*, 17(1), 191–99.

Oron, A. P., and Berkowitz, B. 1998. Flow in rock fractures: the local cubic law assumption re-examined. *Water Resources Research*, 34(11), 2811–25.

Phillips, O. M. 1991. *Flow and Reactions in Permeable Rocks*, Cambridge University Press, Cambridge and New York.

Piggott, A. R., and Elsworth, D. 1992. Analytical models for flow through obstructed domains. *Journal of Geophysical Research*, 97(B2), 2085–93.

Pyrak-Nolte, L. J., Myer, L. R., Cook, N. G. W., and Witherspoon, P.A. 1987. Hydraulic and mechanical properties of natural fractures in low permeability rock, in *Proceedings of the 6th International Congress of Rock Mechanics*, G. Herget and S. Vongpaisal, eds., Balkema, Rotterdam, pp. 225–31.

Ramsay, J. G. 1980. The crack–seal mechanism of rock deformation. *Nature*, 284, 135–39.

Renshaw, C. E. 1995. On the relationship between mechanical and hydraulic apertures in rough-walled fractures. *Journal of Geophysical Research*, 100(B12), 24,629–36.

Reynolds, O. 1886. On the theory of lubrication. *Philosophical Transactions of the Royal Society of London*, 177, 157–234.

Sanematsu, P., Shen, Y., Thompson, K., Yu, T., *et al.* 2015. Image-based Stokes flow modeling in bulk proppant packs and propped fractures under high loading stresses. *Journal of Petroleum Science and Enginering*, 135, 391–402.

Schlichting, H. 1968. *Boundary-Layer Theory*, 6th ed. McGraw-Hill, New York.

Silliman, S. E. 1989. An interpretation of the difference between aperture estimates derived from hydraulic and tracer tests in a single fracture. *Water Resources Research*, 25(10), 2275–83.

Singh, H., and Cai, J. 2019. A feature-based stochastic permeability of shale. *Transport in Porous Media*, 126, 527–60.

Sisavath, S., Al-Yaarubi, A., Pain, C. C., and Zimmerman, R. W. 2003. A simple model for deviations from the cubic law for a fracture undergoing dilation or closure. *Pure and Applied Geophysics*, 160, 1009–22.

Skjetne, E., Hansen, A., and Gudmundsson, J. S. 1999. High-velocity flow in a rough fracture. *Journal of Fluid Mechanics*, 383, 1–28.

Thompson, B. W. 1968. Secondary flow in a Hele-Shaw cell. *Journal of Fluid Mechanics*, 31, 379–95.

Thorpe, M. F. 1992. The conductivity of a sheet containing a few polygonal holes and/or superconducting inclusions. *Proceedings of the Royal Society of London Series A*, 437, 215–27.

Timothy, J. J., and Meschke, G. 2017. The intrinsic permeability of microcracks in porous solids: analytical models and Lattice Boltzmann simulations. *International Journal for Numerical and Analytical Methods in Geomechanics*, 41, 1138–54.

Tokan-Lawal, A., Prodanović, M., and Eichhubl, P. 2015. Investigating flow properties of partially cemented fractures in Travis peak formation using image-based pore-scale modeling. *Journal of Geophysical Research Solid Earth*, 120(8), 5453–66.

Walsh, J. B. 1981. The effect of pore pressure and confining pressure on fracture permeability. *International Journal of Rock Mechanics and Mining Sciences*, 18(5), 429–35.

Wang, J., and Elsworth, D. 2018. Role of proppant distribution on the evolution of hydraulic fracture conductivity. *Journal of Petroleum Science and Engineering*, 166, 249–62.

White, F. M. 2006. *Viscous Fluid Flow*, 3rd ed., McGraw-Hill, New York.

Yeo, I. W., de Freitas, M. H., and Zimmerman, R. W. 1998. Effect of shear displacement on the aperture and permeability of a rock fracture. *International Journal of Rock Mechanics and Mining Sciences*, 35(8), 1051–70.

Zimmerman, R. W. 1996. Effective conductivity of a two-dimensional medium containing elliptical inclusions. *Proceedings of the Royal Society of London Series A*, 452, 1713–27.

Zimmerman, R. W. 2018. *Fluid Flow in Porous Media*, World Scientific, Singapore and London.

Zimmerman, R. W., and Bodvarsson, G. S. 1996. Hydraulic conductivity of rock fractures. *Transport in Porous Media*, 23(1), 1–30.

Zimmerman, R. W., Kumar, S., and Bodvarsson, G. S. 1991. Lubrication theory analysis of the permeability of rough-walled fractures. *International Journal of Rock Mechanics and Mining Sciences*, 28(4), 325–31.

Zimmerman, R. W., Chen, D. W., and Cook, N. G. W. 1992. The effect of contact area on the permeability of fractures. *Journal of Hydrology*, 139(1–4), 79–96.

3

Effect of Stress on Fracture Transmissivity

3.1 Introduction

The hydraulic transmissivity of a rock fracture is controlled and determined by the geometry of the pore space. This geometry is in turn controlled by the morphology of the two opposing rock surfaces, modified by any stress-imposed deformation that may have occurred. If a fracture is subjected to a compressive normal stress, the two opposing rock faces will move closer together, moving normal to the nominal fracture plane, on average. As the two fracture surfaces move closer together, the contact regions between the two faces will grow in size, and new contact regions may be created. Each of these geometric changes – the decrease in mean aperture and the increase in the number and areal extent of the contact regions – will cause the transmissivity to decrease.

Shear stresses acting parallel to the nominal fracture plane will cause the two fracture surfaces to be displaced with respect to one another in a direction aligned with the direction of the shear stress. This motion will have a complicated effect on the aperture field. Initially, the mean aperture will increase as the surfaces must move apart in order to ride over the asperities. As the shear deformation progresses, new contact regions will be formed, and the deformed geometry, which may initially have been isotropic, will become anisotropic, leading to anisotropy in the transmissivity.

Since the transmissivity of a fracture is controlled by the geometry of its aperture field, it follows that the variation of transmissivity with stress is intimately related to the manner in which the fracture deforms in response to applied stresses. This chapter will describe models and present data on the deformation and transmissivity of fractures that are subjected to normal and shear loads. Section 3.2 discusses the effect of normal stress on fracture deformation and introduces the concepts of normal stiffness and normal compliance. Some simple models for relating the normal stiffness to the fracture morphology are presented in Section 3.3. The "row of voids" model for coupling the normal stiffness and the permeability of a fracture is presented in Section 3.4. Data are presented in Section 3.5 to demonstrate the three regimes of the variation of the transmissivity as a function of normal stress, along with a simple conceptual model to explain this behavior. Finally, Section 3.6 discusses the effect of shear stress and shear deformation on fracture transmissivity and the development of hydraulic anisotropy as a function of shear displacement.

Fluid Flow in Fractured Rocks, First Edition. Robert W. Zimmerman and Adriana Paluszny.

Companion website: www.wiley.com/go/zimmerman/fluidflowinfracturedrocks

3.2 The Effect of Normal Stress on Fracture Deformation

Consider a rock specimen that contains a through-going fracture and is subjected to uniaxial compression perpendicular to the plane of the fracture. The total change in length of this specimen will consist of the deformation that would occur if the fracture were not present, plus an excess deformation, known as the *joint closure* or *fracture closure*, that can be attributed to the fracture. (The joint closure is traditionally denoted by δ, which should not be confused with the use of this Greek letter to quantify the fracture aperture variation in Chapter 2. Note also that the joint closure represents an average value over the fracture plane, and is not intended to represent a local value.) This situation can be clearly observed from the measurements of Goodman (1976), who first measured the deformation of an intact cylindrical core of granodiorite under uniaxial compression. He then created an artificially induced tensile fracture, reassembled the core, and measured the deformation of the core that now contained a single *mated* fracture. Finally, after slightly rotating the upper and lower halves of the core relative to each other, he measured the deformation of the core containing an *un-mated* fracture. Both of the fractured specimens exhibited greater axial deformation than did the intact core (Fig. 3.1a).

Using the subscript o to denote the unfractured specimen, the axial displacement of the fractured specimen can be written, by definition, as $\Delta L = \Delta L_o + \delta$. The fracture closure, δ, computed from $\delta = \Delta L - \Delta L_o$, is plotted in Fig. 3.1b. The un-mated fracture exhibited a much greater fracture closure, essentially because the *initial* aperture field of the un-mated fracture was much larger than that of the mated fracture. In both cases, the joint closure was a highly nonlinear function of the applied normal stress, leveling off to some asymptotic value at high stresses. Goodman was able to fit the dependence of fracture closure on the

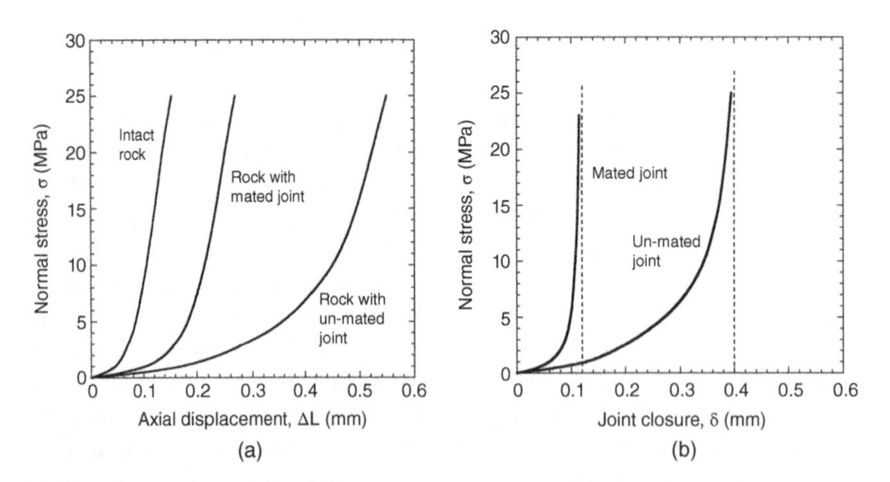

Figure 3.1 Measurements of Goodman (1976) on fracture closure of a granodiorite core: (a) axial displacement of intact core, core with mated fracture, and core with un-mated fracture; (b) fracture closure, computed by subtracting the axial displacement of the intact core from the axial displacement of the fractured core. Source: Goodman (1976)/West Publishing Company.

normal stress with the following empirical relation,

$$\sigma = \sigma_o \left[1 + \left(\frac{\delta}{\delta_m - \delta}\right)^t\right], \quad \text{for } \sigma \geq \sigma_o, \tag{3.1}$$

where σ_o is some initial, low "seating stress," t is a dimensionless empirically derived exponent, and δ_m is the maximum possible fracture closure, which is approached asymptotically as the normal stress increases. The best-fitting values of values of t, for both the mated and un-mated fractures, were found to be very close to 0.6.

Bandis *et al.* (1983) made extensive measurements of joint closure on a variety of natural, unfilled joints in dolerite, limestone, siltstone, and sandstone and observed some hysteresis and a small permanent deformation during cycles of loading and unloading, which diminished rapidly with successive cycles. Barton *et al.* (1985) suggested that the hysteresis was a laboratory artifact and that *in situ* fractures probably behave in a manner similar to the third or fourth laboratory loading cycle, although some other researchers interpret the experimentally observed hysteresis as being reflective of actual fracture behavior (*e.g.*, Haghi and Chalaturnyk, 2022). Bandis *et al.* (1983) fit their measured fracture closures to a function of the form

$$\sigma = \frac{\kappa_o \delta}{1 - (\delta/\delta_m)} = \frac{\kappa_o \delta_m \delta}{\delta_m - \delta}, \tag{3.2}$$

where κ_o [Pa/m] is a fitting parameter that, in some sense, reflects the stiffness of the fracture. This equation can be inverted to yield the fracture closure as a function of normal stress:

$$\delta = \delta_m \left(\frac{\sigma}{\sigma + \kappa_o \delta_m}\right). \tag{3.3}$$

It is useful to define a *fracture normal stiffness*, $\kappa_n = d\delta/d\sigma$, along with a *fracture normal compliance*, $\beta_n = 1/\kappa_n$. The fracture normal stiffness has units of [Pa/m] and can be thought of as an elastic spring constant that is distributed over the plane of the fracture. According to Bandis's equation, the normal stiffness of the fracture is given by

$$\kappa_n = \frac{d\sigma}{d\delta} = \frac{\kappa_o}{(1 - \delta/\delta_m)^2} = \kappa_o \left(1 + \frac{\sigma}{\kappa_o \delta_m}\right)^2, \tag{3.4}$$

which shows that κ_o is the normal stiffness at low confining stress and that the stiffness becomes infinite as the normal stress increases and $\delta \to \delta_m$. The function proposed by Goodman approaches eq. (3.4) asymptotically, if $t = 1$ and $\sigma \gg \sigma_o$.

For fracture surfaces that are unmated, such as those that would result from previous shear displacement, Bandis *et al.* (1983) found that the normal stress could be fitted with an equation of the form

$$\ln(\sigma/\sigma_o) = J\delta, \tag{3.5}$$

where σ_o is a small initial stress at which the joint closure is taken to be zero, and J is a constant having dimensions of $[\mathrm{m}^{-1}]$. The normal stiffness associated with this stress-closure relationship is

$$\kappa_n = \frac{d\sigma}{d\delta} = J\sigma, \tag{3.6}$$

which increases linearly with stress.

Normal stiffnesses that increase linearly with stress were obtained by Lang *et al.* (2016) from numerical simulations of the normal compression of a fracture whose surfaces were isotropic and self-affine, with approximately Gaussian height distributions. Morris *et al.* (2017) assumed the elastic behavior of the rock and used a boundary element method to simulate the normal deformation of rough fractures whose surfaces followed a self-affine distribution. They also found that the stiffness increases linearly with stress and observed that normal stiffness scales with fracture length. Zou *et al.* (2020) measured the stiffness of several fractures in Miluo granite samples from Hunan, China, and again found that the stiffness increased roughly linearly with normal stress. They were able to fit the measured stiffness data using an elastic-plastic contact model that accounted for multi-scale fracture surface roughness. The numerical values of fracture normal stiffness tend to lie in the range of 10^{10}–10^{12} Pa/m.

3.3 Models for the Normal Stiffness of Rock Fractures

Many aspects of the normal closure of an initially mated fracture can be qualitatively explained by the conceptual model developed by Myer (2000), in which a fracture is represented by a collection of collinear, thin elliptical voids (Fig. 3.2a). In this model, the voids are not intended to represent discrete micro-fractures but rather the regions of the fracture that are still "open" after two undulating fracture surfaces are brought into contact by the application of a normal stress. (Although this model is often referred to as a "row of cracks" model, this term can be misleading since the open voids do not represent individual cracks and the region between the voids is not actually "welded" together. Therefore, these open regions will be referred to as "elliptical voids" in this section.) The voids are assumed to have length $2a$, the spacing between the centers of adjacent ellipses is 2λ, and the fractional contact area is given by $c = 1-(a/\lambda)$. The maximum aperture of each void, which is the aperture at the void's mid-point, is $2a\alpha$, where the aspect ratio α is allowed to vary from one void to the next. The intact rock on both sides of

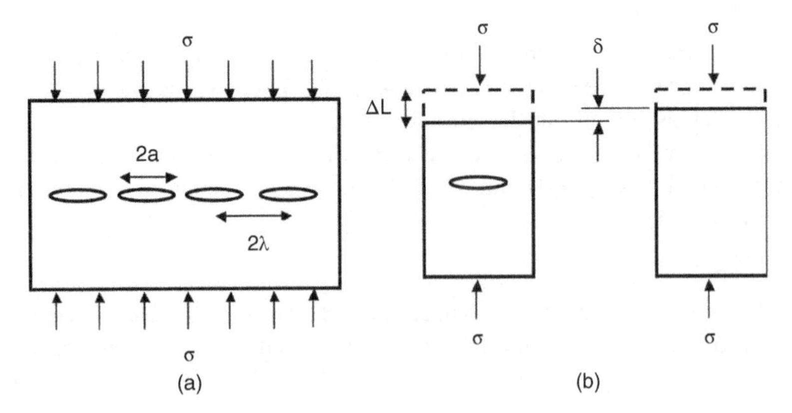

Figure 3.2 (a) Schematic model of a fracture as an array of two-dimensional elliptical voids of length $2a$ and periodicity 2λ. (b) Unit cell of fractured and intact rock, showing definition of δ. Source: Adapted from Myer (2000).

the fracture plane is assumed to be linearly elastic, with elastic modulus E_o and Poisson's ratio ν_0.

According to the elasticity solution for this geometry that was presented by Sneddon and Lowengrub (1969), the incremental fracture closure due to a small increase in normal stress is (see Fig. 3.2b)

$$\delta = \frac{8\lambda(1-\nu_0^2)\sigma}{\pi E_o} \ln \sec\left(\frac{\pi a}{2\lambda}\right) = \frac{8a(1-\nu_0^2)\sigma}{\pi E_o(1-c)} \ln \sec\left[\frac{\pi}{2}(1-c)\right]. \tag{3.7}$$

It follows that the normal compliance of this fracture is given by Baik and Thompson (1984)

$$\beta_n = \frac{1}{\kappa_n} = \frac{d\delta}{d\sigma} = \frac{8a(1-\nu_0^2)}{\pi E_o(1-c)} \ln \sec\left[\frac{\pi}{2}(1-c)\right]. \tag{3.8}$$

This model reveals an interesting scaling law: the fracture normal stiffness scales like $\kappa_n \propto E_0/a$, which implies that smaller "channel" sizes lead to stiffer fractures.

At low stresses, the fractional contact area c is small, and the normal compliance will be large. For very small values of c, the compliance behaves asymptotically as (see Problem 3.1)

$$\beta_n \approx \frac{8a(1-\nu_0^2)}{\pi E_o} \ln\left(\frac{2}{\pi c}\right) \approx \frac{8a(1-\nu_0^2)}{\pi E_o} \ln(1/c). \tag{3.9}$$

The second approximation contains only the term that becomes singular as the fractional contact area goes to zero. As would be expected, the compliance becomes infinite as the fractional contact area vanishes, although it does so at a very weak, logarithmic rate.

As the normal stress increases, voids with smaller aspect ratios close up. Although this disturbs the periodicity of the array, this process can still be modeled approximately with the above equations, by allowing the contact fraction c to increase, but now interpreting λ as the *mean* spacing between adjacent voids. Expansion of eq. (3.8) for large contact fractions, which corresponds to *small* values of $1-c$, shows that, as the contact fraction increases (see Problem 3.2),

$$\beta_n \approx \frac{\pi a(1-\nu_0^2)}{E_o}(1-c). \tag{3.10}$$

As the contact area increases, the compliance decreases and the stiffness increases, in qualitative accordance with experimental observations. At the limit of complete contact, the fracture compliance vanishes.

Another geometrically simple model that can be used to estimate the normal compliance of a fracture is that of a fracture plane that consists of a dispersion of isolated circular contact regions of radius a (Baik and Thompson, 1984; Markov *et al.*, 2019). The normal compliance of such a fracture is given by (Baik and Thompson, 1984)

$$\beta_n = \frac{\pi a(1-\nu_0^2)}{E_o}\left[\frac{1}{c} - \frac{1.413}{\sqrt{c}} + 0.676\sqrt{c} - \dots\right]. \tag{3.11}$$

For very low contact fractions, the circular contact model predicts that the compliance will vary as $1/c$, in contrast to the $\ln(1/c)$ behavior that occurs when the contact regions are parallel strips.

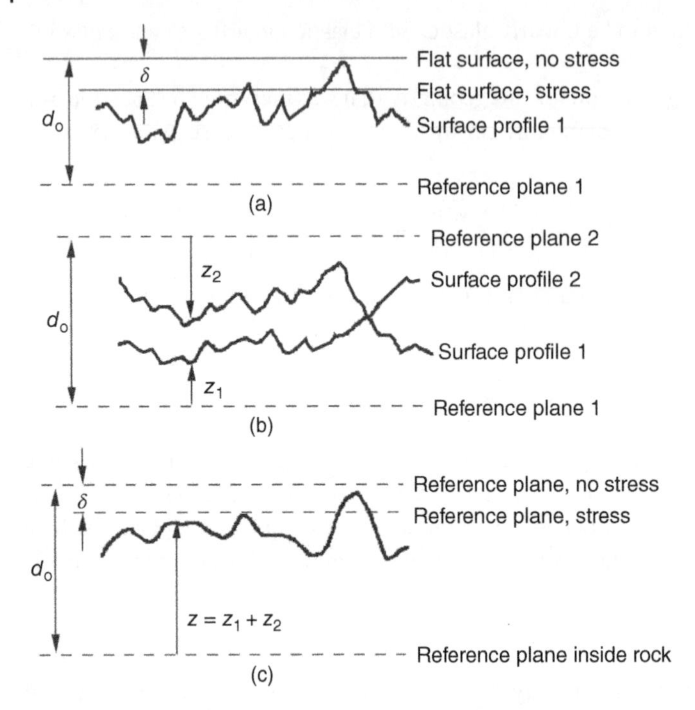

Figure 3.3 (a) Single rough profile in contact with a smooth surface; (b) two rough surfaces; (c) composite profile; after Cook (1992). Source: Adapted from Cook (1992).

A more complicated, and perhaps more physically realistic, conceptual model for the normal stiffness of a rock fracture is to treat the fracture surface as a rough elastic surface and use Hertzian contact theory (Timoshenko and Goodier, 1970, pp. 409–16) to analyze the deformation of the contacting asperities. This approach was taken by Greenwood and Williamson (1966) and extended by Brown and Scholz (1985). Greenwood and Williamson considered a single, rough elastic surface whose asperities each have a radius of curvature R [m], with a distribution of *peak* heights $\phi(Z^*)$, where the height Z of an asperity is measured relative to a reference plane that is parallel to the nominal fracture plane and which can conveniently be located entirely within the rock, *i.e.*, below the lowest troughs of the fracture surface. (The symbol ϕ used in this context is unrelated to the rock porosity, which is generally also denoted by this symbol). A value of Z^* is associated with each local peak, and there are assumed to be η peaks per unit area of fracture in the undeformed (zero stress) state. The height of the highest peak, measured from the reference plane, is initially equal to d_o (Fig. 3.3a).

If such a surface is pressed against a smooth elastic surface of area A, the density of contacts, n, will be given by

$$n = \frac{N}{A} = \eta \int_{d_o - \delta}^{\infty} \phi(Z^*) dZ^*. \tag{3.12}$$

As the distribution function $\phi(Z^*)$ vanishes for $Z^* > d_0$, by construction, the contact density is zero when the joint closure δ is zero. In the hypothetical situation in which all asperities were pressed flat against the upper flat surface, δ would equal d_0, so the integral in eq. (3.12) would approach unity. Since n is the number of contacts per unit area, and η is the number of peaks per unit area, the fraction of asperity peaks that are in contact, n/η, would reach unity.

The fractional contact area of asperities is given by Cook (1992)

$$c = A_{contact}/A = \pi R\eta \int_{d_o-\delta}^{\infty} (Z^* - d_o + \delta)\phi(Z^*)dZ^*, \tag{3.13}$$

and the average normal stress acting over the surface is

$$\sigma = \frac{2\eta R^{1/2} E_0}{3\left(1 - \nu_0^2\right)} \int_{d_o-\delta}^{\infty} (Z * -d_o + \delta)^{3/2}\phi(Z^*)dZ^*, \tag{3.14}$$

where, as before, the intact rock on both sides of the fracture plane is assumed to have elastic modulus E_0 and Poisson's ratio ν_0.

Swan (1983) measured the topography of 10 different surfaces of Offerdale slate and showed that the peak heights of asperities followed a Gaussian distribution. Greenwood and Williamson (1966) had already shown that the upper quartile of a Gaussian distribution could be approximated by an exponential distribution of the form

$$\phi(Z^*) = \frac{1}{s}\exp(-Z^*/s), \tag{3.15}$$

where s is the mean, as well as the standard deviation, of the exponential distribution. Equations (3.12)–(3.14) lead in this case to

$$\ln\{\sigma/[(\pi Rs)^{1/2} sE'\eta]\} = (\delta - d_0)/s, \tag{3.16}$$

where $E' = E_0/2\left(1 - \nu_0^2\right)$. This expression has the same form as the empirical relation found by Bandis *et al.* (1983) for fractures with unmated surfaces: $\ln(\sigma/\sigma_0) = J\delta$. Comparison of eqs. (3.5) and (3.16) shows that, according to the model of Swan, Greenwood, and Williamson, the coefficients that appear in the Bandis equation are given by

$$J = 1/s, \quad \sigma_0 = (\pi Rs)^{1/2}(sE'\eta)\exp(-d_0/s). \tag{3.17}$$

Comparison of eqs. (3.6) and (3.17) shows that the normal stiffness is equal to σ/s and therefore increases with stress at a rate that is inversely proportional to the "roughness" of the fracture.

3.4 "Row of Elliptical Voids" Model for Fracture Transmissivity

If a cross-section is taken through a fracture perpendicular to the nominal plane of the fracture, it will reveal a sequence of open regions separated by regions of contact between the two opposing faces of rock. Pyrak-Nolte *et al.* (1987) made casts of a quartz monzonite fracture by injecting molten Wood's metal, which has a melting point of 70 °C, into the fracture under a fixed normal stress and allowing it to harden, thereby forming a cast of the interconnected fracture void space (Fig. 3.4a; top). Myer (2000) took transects through this image, showing the sequence of alternating open and closed regions (Fig. 3.4a; bottom). Each of the open regions can be thought of as a cross-section of a flow channel.

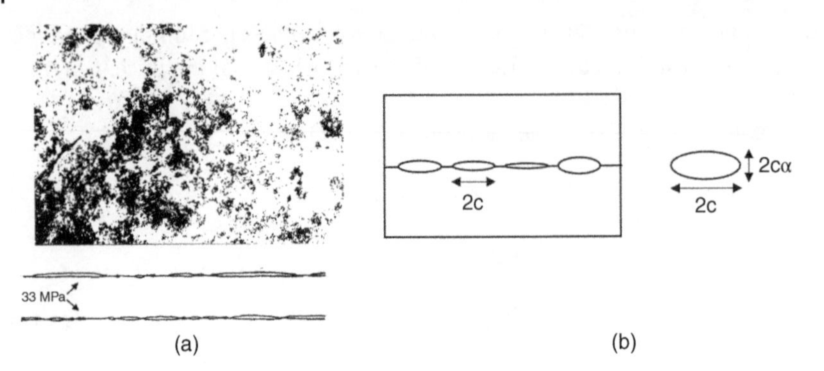

Figure 3.4 (a) Cast of the void space in a quartz monzonite fracture (Pyrak-Nolte *et al.*, 1987), showing regions of contact (black) and open flow paths (white). Two transects through the fracture, showing alternating open and closed regions, are shown below (Myer, 2000). (b) Parallel channels having an initial distribution of aspect ratios, α.

The flow channels in an actual fracture plane will generally form an irregular "random" network, with series and parallel arrangements being the two extreme cases. However, a simple coupled hydro-mechanical model of fracture behavior under normal stress can be developed by assuming that the channels are arranged in parallel (Tsang and Witherspoon, 1981; Zimmerman, 2008). Based on the row of voids conceptual model of Myer (2000), a further simplification can be made by considering the case of large contact fractions, c, which correspond to small values of a/λ (Zimmerman, 2008). In this regime, $\cos(\pi a/2\lambda) \approx 1-(\pi^2 a^2/8\lambda^2)$, and so $\ln[\cos(\pi a/2\lambda)] \approx -\pi^2 a^2/8\lambda^2$. Hence, for large values of c, the fracture compliance is approximately given by

$$\beta_n = \frac{(1-\nu_o^2)\,\pi a^2}{\lambda E_o}. \tag{3.18}$$

The fracture contains one channel for each distance λ, so the number of channels contained in a region having width w in the transverse direction will be $N = w/\lambda$. The fracture compliance can therefore also be written as

$$\beta_n = \frac{(1-\nu_o^2)\pi N a^2}{w E_o}. \tag{3.19}$$

Before the application of normal stress, the initial aspect ratios are distributed according to some distribution function $n(\alpha^i)$, where the subscript i refers to the initial, stress-free state. The fracture compliance can therefore be expressed as

$$\beta_n = \frac{(1-\nu_o^2)\pi a^2}{w E_o} \int_0^\infty n(\alpha^i) d\alpha^i, \tag{3.20}$$

where the integral extends over all possible aspect ratios, from 0 to ∞, although in practice $n(\alpha^i)$ will rapidly go to zero well before α^i reaches 1.

As the normal load on the fracture increases, channels will close up, starting with those channels that have small initial aspect ratios. At a stress σ, all channels whose initial aspect ratio was less than some critical value α^* will be closed. The stress needed to close a channel whose initial aspect ratio is α^i equal to $\alpha^i E_o/2(1-\nu_o^2)$ (Walsh, 1965), so the critical value of α^i at any given stress σ is

$$\alpha^*(\sigma) = \frac{2(1-\nu_o^2)\sigma}{E_o}. \tag{3.21}$$

At a given normal stress σ, the integral in eq. (3.20) should be taken over only those channels whose initial aspect ratios were greater than α^*, *i.e.*,

$$\beta_n = \frac{(1-\nu_o^2)\pi a^2}{wE_o}\int_{\alpha*}^{\infty} n(\alpha^i)d\alpha^i. \tag{3.22}$$

Differentiation of this equation with respect to σ yields

$$\frac{d\beta_n}{d\sigma} = \frac{-(1-\nu_o^2)\pi a^2}{wE_o} n(\alpha^*)\frac{d\alpha^*}{d\sigma} = \frac{-2(1-\nu_o^2)^2\pi a^2}{wE_o^2} n(\alpha^*). \tag{3.23}$$

The aspect ratio distribution function is therefore related to the derivative of the fracture compliance by

$$n(\alpha^i) = \frac{-wE_o^2}{2(1-\nu_o^2)^2\pi a^2}\frac{d\beta_n}{d\sigma}, \tag{3.24}$$

where the derivative must be evaluated at $\sigma = \alpha^i E_o/2(1-\nu_o^2)$. This expression is very similar to the one derived by Morlier (1971) for a three-dimensional rock randomly permeated with thin oblate spheroidal micro-cracks, under hydrostatic loading.

The total hydraulic conductivity of the fracture will be the sum of the conductivities of the individual channels. The volumetric flow through a channel of elliptical cross-section, with semi-major axis a and aspect ratio α, is given by (Boussinesq, 1868; Bernabe *et al.*, 1982; White, 2006)

$$Q_{channel} = -\left(\frac{dP}{dz}\right)\frac{\pi a^4}{4\mu}\left(\frac{\alpha^3}{1+\alpha^2}\right), \tag{3.25}$$

in which α is the "current" aspect ratio at stress σ, and not the initial (zero-stress) aspect ratio. Since fracture compliances tend to be negligible at stresses greater than about 100 MPa, and E_o will be on the order of 10–100 GPa, eq. (3.21) shows that the initial aspect ratio of any closable channel must be less than 0.01. Consequently, the α^2 term in the denominator of eq. (3.25) can be ignored.

Elliptical voids of low aspect ratio close up linearly with stress (Sneddon and Lowengrub, 1969; Zimmerman, 1991), and so the aspect ratio at stress σ is related to the initial aspect ratio α^i by

$$\alpha(\sigma) = \alpha^i\left[1 - \frac{2(1-\nu_o^2)\sigma}{\alpha^i E_o}\right]. \tag{3.26}$$

This equation holds until $\sigma = \alpha^i E_o/2(1-\nu_o^2)$, beyond which the aspect ratio is zero for all higher stresses. The total flux through the fracture at some given stress σ will be the sum of the fluxes through the individual channels, summed over all channels that are still open, which are those channels whose initial aspect ratios are *greater* than α^*:

$$Q_{total}(\sigma) = -\left(\frac{dP}{dz}\right)\frac{\pi a^4}{4\mu}\int_{\alpha^*}^{\infty}\alpha^3 n(\alpha^i)d\alpha^i. \tag{3.27}$$

The variable over which the integration is taken is the *initial* aspect ratio, α^i, whereas the aspect ratio that controls the conductivity is the *current* aspect ratio, α.

Comparison of eqs. (3.21) and (3.26) shows that $\alpha(\sigma) = \alpha^i - \alpha^*$, which allows eq. (3.27) to be written, after dropping the subscript "total" for simplicity, as

$$Q(\sigma) = -\left(\frac{dP}{dz}\right)\frac{\pi a^4}{4\mu}\int_{\alpha^*}^{\infty}(\alpha^i - \alpha^*)^3 n(\alpha^i)d\alpha^i. \tag{3.28}$$

Use of eq. (3.24) to express the aspect ratio distribution in terms of the fracture compliance, results in an expression that relates the flow through the fracture to its normal compliance:

$$Q(\sigma) = -\left(\frac{dP}{dz}\right)\frac{wE_0^2a^2}{8\mu\left(1 - \nu_0^2\right)^2}\int_{\alpha^*}^{\infty}(\alpha^i - \alpha^*)^3\frac{d\beta_n}{d\sigma}d\alpha^i, \tag{3.29}$$

where, as before, σ must be related to α^i by eq. (3.21). A more convenient form of eq. (3.29) can be found by integrating over σ instead of α^i, where the relation $\sigma = \alpha^i E_0/2\left(1 - \nu_0^2\right)$ is used to transform the integrand and the limits of integration:

$$Q(\sigma) = -\left(\frac{dP}{dz}\right)\frac{2\left(1 - \nu_0^2\right)^2 wa^2}{\mu E_0^2}\int_{\sigma}^{\infty}(\sigma' - \sigma)^3\frac{d\beta_n}{d\sigma'}d\sigma'. \tag{3.30}$$

In practice, the integral only needs to extend out to a stress sufficiently large that the fracture compliance has fallen to zero, so that $d\beta_n/d\sigma' = 0$ in the integrand. The transmissivity can be obtained from eq. (3.30) through the definition $Q = -(T/\mu)(dP/dz)$:

$$T(\sigma) = \frac{-2\left(1 - \nu_0^2\right)^2 wa^2}{E_0^2}\int_{\sigma}^{\infty}(\sigma' - \sigma)^3\frac{d\beta_n}{d\sigma'}d\sigma', \tag{3.31}$$

where w is the width of the fractured specimen in the direction normal to the pressure gradient, *i.e.*, the horizontal direction in Fig. 3.2a. The compliance decreases with stress, and so $d\beta_n/d\sigma$ is negative, and the transmissivity is, of course, positive.

To test this model, consider the data of Iwai (1976), who measured the mechanical deformation and transmissivity of a Sierra white Cretaceous granite from Raymond, California. The deformation of the fractured core with stress is shown in Fig. 3.5a, and the variation of transmissivity is shown in Fig. 3.5b, during the first loading cycle. To remove experimental artifacts related to the seating of the specimen within the testing apparatus, the transmissivity can be normalized against its value at some low reference stress, such as $\sigma_o = 1.45$ MPa. According to eq. (3.31),

$$\frac{T(\sigma)}{T(\sigma_o)} = \frac{\int_{\sigma}^{\infty}(\sigma' - \sigma)^3\frac{d\beta_n}{d\sigma'}d\sigma'}{\int_{\sigma_o}^{\infty}(\sigma' - \sigma)^3\frac{d\beta_n}{d\sigma'}d\sigma'}. \tag{3.32}$$

This normalized transmissivity therefore depends *only* on the normal compliance of the fracture, and any explicit reference to other parameters, such as the length of the channels or the elastic moduli of the intact rock, does not appear. The fracture compliance β_n that is required in eq. (3.32) is found by fitting a curve through the deformation data shown in Fig. 3.5a and then differentiating it with respect to stress. The predicted normalized transmissivity (Fig. 3.5b) matches the data reasonably well over two-orders of magnitude variation in transmissivity.

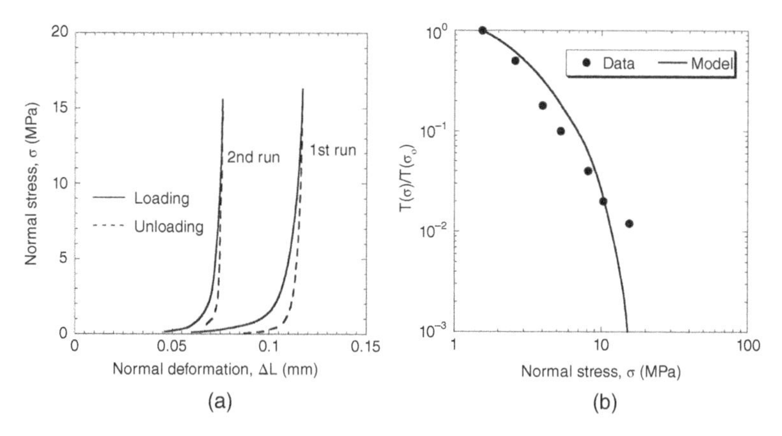

Figure 3.5 (a) Deformation under normal loading of a block of Sierra white granite, containing a through-going fracture. Source: Adapted from Iwai (1976). (b) Variation of transmissivity with stress, showing the comparison between the measured values and the values predicted from eq. (3.32).

Figure 3.5b shows a hint of the "leveling off" of the transmissivity that often occurs at high stresses. In the context of the present model, this can be attributed to the presence of a few non-closable channels of high aspect ratio. For example, consider a granite with $E_0 \approx$ 100 GPa. At a stress of 20 MPa, eq. (3.26) shows that a channel with an initial aspect ratio of, say, 0.1, will essentially have its aspect ratio, and consequently its transmissivity, unchanged under a stress of 20 MPa. Therefore, any channels having initial aspect ratios of 0.1 or larger will give rise to a contribution to the overall transmissivity that does not sensibly vary with stress. This phenomenon is discussed further in Section 3.5.

Based on a Hertzian contact model for an irregular fracture surface such as shown in Fig. 3.3, the following simple expression for the variation of fracture transmissivity with normal stress can be derived (Walsh and Grosenbaugh, 1979; Walsh, 1981):

$$\frac{T(\sigma)}{T(\sigma_0)} = [1 - 2^{3/2} (s/h_0) \ln(\sigma/\sigma_0)]^3, \tag{3.33}$$

where s is the standard deviation of the asperity height distribution, and σ_0 is some reference stress at which the mean aperture is h_0. Walsh (1981) showed that this equation provides a good fit to several data sets on the transmissivity of fractures in Barre granite.

Equation (3.31), although based on a specific, simplified micromechanical model, implies that the normal compliance (or stiffness) of a fracture will be closely related to its hydraulic transmissivity. In fact, all of the data and models discussed in this section, and Section 3.2, show that as the normal stress increases on a given fracture, the stiffness increases, and the transmissivity decreases. This seems to imply that transmissivity should be a decreasing function of stiffness. The relationship between normal stiffness and transmissivity has been investigated by Pyrak-Nolte and coworkers (Pyrak-Nolte and Morris, 2000; Petrovitch *et al.*, 2013; Pyrak-Nolte and Nolte, 2016). Using ideas from percolation theory, and guided by the results of extensive numerical simulations of deformation and flow through numerous fracture aperture distributions generated using a stratified percolation approach, Pyrak-Nolte and Nolte (2016) developed a universal scaling law that collapsed essentially all of their simulated transmissivity data, over fifteen orders of magnitude of variation, onto one curve.

3.5 Relation Between Transmissivity and Mean Aperture During Normal Compression

As the normal stress on a fracture increases, the mean aperture decreases, causing the transmissivity to decrease. Although transmissivity is proportional to mean aperture cubed, it also depends on the variance of the aperture, and the amount of contact area, through relations such as eqs. (2.45) and (2.52). Consequently, the transmissivity of a fracture that is deforming under a normal load should not be expected to be *precisely* proportional to the cube of the mean aperture.

For example, consider the measurements of Witherspoon *et al.* (1980), in which the transmissivity of a tensile fracture in marble was measured while compressing it under a normal load. As seen in Fig. 3.6a, the graph of the hydraulic conductivity (defined, as in Chapter 2, as T/w) *vs.* mean aperture displays three regimes. At low stresses (region I), the aperture is large, the relative roughness σ_h/h_m is therefore low, and the fracture approximates, to some extent, the parallel plate model. Consequently, the transmissivity will vary with the cube of the mean aperture.

As the normal stress increases (region II), the mean aperture of the open areas of the fracture will decrease, while the "roughness" σ_h remains nearly constant (Renshaw, 1995). Hence, according to eq. (2.45), the transmissivity will decrease more severely than the cube of the mean aperture. In other regions of the fracture, the two opposing faces will come into contact, causing an additional decrease in transmissivity, according to, for example, eq. (2.52). Both of these effects cause the transmissivity to decrease more rapidly than the cube of the mean aperture. In this regime, the graph of transmissivity *vs.* mean aperture can be approximated by a power-law with an exponent that is greater than 3 (Pyrak-Nolte

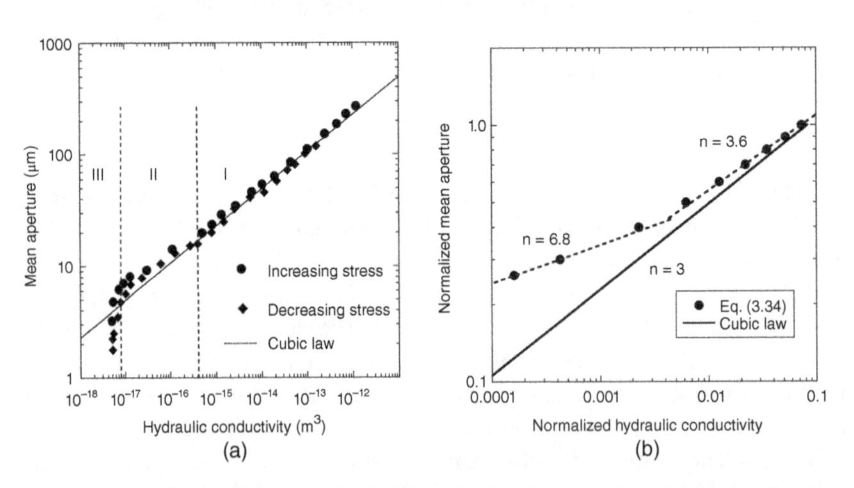

Figure 3.6 (a) Hydraulic conductivity (the transmissivity per unit width in the direction normal to the pressure gradient and the flow) of a marble fracture as a function of mean aperture, as measured by Witherspoon *et al.* (1980). Data points are taken from three different loading cycles; for clarity, not all points are shown. Source: Adapted from Witherspoon *et al.* (1980). (b) Hydraulic conductivity *vs.* mean aperture, according to the sinusoidal aperture model of Sisavath *et al.* (2003); all values are normalized against the values at zero normal stress. Source: Adapted from Sisavath *et al.* (2003).

et al., 1987), and often as high as 8–10, although it should be noted that these power-laws hold over a narrow range of mean apertures, usually less than one order of magnitude.

Sisavath *et al.* (2003) proposed a simple model to explain the transition between the behaviors observed in regions I and II. They computed the second-order perturbation solution to the Stokes equations for a two-dimensional fracture having a sinusoidal aperture variation, described by $h(x) = h_m[1 + \delta \sin(2\pi x/\lambda)]$, and found the transmissivity to be given by

$$T = \frac{wh_m^3(1-\delta^2)^{5/2}}{12[1+(\delta^2/2)]}\left[1+\frac{3\pi^2(1-\delta^2)\delta^2}{5[1+(\delta^2/2)]}\left(\frac{h_m}{\lambda}\right)^2\right]^{-1}. \tag{3.34}$$

They then assumed that, as the normal stress increases, the two surfaces of the fracture move toward each other, causing h_m to decrease, while σ_h remains constant. For a fracture having a sinusoidal aperture variation, the standard deviation of the aperture is given by $\sigma_h = h_m\delta/\sqrt{2}$ (Zimmerman *et al.*, 1991). Hence, in order to use eq. (3.34) to model the process in which h_m decreases while σ_h remains constant, δ must increase according to $\delta = \sqrt{2}\sigma_h/h_m$. Although eq. (3.34) does not explicitly show power-law behavior in this situation, for plausible values of the wavelength and the initial relative roughness, the predicted relationship between T and h_m can be approximated very well by two power-law functions, having slopes of around 3 at high mean apertures, in region I, and much higher slopes for smaller mean apertures, in region II.

As an example, consider the case in which the initial stress-free aperture profile is characterized by $\lambda/h_m = 5$ and $\delta = 0.2$, which implies $\sigma_h = 0.2h_m/\sqrt{2}$. As h_m decreases due to an increase in normal stress, and δ is increased (mathematically) according to $\delta = \sqrt{2}\sigma_h/h_m$ so as to hold σ_h constant, the transmissivity will decrease, according to eq. (3.34). The predicted hydraulic conductivity, normalized against the initial stress-free value, is plotted in Fig. 3.6b, as a function of the normalized mean aperture. Initially, the transmissivity drops off as $(h_m)^{3.6}$, which is nearly indistinguishable from the "cubic law," but then transitions to a regime in which it drops off much more rapidly, as $(h_m)^{6.8}$.

As the normal stress increases further, the mean aperture continues to decrease, but the transmissivities measured by Witherspoon *et al.* (1980) actually stabilized at some small but nonzero residual value (Fig. 3.6a; region III). As discussed in the previous section, the existence of this "residual" transmissivity can be explained by the presence of some flow channels whose aspect ratios are sufficiently large so as to cause these channels to be nearly "rigid." For example, the channels observed by Pyrak-Nolte *et al.* (1987) had widths of less than 100 μm (Myer, 2000). The mean aperture of those channels remaining open at high stresses was on the order of 10 μm, and so the aspect ratios of these channels were on the order of 0.1. If the Young's modulus of the intact rock is on the order of a few tens of GPa, the apertures of such channels will decrease less than 1% under a normal stress of a few tens of MPa. Hence, these channels will be essentially rigid under the levels of normal stress that might occur in the subsurface, resulting in a residual fracture transmissivity that is nearly independent of stress. This qualitative explanation is supported by several sets of experimental data on fractures in basalt, granite, granodiorite, marble, and quartzite that were collated from the literature and analyzed by Renshaw (1995).

A model that is able to *quantitatively* match the data of Witherspoon *et al.* (1980) in regime III has been developed by Yu (2015), based on the assumption of flow through parallel channels, and an aperture field that is initially normally distributed.

3.6 Effect of Shear Deformation on Fracture Transmissivity

Normal stresses invariably cause the fracture aperture to decrease, thereby decreasing the transmissivity. Shear stresses, on the other hand, cause tangential displacement of the two fracture surfaces. As the two surfaces move nominally parallel to each other, in some situations the asperities will be sheared off (Belem *et al.*, 2007; Gui *et al.*, 2017). More commonly, the fracture will dilate, as one surface moves away from the opposing surface in order to ride over the asperities. In this situation, the transmissivity will increase, often by a substantial amount.

Olsson and Brown (1993) measured the hydraulic transmissivity of a fracture in Austin chalk, while increasing the shear displacement, and holding the normal stress constant. The flow geometry was annular-radial, with fluid entering the fracture plane through a borehole of radius of 24.0 mm, and leaving at the outer boundary, which had a radius of 60.3 mm. When the fracture was subjected to 3.5 mm of shear displacement at constant normal stress of 4.3 MPa, the joint dilated (*i.e.*, *negative* joint closure) at a rate of about 50 μm per millimeter of shear displacement, and the transmissivity increased by about two orders of magnitude. Qualitatively similar results were found by Yeo *et al.* (1998) for a red Permian sandstone fracture from the North Sea, and by Chen *et al.* (2000) for a granitic fracture from Olympic Dam mine in Central Australia.

Esaki *et al.* (1999) subjected an artificially split fracture in a granite from Nangen, Korea to shear displacements up to 20 mm, under various levels (1, 5, 10, and 20 MPa) of normal stress. Flow was measured from a central borehole to the outer boundaries of a rectangular fracture plane of dimensions 100 mm × 120 mm. Due to the radial flow geometry, the measured transmissivity therefore reflected some sort of average value of the transmissivity in the direction of shear, and the transmissivity in the direction perpendicular to the shear. Bearing this in mind, the measured transmissivity typically decreased very slightly for the first 0.5–1.0 mm or so of shear, until the peak shear stress was reached. It then increased by about two orders of magnitude as the shear offset increased to about 5 mm. After this point, when the shear stress had reached its residual value, the transmissivity essentially leveled off (Fig. 3.7a). When the shear displacement was reversed, the transmissivity decreased, but not to its original value, leaving an excess residual transmissivity at zero shear displacement. The hysteresis was larger at larger values of the normal stress.

For a fracture undergoing shear displacement, it was found by Auradou *et al.* (2005) that the transmissivity in the direction parallel to the shear (T_{II}) decreases significantly, whereas the transmissivity in the direction perpendicular to the shear ($T_{\perp}$) either increases or remains constant, as the shear displacement increases. In general, the variance of the aperture field increases with increasing shear displacement. Auradou *et al.* (2005) proposed the following bounds for the two resultant transmissivities, in terms of the normalized variance:

$$1 - 6(\sigma_h/h_m)^2 + O(\sigma_h/h_m)^4 \leq \frac{T_{\mathrm{II}}}{T_o} \leq 1, \tag{3.35}$$

$$1 \leq \frac{T_{\perp}}{T_o} \leq 1 + 3(\sigma_h/h_m)^2, \tag{3.36}$$

where T_o is the transmissivity before any shear has occurred, σ_h is the "current" standard deviation of the fracture aperture, h_m is the "current" mean fracture aperture (*i.e.*, after shear has occurred), and $O(\sigma_h/h_m)^4$ denotes additional terms "on the order of" $(\sigma_h/h_m)^4$.

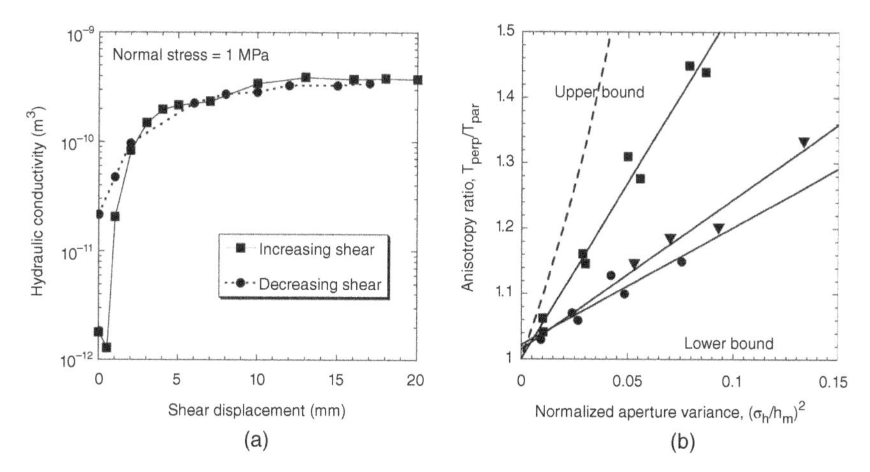

Figure 3.7 (a) Transmissivity of a granitic joint as a function of shear displacement, measured by Esaki *et al.* (1999) under a normal stress of 1 MPa. Source: Adapted from Esaki *et al.* (1999). (b) Shear-induced anisotropy ratio of tensile fractures in granite, as a function of the normalized aperture variance, $(\sigma_h/h_m)^2$, as measured by Auradou *et al.* (2005). Squares: shear displacement in the y-direction, $h_m = 1$ mm. Circles: shear displacement in the x-direction, $h_m = 1$ mm. Triangles: 2 mm displacement in x-direction, for various values of 1.25 mm $< h_m <$ 2.0 mm. Linear fits to data are drawn mainly to aid visualization. See text for additional details. Source: Adapted from Auradou *et al.* (2005).

Numerical simulations of flow through self-affine fractures, conducted by Auradou *et al.* (2005) using the Lattice Boltzmann method, as well as the transmissivities that were inferred from flow experiments performed on transparent epoxy casts of fractured granite blocks extracted from a quarry in Lanhelin (Brittany), were found to be consistent with these bounds, for shear displacements as large as about 2 mm. Since $T_\perp$ generally increased with shear, and T_{II} decreased, the shear-induced anisotropy ratio increased with shear (Fig. 3.7b). The "upper bound" of the anisotropy ratio $T_\perp/T_{II}$ plotted in Fig. 3.7b is computed by taking the upper bound for $T_\perp$ as given by eq. (3.36), and the lower bound for T_{II} as given by eq. (3.35). The "lower bound" for $T_\perp/T_{II}$, computed by taking the lower bound for $T_\perp$ from eq. (3.36), and the upper bound for T_{II} from eq. (3.35), is unity.

Problems

3.1 Carefully expand eq. (3.8) for small values of the contact fraction c, and thereby derive the asymptotic expression given by eq. (3.9).

3.2 Carefully expand eq. (3.8) for large values of the contact fraction c, and thereby derive the asymptotic expression given by eq. (3.10).

3.3 Consider the first loading cycle performed by Iwai (1976) on the Sierra white granite, as shown in Fig. 3.5a. The thickness of the fractured specimen in the direction normal to the fracture plane was 15.5 cm (corresponding to L_o in the terminology of Section 3.2), and the Young's modulus of the intact rock was 44.4 GPa. Extract the joint closure δ from this graph, and fit the closure *vs.* stress data with the Bandis *et al.* (1983) model, eq. (3.3).

3.4 Fit the normalized transmissivity data shown in Fig. 3.5b to the Walsh–Grosenbaugh model, eq. (3.33). The parameter s/h_0 can be treated as a fitting parameter. Does the best-fitting value found for s/h_0 seem to be physically plausible?

References

Auradou, H., Drazer, G., Hulin, J. P., and Koplik, J. 2005. Permeability anisotropy induced by the shear displacement of rough fracture walls. *Water Resources Research*, 41, W09423.

Baik, J.-M., and Thompson, R. B. 1984. Ultrasonic scattering from imperfect interfaces: a quasi-static model. *Journal of Nondestructive Evaluation*, 4(3,4), 177–96.

Bandis, S. C., Lumsden, A. C., and Barton, N. R. 1983. Fundamentals of rock joint deformation. *International Journal of Rock Mechanics and Mining Sciences*, 20(6), 249–68.

Barton N., Bandis, S., and Bakhtar, K. 1985. Strength, deformation and conductivity coupling of rock joints. *International Journal of Rock Mechanics and Mining Sciences*, 22(3), 121–40.

Belem, T., Mountaka, S., and Homand, E. F. 2007. Modeling surface roughness degradation of rock joint wall during monotonic and cyclic shearing. *Acta Geotechnica* 2(4), 227–48.

Bernabe, Y., Brace, W. F., and Evans, B. 1982. Permeability, porosity, and pore geometry of hot-pressed calcite. *Mechanics of Materials*, 1(3), 173–83.

Boussinesq, J. 1868. Mémoire sur l'influence des frottements dans les mouvements réguliers des fluids [Study of the influence of friction on the laminar motion of fluids]. *Journal de Mathématiques Pures et Appliqués*, 2(13), 377–424.

Brown, S. R., and Scholz, C. H. 1985. Closure of random elastic surfaces in contact. *Journal of Geophysical Research*, 90, 5531–45.

Chen, Z., Narayan, S. P., Yang, Z., and Rahman, S. S. 2000. An experimental investigation of hydraulic behaviour of fractures and joints in granitic rock. *International Journal of Rock Mechanics and Mining Sciences*, 37(7), 1061–71.

Cook, N. G. W. 1992. Natural joints in rock: mechanical, hydraulic and seismic behaviour and properties under normal stress. *International Journal of Rock Mechanics and Mining Sciences*, 29(3), 198–223.

Esaki, T., Du, S., Mitani, Y., Ikusada, K., *et al.* 1999. Development of a shear-flowtest apparatus and determination of coupled properties for a single rock joint. *International Journal of Rock Mechanics and Mining Sciences*, 36(5), 641–50.

Goodman, R. E. 1976. *Methods of Geological Engineering in Discontinuous Rocks*, West Publishing, New York.

Greenwood, J. A., and Williamson, J. 1966. Contact of nominally flat surfaces. *Proceedings of the Royal Society London Series A*, 295(1442), 300–19.

Gui, Y., Xia, C., Ding, W., Qian, X., and Du, S. 2017. A new method for 3D modeling of joint surface degradation and void space evolution under normal and shear loads. *Rock Mechanics and Rock Engineering*, 50, 2827–36.

Haghi, A. H., and Chalaturnyk, R. 2022. Experimental characterization of hydrodynamic properties of a deformable rock fracture. *Energies*, 15, 6769.

Iwai, K. 1976. *Fundamental Studies in Fluid Flow through a Single Fracture*, Ph.D. dissertation, University of California, Berkeley.

Lang, P. S., Paluszny, A., and Zimmerman, R. W. 2016. Evolution of fracture normal stiffness due to pressure dissolution and precipitation. *International Journal of Rock Mechanics and Mining Sciences*, 88, 12–22.

Markov, A., Abaimov, S., Sevostianov. I., Kachanov, M., *et al.* 2019. The effect of multiple contacts between crack faces on crack contribution to the effective elastic properties. *International Journal Solids and Structures*, 163, 75–86.

Morlier, P. 1971. Description de l'état de fissuration d'une roche à partir d'essais non-destructifs simples [Description of the state of rock fracturization from simple non-destructive tests]. *Rock Mechanics*, 3(3), 125–38.

Morris, J. P., Jocker, J., and Prioul, R. 2017. Numerical investigation of alternative fracture stiffness measures and their respective scaling behaviours. *Geophysical Prospecting*, 2017, 65, 791–807

Myer, L. R. 2000. Fractures as collections of cracks. *International Journal of Rock Mechanics and Mining Sciences*, 37(1, 2), 231–43.

Olsson, W. A., and Brown, S. R. 1993. Hydromechanical response of a fracture undergoing compression and shear, *International Journal of Rock Mechanics and Mining Sciences*, 30(7), 845–51.

Petrovitch, C. L., Nolte, D. D., Pyrak-Nolte, L. J. 2013. Scaling of fluid flow versus fracture stiffness. 2013. *Geophysical Research Letters*, 40(10), 2076–80.

Pyrak-Nolte, L. J., and Morris, J. P. 2000. Single fractures under normal stress: The relation between fracture specific stiffness and fluid flow. *International Journal of Rock Mechanics and Mining Sciences*, 37, 245–62.

Pyrak-Nolte, L. J., and Nolte, D. D. 2016. Hydraulic and approaching a universal scaling relationship between fracture stiffness and fluid flow. *Nature Communications*, 7, 10663.

Pyrak-Nolte, L. J., Myer, L. R., Cook, N. G. W., and Witherspoon, P. A. 1987. Hydraulic and mechanical properties of natural fractures in low permeability rock, in *Proceedings of the 6th International Congress of Rock Mechanics*, G. Herget and S. Vongpaisal, eds., Balkema, Rotterdam, pp. 225–31.

Renshaw, C. E. 1995. On the relationship between mechanical and hydraulic apertures in rough-walled fractures. *Journal of Geophysical Research*, 100(B12), 24, 629–36.

Sisavath, S., Al-Yaarubi, A., Pain, C. C., and Zimmerman, R. W. 2003. A simple model for deviations from the cubic law for a fracture undergoing dilation or closure. *Pure and Applied Geophysics*, 160, 1009–22.

Sneddon, I. N., and Lowengrub, M. 1969. *Crack Problems in the Classical Theory of Elasticity*, Wiley, New York.

Swan, G. 1983. Determination of stiffness and other joint properties from roughness measurements. *Rock Mechanics and Rock Engineering*, 16, 19–38.

Timoshenko, S. P. and Goodier, J. N. 1970. *Theory of Elasticity*, 3rd ed., McGraw-Hill, New York.

Tsang, Y. W., and Witherspoon, P. A. 1981. Hydro-mechanical behavior of a deformable rock fracture subject to normal stress. *Journal of Geophysical Research*, 86(B10), 9287–98.

Walsh, J. B. 1965. The effect of cracks on the compressibility of rock. *Journal of Geophysical Research*, 70(2), 381–89.

Walsh, J. B. 1981. The effect of pore pressure and confining pressure on fracture permeability. *International Journal of Rock Mechanics and Mining Sciences*, 18(5), 429–35.

Walsh, J. B., and Grosenbaugh, M. A., 1979. A new model for analyzing the effect of fractures on compressibility. *Journal of Geophysical Research*, 84(7), 3532–36.

White, F. M. 2006. *Viscous Fluid Flow*, 3rd ed., McGraw-Hill, New York.

Witherspoon, P. A., Wang, J. S. Y., Iwai, K., and Gale, J. E. 1980. Validity of cubic law for fluid flow in a deformable rock fracture, *Water Resources Research*, 16(6), 1016–24.

Yeo, I. W., de Freitas, M. H., and Zimmerman, R. W. 1998. Effect of shear displacement on the aperture and permeability of a rock fracture. *International Journal of Rock Mechanics and Mining Sciences*, 35(8), 1051–70.

Yu, C. 2015. A simple statistical model for transmissivity characteristics curve for fluid flow through rough-walled fractures. *Transport in Porous Media*, 108, 649–57.

Zimmerman, R. W. 1991. *Compressibility of Sandstones*, Elsevier, Amsterdam.

Zimmerman, R. W. 2008. A simple model for coupling between the normal stiffness and the hydraulic transmissivity of a fracture, in *Proceedings of the 42nd US Rock Mechanics Symposium*, San Francisco, paper ARMA 2008-314.

Zimmerman, R. W., Kumar, S., and Bodvarsson, G. S. 1991. Lubrication theory analysis of the permeability of rough-walled fractures. *International Journal of Rock Mechanics and Mining Sciences*, 28(4), 325–31.

Zou, L. C., Li, B., Mo, Y. Y., and Cvetkovic, V. 2020. A high-resolution contact analysis of rough-walled crystalline rock fractures subject to normal stress. *Rock Mechanics and Rock Engineering* 53, 2141–55.

4

Fluid Flow Through Fractures at Moderate to High Reynolds Numbers

4.1 Introduction

As shown in Chapter 2, fluid flow through a rock fracture at low Reynolds numbers is governed by a Darcy-like equation in which the flowrate is linearly proportional to the pressure gradient. The Reynolds number, as defined by eq. (2.21), is essentially a dimensionless ratio of the inertial forces to the viscous forces. The order-of-magnitude analysis of the Navier–Stokes equations that was presented in Section 2.3, supported by analytical solutions for flow through a few simplified fracture geometries (Hasegawa and Izuchi, 1983; Oron and Berkowitz, 1998) and by numerical simulations (Skjetne *et al.*, 1999), implies that fluid flow through a rock fracture will exhibit Darcy-like linear behavior for Reynolds numbers less than 1, and will start exhibiting appreciable deviations from linearity when *Re* reaches about 10.

Most subsurface flow scenarios of engineering or scientific interest occur in the regime $Re < 1$. However, there are a few important situations in which higher Reynolds numbers occur during fracture flow. For example, consider fluid flowing under quasi-steady-state conditions into a vertical wellbore through a parallel plate fracture that lies in a horizontal plane, such as may occur in an engineered geothermal system. If Q [m^3/s] is the total volumetric flowrate and h is the aperture of the fracture, conservation of mass implies that the velocity at a distance r from the center of the borehole is given by $v = Q/A = Q/2\pi rh$. Hence, the velocity increases inversely with the radius, which can lead to large Reynolds numbers near the borehole. Vertical flow of magma through a dike often occurs under conditions of moderate to high Reynolds numbers (Emerman *et al.*, 1986). High Reynolds numbers are also sometimes reached during the flow of low-viscosity "slick water" fracturing fluids in a hydraulic fracture (Dontsov and Peirce, 2017). Non-Darcy flow in rock fractures therefore deserves to be investigated.

Section 4.2 contains a detailed discussion of the analytical solution for flow through a fracture having a sinusoidal aperture in order to develop an estimate of the "critical" Reynolds number at which Darcy-like flow is expected to break down. The two "non-Darcy" flow regimes that occur as the Reynolds number becomes greater than 1, namely the weak inertia regime and the Forchheimer regime, are discussed from a theoretical viewpoint in Section 4.3. The first Navier–Stokes simulations of fluid flow through a real rock fracture profile, conducted by Al-Yaarubi (2003), are discussed in some detail in Section 4.4, with specific focus on verifying the existence of weak inertia and

Fluid Flow in Fractured Rocks, First Edition. Robert W. Zimmerman and Adriana Paluszny.

Companion website: www.wiley.com/go/zimmerman/fluidflowinfracturedrocks

Forchheimer regimes. Additional data sets on flow through fractures in the non-Darcy regime are discussed in Section 4.5 and analyzed in the framework of the Forchheimer equation. Since non-Darcy flow is more likely to occur for gases than for liquids, Section 4.6 is devoted to studying the flow of a highly compressible gas through a rock fracture.

4.2 Approximate Analytical Solution for a Sinusoidal Fracture Aperture

A useful and instructive starting point for understanding the way in which a Darcy-like linear relationship between pressure gradient and flowrate will be expected to break down is the second-order perturbation solution derived by Hasegawa and Izuchi (1983) for flow through a channel bounded by one flat wall and one sinusoidal wall (Fig. 4.1a). The aperture of this fracture is described by

$$h(x) = h_m[1 + \delta \sin(2\pi x/\lambda)], \tag{4.1}$$

in which h_m [m] is the mean aperture, λ [m] is the wavelength, and δ [-] is the dimensionless relative amplitude of the aperture variation. Note that this parameter δ should not be confused with the use of the same symbol in Chapter 3 to denote the fracture normal deformation.

Hasegawa and Izuchi computed a few terms in the perturbation solution for the pressure and velocity fields, using Re and h_m/λ as the two "small" perturbation parameters. To second order in each of these parameters, the transmissivity was found to be given by

$$T = \frac{wh_m^3(1-\delta^2)^{5/2}}{12[1+(\delta^2/2)]}\left\{1 - \frac{3\pi^2(1-\delta^2)\delta^2}{5[1+(\delta^2/2)]}\left[1 + \frac{13(1-\delta^2)^5}{8085[1+(\delta^2/2)]^2}Re^2\right]\left(\frac{h_m}{\lambda}\right)^2\right\}, \tag{4.2}$$

where $Re = \rho U_x h/\mu$. In order to isolate the effect of Re, eq. (4.2) can be written as

$$T = \frac{wh_m^3(1-\delta^2)^{5/2}}{12[1+(\delta^2/2)]}\left(\left\{1 - \frac{3\pi^2(1-\delta^2)\delta^2}{5[1+(\delta^2/2)]}\left(\frac{h_m}{\lambda}\right)^2\right\} - \left\{\frac{39\pi^2(1-\delta^2)^6\delta^2}{40,425[1+(\delta^2/2)]^3}\left(\frac{h_m}{\lambda}\right)^2\right\}Re^2\right). \tag{4.3}$$

As mentioned in Chapter 2, it is *not* instructive to use the case $\delta = 0$, *i.e.*, the parallel plate geometry, as a basis of comparison since, for this case, the Re term drops out of the expression for transmissivity. In order to gain some understanding of the predictions of this model for a rough-walled, albeit highly idealized, sinusoidal geometry, a specific case such as $h_m/\lambda = 0.5$ and $\delta = 0.5$ can be considered. In this case, eq. (4.3) takes the form

$$T = \frac{0.326wh_m^3}{12}(1 - 9.88 \times 10^{-5}Re^2). \tag{4.4}$$

The pre-factor 0.326 shows that the roughness of this fracture has reduced the transmissivity to about one-third of the transmissivity of a parallel plate fracture having the same mean aperture. The inertial effects, as embodied in the Re^2 term, will reduce the effective transmissivity further. However, the deviation from Darcy-like behavior is second-order in Re, which implies that these nonlinear effects will be negligible for $Re < 1$. Moreover, the coefficient that multiplies Re is very small. For this specific geometry, eq. (4.4) indicates

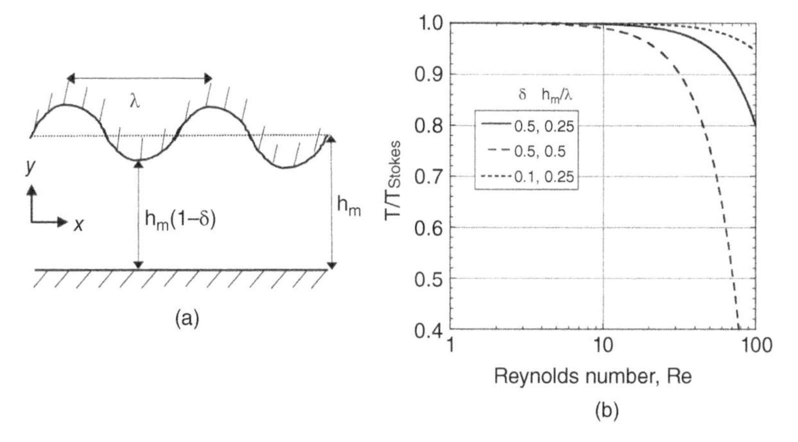

Figure 4.1 (a) Fracture bounded by one smooth wall and one sinusoidal wall, in which the mean aperture is h_m, the wavelength is λ, and the relative amplitude of the aperture variation is δ. (b) Transmissivity as a function of Reynolds number, for several values of h_m/λ and δ. To isolate the effect of *Re*, the transmissivities have been normalized against T_{Stokes}, which can be obtained from eq. (4.2) by setting $Re = 0$.

that inertial effects will cause a 1% deviation from Darcy-like behavior when *Re* reaches about 10, and a 10% deviation from Darcy-like behavior when *Re* reaches about 30.

More generally, the Hasegawa–Izuchi model predicts that deviations from linearity will begin (see Fig. 4.1b) at lower Reynolds numbers if the aperture variations are greater (*i.e.*, larger values of δ), and/or if the wavelength is smaller (*i.e.*, larger values of h_m/λ). Larger values of δ and larger values of h_m/λ can both be thought of as corresponding to "rougher" fractures. Bearing in mind the highly idealized geometry, and the approximate nature of the perturbation solution, these results should probably be interpreted as providing only qualitative guidelines for the onset of nonlinearity.

Finally, it is worth noting that the deviations from Darcy-like behavior that begin to emerge as *Re* increases above about 1 are not due to, or indicative of, turbulence (see Bear, 1988). These deviations arise from the inertial terms in the Navier–Stokes equations and can be attributed to the curvature of the local flowlines. Any such curvature will cause the velocity gradient in eq. (2.3) to no longer be perpendicular to the velocity vector, which in turn causes the nonlinear term in the Navier–Stokes equation to become nonzero. Consequently, the local velocities will no longer be directly proportional to the magnitude of the imposed pressure gradient, and hence the total flowrate will not follow a Darcy-like linear law. This nonlinear behavior begins to occur at Reynolds numbers below the values that would be needed to create true turbulence, which is a type of flow that is both spatially and temporally chaotic.

4.3 Weak Inertia Regime and Forchheimer Regime

Although the numerical values of the coefficients that appear in eq. (4.4) depend on the geometrical parameters δ and h_m/λ, the fact that the initial deviations from Darcy-like behavior are second-order in *Re* is "universal" and applies to any fracture geometry that exhibits "left-right" symmetry. This is also expected to be true for fractures that have no true plane of symmetry but are "symmetrical" in some statistical sense. For any such symmetric geometry, changing the direction of the macroscopic pressure gradient, which mathematically is equivalent to changing the *sign* of the pressure gradient, will necessarily have the effect of merely changing the sign of the local velocities and ultimately will change the sign of the macroscopic flux. Therefore, the flux must be an *odd* function of the flowrate. Any mathematical solution for the flowrate that results in an analytical function for the pressure gradient as a function of flowrate can therefore only contain *odd* powers of Q in its power series expansion. When expressed in terms of the Reynolds number, which is essentially a dimensionless flowrate, the pressure gradient can contain only terms such as Re and Re^3, *etc.* and cannot contain a term such as Re^2. The transmissivity, which is essentially the ratio of the flowrate to the pressure gradient, will therefore only contain even powers of the Reynolds number, such as Re^0 and Re^2, *etc.* This result was proven by Mei and Auriault (1991) for three-dimensional porous media, but their arguments can easily be adapted to a fracture, which can be thought of as a porous medium having a very specific type of geometry.

This regime of "small but non-negligible" Reynolds numbers was referred to by Mei and Auriault as the "weak inertia regime," since in this regime, inertia forces cannot be neglected but are still small relative to viscous forces. If perturbation solutions, such as those derived by Hasegawa and Izuchi (1983) and Mei and Auriault (1991), could be extended to higher orders, the result would presumably be a slowly convergent series that contains all higher odd powers of Re. Such solutions have not yet been obtained for any fracture geometries, but would probably be of little practical use due to their slow convergence. The important point is that for Re on the order of about 1, the apparent transmissivity will begin to diverge below the "Darcy transmissivity," initially according to a term proportional to Re^2.

For flow between two parallel, smooth-walled plates, the Darcy-like relation between flowrate and pressure drop will break down at some specific flowrate at which the flow *abruptly* switches from laminar to turbulent. This generally occurs at a Reynolds number of around 2000 (Sano and Tamai, 2016). In turbulent flow, the pressure drop will be proportional to the square of the flowrate rather than to the flowrate to the first power (Schlichting, 1968, pp. 578–80). When fluid flows through a rough-walled fracture, as the Reynolds number increases, turbulence will eventually occur, at first locally, in different parts of the fracture, and gradually spreading to the entire fracture plane (Zhang *et al.*, 2017). Eventually, it would be expected that at high Reynolds numbers, the overall (macroscopic) pressure drop will be proportional to the square of the flowrate. Hence, the relationship between pressure drop and flowrate will at first be linear for $Re < 1$, will begin to include a term that is cubic in Re as Re rises above 1, and will eventually be expected to be dominated by a term that is *quadratic* in Re. The various ranges of behavior that are expected to occur for fluid flow through a rough-walled fracture are categorized and quantified below, following the discussions given by Mei and Auriault (1991) and Zimmerman *et al.* (2004).

At sufficiently low flowrates, the flow can be described by a linear relation between flowrate and pressure drop:

$$-\frac{dp}{dx} = \frac{\mu Q}{T_0}, \tag{4.5}$$

in which T_o [m^4] is a constant. If flow in this regime is represented by a Darcy-type law, it would be found that the transmissivity T is independent of *Re*, *i.e.*,

$$T \equiv \frac{-\mu Q}{\nabla p} = T_o = \text{constant}, \tag{4.6}$$

where ∇p is written for dp/dx.

At somewhat higher flowrates, the precise value of which seems to depend on the small-scale roughness of the fracture walls (Wang *et al.*, 2016), a transitional regime occurs, in which an *additional* pressure drop emerges that is proportional to the *cube* of the flowrate:

$$-\frac{dp}{dx} = \frac{\mu Q}{T_o} + aQ^3, \tag{4.7}$$

where a is some constant that has units of [kg s/m^{11}]. If flow data in this regime were analyzed using a Darcy-type law, the "apparent" transmissivity T would be found to vary with flowrate, according to

$$T \equiv \frac{-\mu Q}{\nabla p} = \frac{T_o}{1 + (aQ^2 T_o/\mu)}. \tag{4.8}$$

This regime is known as the "weak inertia regime." Due to the additional non-Darcy pressure drop, the apparent transmissivity in this regime will be less than the "Darcy transmissivity," T_0. In this regime, the transmissivity can be written in dimensionless form as (see Problem 4.1):

$$T = \frac{T_o}{1 + \alpha Re^2}, \tag{4.9}$$

in which α is a dimensionless *weak inertia parameter*.

At yet higher flowrates, the additional non-Darcy pressure drop will be proportional to the *square* of the flowrate, rather than the *cube* of the flowrate:

$$-\frac{dp}{dx} = \frac{\mu Q}{T_o} + bQ^2, \tag{4.10}$$

where b is another constant, having units of [kg/m^8]. This regime is known as the "strong inertia regime," and eq. (4.10) is generally known as the Forchheimer equation (Bear, 1988; Zimmerman, 2018), although it seems to have first been proposed by Dupuit (1863); see the discussion given by Lage and Antohe (2000). In this regime, the apparent transmissivity is again less than the Darcy transmissivity and is given, after rearranging eq. (4.10), by

$$T \equiv \frac{-\mu Q}{\nabla p} = \frac{T_o}{1 + (bQT_o/\mu)}. \tag{4.11}$$

This relation can be written in dimensionless form as (see Problem 4.2):

$$T = \frac{T_o}{1 + \beta Re}, \tag{4.12}$$

in which β is a dimensionless parameter that can be referred to as the *dimensionless Forchheimer coefficient*.

Strictly speaking, the quadratic term in eq. (4.10) should be written as $b|Q|Q$, where $|Q|$ is the absolute value of Q. In this form, the equation correctly predicts that changing the sign of the flowrate will have the effect of changing the sign of the pressure gradient. Provided that this obvious fact is borne in mind, it is convenient to ignore the absolute value signs and use the less cumbersome expression given in eq. (4.10).

The Darcy regime and Forchheimer regime have for many years been known to occur in both porous media and fractures (Forchheimer, 1901; Geertsma, 1974; Barree and Conway, 2004; Zhou *et al.*, 2019). Although the existence of these two regimes is essentially an empirical fact, both can be heuristically rationalized. At low flowrates, the flow is governed on a microscopic scale by the Stokes equations, which contain no inertia terms and are consequently linear. Hence, it is to be expected that the mean flowrate and mean pressure drop will be linearly related. At high flowrates, inertial effects (*i.e.*, kinetic energy) become dominant, and since kinetic energy is proportional to velocity squared, it is therefore to be expected that the pressure drop will be proportional to the flowrate squared. This is consistent with eq. (4.10), since at very high flowrates, the quadratic term will dominate the linear term.

The existence of the weak inertia regime can be explained by the arguments given above, which were based on the pressure gradient being an analytical function of flowrate that can contain only *odd* powers. Although the transition between the weak inertia regime and the Forchheimer regime will occur gradually as Re increases, there can be no analytically derived mathematical equation that bridges these two regimes, because, as pointed out above, analytical expressions for the flowrate as a function of pressure gradient will never contain Re^2 or Q^2 terms. Indeed, the "correct" version of eq. (4.10), which contains the absolute value term $|Q|$, is *not* an analytical function of Q, since the term $|Q|$ is not differentiable at $Q = 0$. Hence, the lowest-order nonlinear term in the relationship between pressure gradient and flowrate must necessarily be a cubic term. From a physical point of view, these incipient nonlinear terms correspond to the growth of small recirculating eddies within the "nooks and crannies" of the rough-walled fracture, near the fracture walls, as has recently been imaged by Kim and Yeo (2022).

The flow regime characterization outlined above is a simplification of the complex flow behavior that occurs in rough-walled rock fractures. More detailed analyses of the various flow regimes, and the transitions between them, have been presented by Panfilov *et al.* (2003) and Lucas *et al.* (2007). Lage and Antohe (2000) have also provided an interesting discussion of these regimes. Nevertheless, as shown by the data presented and discussed below, the three regimes outlined above are probably adequate for most engineering purposes.

4.4 Verification of the Weak Inertia and Forchheimer Regimes

The first Navier–Stokes simulations of fluid flow in an actual fracture profile were performed by Al-Yaarubi (2003). These simulations were described by Zimmerman *et al.* (2004), from which the following discussion is taken. Replicas of the surface of a natural fracture in a red Permian sandstone were made using an epoxy adhesive known as Araldite, following the process described by Yeo *et al.* (1998). The surface profiles of both opposing surfaces were measured using a Talysurf profilometer, a device typically used for tribological studies of automotive parts. This profilometer utilizes a diamond stylus of 2 mm radius, positioned at the end of a cantilevered arm. The vertical motion of the

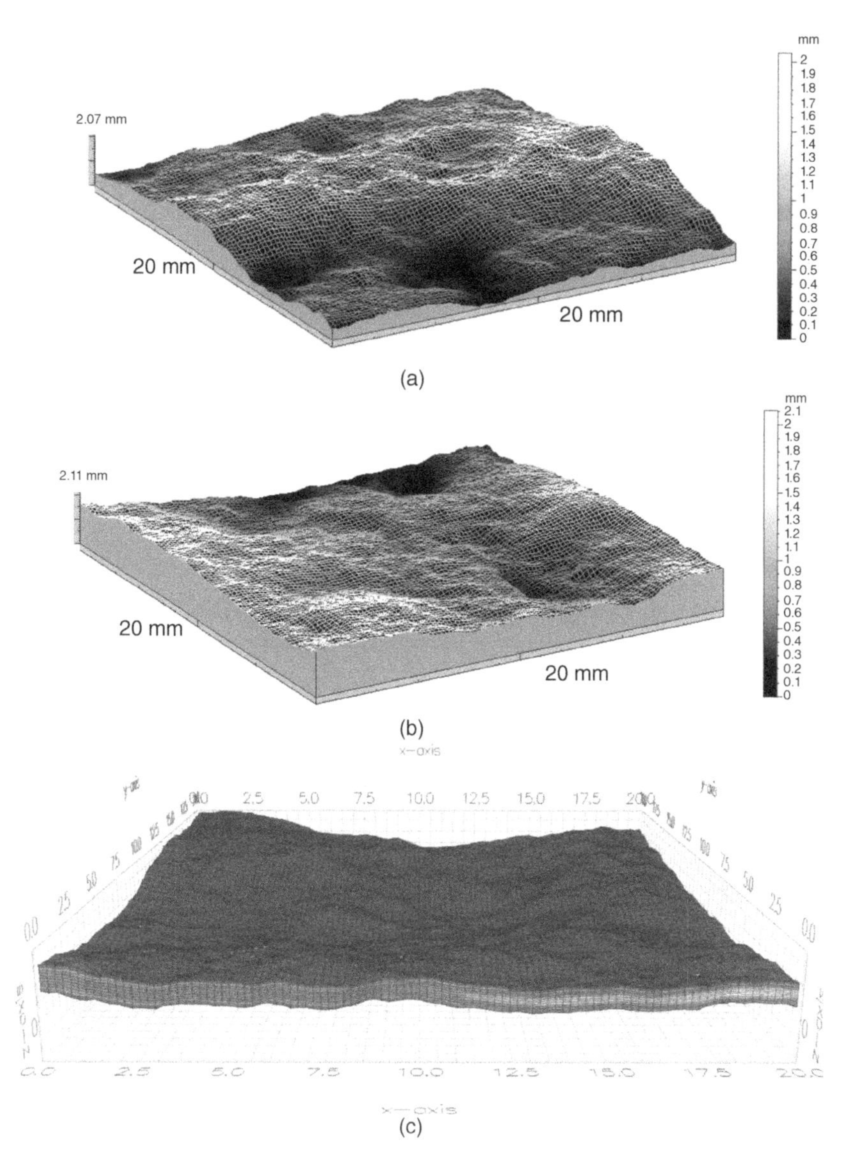

Figure 4.2 (a) Upper and (b) lower surfaces of a 2 cm × 2 cm region of the fracture. The gray scale represents elevations above some nominal plane, with dark = 0 mm and light = 2.1 mm. (c) Finite element grid used in simulations. Side view shows velocity magnitudes, with the same range of gray scale as in (a) and (b), with dark = zero velocity and light = highest velocity.

stylus was measured with a laser, to within an accuracy of ±10 nm, as the stylus traversed the surface. Surface elevations of the fracture surface were thereby measured to within a vertical accuracy of much better than 1 μm, every 20 mm in both the x and y directions. Measurements were made of several 2 cm × 2 cm regions of the fracture surfaces, leading to data files containing about 10^6 profile height values for each of the two surfaces (Fig. 4.2).

The surface data were then used as input to a mesh generator to generate a computational finite element mesh, having a spacing of 200 mm in the x and y directions (Fig. 4.2), which is coarser than the grid on which the surface data were originally measured. Comparison of the discretized boundary as represented in the mesh and the original surface profile data showed that the root-mean-squared discrepancy between the measured profiles and the computational grid was less than 1% of the mean aperture. Hence, the small-wavelength roughness components that were lost in this smoothing process were of small amplitude. This is consistent with the findings of Brown and Scholz (1985), who found that the roughness amplitudes of fractures scale with the wavelength. The grid contained ten (or fifteen) layers of elements in the transverse (z) direction, giving a total of 100,000 (or 150,000) elements.

The finite element code FLUIDITY (Pain, 2000) was used to solve the flow equations in the 2 cm × 2 cm region. FLUIDITY solves the full Navier–Stokes equations, including the inertial terms. The boundary conditions were taken to be uniform pressures p_1 and p_2 on two opposing faces, $x = 0$ and $x = 2$ cm, with zero normal velocities along the two lateral faces, $y = 0$ and $y = 2$ cm. All velocity components were taken to vanish along the upper and lower boundaries of the flow region, which correspond to the two rock walls. The total flux Q through the fracture was found by integrating the normal component of the velocity across the outlet face at $x = 2$ cm. Finally, the transmissivity T was computed from its defining equation, $Q = -(T/\mu)(\Delta p/\Delta x)$.

The transmissivities of two 2 cm × 2 cm regions of fracture, one of which is shown in Fig. 4.2, were also measured in the laboratory using the aforementioned fracture surface casts. The top and bottom halves of the fracture casts were connected to V-shaped channel extensions, thereby allowing the fluid to be spread uniformly across the inlet and the outlet regions. Four pressure-monitoring ports of 0.75 mm diameter, two at each end of the fracture, were drilled through the upper half of the fracture, terminating at the fracture plane. The two fracture halves were then brought together until they contacted each other at a few contact points, after which the sides of the device were sealed. The mean aperture of the assembled fracture apparatus was then determined by a procedure described in detail by Al-Yaarubi (2003). The inlet to the fracture system was connected to a syringe pump, from which distilled water was injected into the fracture. The outlet was connected to a graduated measuring cylinder to collect and measure the outflow fluid during a specific period of time. The four pressure ports on the top of the device were attached to liquid manometers to measure the liquid heads at the inlet and outlet of the fracture.

Flow experiments and simulations were conducted on the fracture cast sample, as shown in the top panels of Fig. 4.2, over a range of flowrates. For both the flow experiments and simulations, the fracture void space had a mean aperture of 148.9 mm and a standard deviation of 55.5 mm. The computed value of the Darcy transmissivity T_o was 0.134×10^{-12} m^4, whereas the best fit to the measured values, in the low-*Re* regime, was 0.136×10^{-12} m^4.

The measured and calculated transmissivities are plotted in normalized form in Fig. 4.3, as functions of *Re*. The filled circles represent the computed values, and the open circles represent the measured values. The large amount of scatter at low flowrates is due to the fact that at low flowrates, and hence at low pressure drops, the absolute errors in reading the manometers and reading the effluent volume create large *relative* errors in the computed transmissivities. At higher flowrates, the relative errors become small – less than

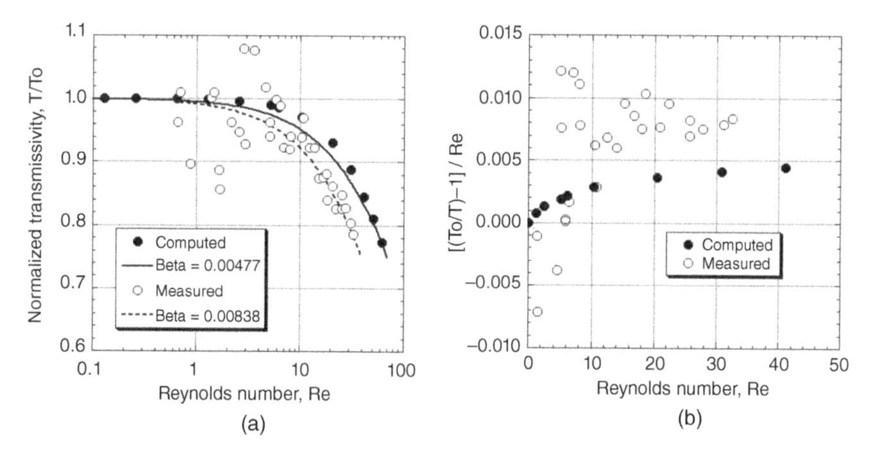

Figure 4.3 (a) Measured and computed transmissivities of the fracture are shown in Fig. 4.2. (a) Normalized transmissivity as a function of Reynolds number, fitted to the Forchheimer equation. (b) Skjetne-type plots of the excess hydraulic resistance, as a function of Reynolds number.

the diameter of the plotted circles – and the data consequently show less scatter. The lines in Fig. 4.3a are Forchheimer curves of the form (4.12), with the dimensionless Forchheimer coefficient β chosen so as to optimize the fit at high Re. Both the computed and measured transmissivities were fit fairly well by the Forchheimer relation, although the best-fitting values of β for the measured data and the computed data differed by nearly a factor of 2. This is probably due to the fact that the computational grid did not include the roughness components with the smallest wavelengths, and so the simulations underestimated the frictional resistance in the inertial regime. This is consistent with the known results for flow in rough-walled pipes (Schlichting, 1968, pp. 578–80), which show that although small-scale roughness is essentially irrelevant at low Reynolds numbers, it greatly *increases* the hydraulic resistance at high Reynolds numbers. This fact, essentially corresponding to an increase in the value of the dimensionless Forchheimer coefficient β as the surface roughness increases, has been demonstrated in rough-walled rock fractures by Briggs *et al.* (2017) and Cunningham *et al.* (2020).

When the transmissivities are plotted as in Fig. 4.3a, the transitional regime between Darcy flow and Forchheimer flow, the so-called weak inertia regime, tends to be obscured. However, its existence can be inferred, at least for the computed transmissivities if not for the measured values, by the fact that in the range $1 < Re < 30$, the Forchheimer equation slightly underestimates the transmissivity. The unavoidable scatter in the measured transmissivity values prevents this transitional regime from being evident in the experimental data plotted in Fig. 4.3a. However, Skjetne *et al.* (1999) showed that the character of the transitional regime could be probed by plotting the excess hydraulic resistance divided by Re. Specifically, in the weak inertia regime, eq. (4.9) can be rearranged into the form

$$\frac{(T_0/T) - 1}{Re} = \alpha Re. \tag{4.13}$$

In the Forchheimer regime, eq. (4.12) can be rewritten as

$$\frac{(T_0/T) - 1}{Re} = \beta. \tag{4.14}$$

Hence, if the ratio $[(T_0/T)-1]/Re$ is plotted against Re, the data should, according to the model of the "weak inertia regime transitioning to the Forchheimer regime," initially fall on a straight line, as in eq. (4.13), and then level off to some constant value, as in eq. (4.14).

The measured and computed transmissivities are plotted in this manner in Fig. 4.3b. The computed values obey the scheme described above quite well: increasing at first and then leveling off as Re increases. Unfortunately, the scatter in the measured values of the transmissivity is enhanced by plotting the data as in Fig. 4.3b, which is essentially equivalent to numerical differentiation, which is known to be an unstable operation. It is therefore difficult to discern any trend for $Re < 10$. However, for $Re > 10$, the data do seem to level off to a constant value, which corresponds to the dimensionless Forchheimer coefficient, β.

Although the numerical and experimental data shown in Fig. 4.3b can be said to confirm the existence of the weak inertia regime, a more useful interpretation of Fig. 4.3a is that the Forchheimer equation is probably sufficiently accurate for modeling the flow over all ranges of Re. This is because, in the range of Reynolds numbers for which the Forchheimer equation fails to capture the weak inertia effects, these effects are too small to be of practical importance. This argument has been put forward by Lage and Antohe (2000) and Zimmerman *et al.* (2004), among others.

4.5 Experimental Data on Fluid Flow at Moderate to High Reynolds Numbers

Despite the recent increased interest in fluid flow through fractures in the non-Darcy regime, few reliable data sets of transmissivity measurements made on actual rock fractures, or casts thereof, are available in the literature. One of the earliest such data sets are the classic measurements made by Iwai (1976) on flow through a tension fracture in granite. Iwai's measured transmissivities are plotted in Fig. 4.4a. As pointed out in the discussion of Al-Yaarubi's data in Section 4.4, transmissivities at low Reynolds numbers inevitably suffer from large scatter due to the larger relative errors incurred when measuring low flowrates and low pressure drops. Nevertheless, the data are moderately well fit by the Forchheimer equation, in the regime $Re > 10$.

Jin *et al.* (2020) measured the flow of cement grout through a set of artificial rock fractures, and found values of the dimensionless Forchheimer coefficients β in the range of 0.002–0.02. Wang *et al.* (2016) used the lattice Boltzmann method to simulate flow through self-affine rock fractures, for Reynolds numbers up to about 50, and found that β decreased from 2.95×10^{-2} to 2.39×10^{-3} as the Hurst exponent of the fracture surfaces increased from 0.5 to 0.8. Their simulations showed that β is strongly influenced by small-wavelength roughness, since simulations through fractures that had their short-wavelength roughness removed invariably exhibited much lower values of β.

Wang *et al.* (2020) conducted flow experiments on a set of replicas of tensile fractures that were artificially induced in granite, at varying levels of shear displacement and under varying normal stresses. Their measured transmissivities could be fit very well by the Forchheimer equation, with β coefficients that were found to lie in the range of 0.002–0.09, decreasing with increasing shear displacement. A typical example of their data is shown in Fig. 4.4b, for their sample G4, under a normal stress of 1 MPa, and at various levels of shear

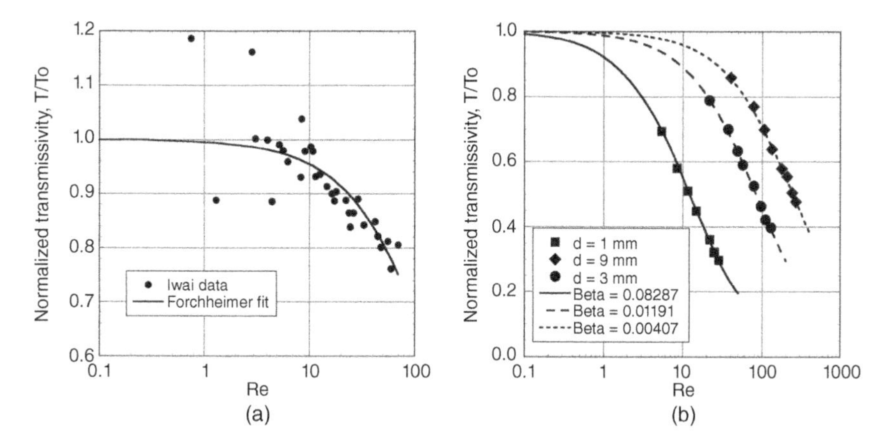

Figure 4.4 (a) Classic data of Iwai (1976) on flow through a tension fracture in granite, fit to a Forchheimer equation, with $\beta = 0.00476$. (b) Data from Wang *et al.* (2020) measured at different shear displacements on replicas of tensile fractures that were artificially induced in granite.

displacement, d. For ease of visualization, the data collected for $d = 5$ mm and $d = 7$ mm are not shown. The transmissivities were fit well by the Forchheimer equation, although it should be acknowledged that in each case the value of T_0 was an extrapolated fitting parameter, as no data were collected in the regime $T/T_0 > 0.9$.

Yu *et al.* (2017) reviewed numerous experimental and numerical studies of nonlinear flow through fractures and focused on the "critical Reynolds number for the onset of nonlinearity," defined as the Reynolds number at which the apparent transmissivity has been reduced to 90% of the value that it has at very low Re. If the transmissivities can be fit by the Forchheimer equation, as in eq. (4.14), then the critical Reynolds number will be inversely related to the dimensionless Forchheimer coefficient, according to $Re_c = 0.1/\beta$. They concluded that the critical Reynolds number increases with increasing shear displacement, at first rapidly, and then gradually, reaching about 25 when the shear displacement reaches about 20 mm. Yu *et al.* (2017) also found that the critical Reynolds number increases with increasing normal stress, reaching about 10 when the normal stress reaches 30 MPa.

4.6 Flow of Compressible Gases Through Fractures

Inertia effects, and consequently non-Darcy behavior, are more liable to occur for gas flow through a fracture, than for liquid flow. To understand why this is the case, consider a parallel plate fracture having aperture h, with the x-axis aligned with the macroscopic pressure gradient. Recalling that $Re = \rho U_x h/\mu$, and that the mean velocity in the x-direction is given by eq. (2.11) as $U_x = Q_x/wh = (h^2/12\mu)(\partial p/\partial x)$, where w is the width of the fracture in the direction normal to the macroscopic pressure gradient, the Reynolds number can be written as

$$Re = \frac{\rho h^3}{12\mu^2}\frac{\partial p}{\partial x}. \tag{4.15}$$

For fluid flowing through a given fracture under a given pressure gradient, the Reynolds numbers for gas flow and for liquid flow will therefore be in the ratio of $Re_G/Re_L = \rho_G\mu_L^2/\rho_L\mu_G^2$. At a temperature of 300 °K, water has a density of about 1000 kg/m^3, and a viscosity of about 0.001 Pa s, whereas air has a density of about 1.18 kg/m^3 and a viscosity of about 1.85×10^{-5} Pa s. Hence, the Reynolds number for air flow would be greater than that for water flow by a factor of about 8. Consequently, inertial effects are more likely to be important for gas flow than for liquid flow.

Consider now the pressure-driven flow of an ideal gas through a parallel plate channel of aperture h, as in the standard laboratory configuration shown in Fig. 2.1. If the x-axis is aligned with the macroscopic pressure gradient, then the local flow velocity is governed by eq. (2.8), which is repeated here:

$$\mu\frac{\partial^2 u_x}{\partial z^2} = \frac{\partial p}{\partial x}. \tag{4.16}$$

If the fluid were incompressible, then u_x would not vary along the flow direction, and consequently the pressure gradient, $\partial p/\partial x$, would also be constant. It would then be possible to develop the cubic law in the form of eq. (2.11). However, the density of a *gas* cannot be considered to be constant as the pressure varies along the length of the fracture, and so a linear pressure profile would not satisfy the conservation of mass. Instead, the analysis proceeds as follows.

If the pressure gradient, $\partial p/\partial x$, is not constant, it is still possible to integrate eq. (4.16) with respect to z, at any fixed value of x, and apply the no-slip boundary conditions, to obtain

$$u_x = -\frac{1}{2\mu}\frac{\partial p}{\partial x}[(h/2)^2 - z^2], \tag{4.17}$$

where $\partial p/\partial x$ must now be considered to be a function of x. This velocity profile can be integrated across the fracture, across the $y-z$ plane, to obtain, as in Section 2.2, the *volumetric* flowrate in the form

$$Q_x = w\int_{-h/2}^{+h/2} u_x dz = -\frac{wh^3}{12\mu}\frac{\partial p}{\partial x}. \tag{4.18}$$

Hence, the flow of a compressible gas through a parallel plate fracture is governed *locally* by the usual cubic law, with a transmissivity given by $T = wh^3/12\mu$. However, the *integrated* form of Darcy's law, for the standard laboratory configuration such as shown in Fig. 2.1, takes a different form for gas flow than for liquid flow, as explained below.

The equation of conservation of mass, in its general form that applies to all one-dimensional flows, whether compressible or incompressible, requires that $\rho Q_x = \dot{m} =$ constant, where $\dot{m} = dm/dt$ is the mass flowrate, with units of [kg/s] (Batchelor, 1967, p. 163). Hence, at each location x along the fracture, the volumetric flowrate will be given by $Q_x = \dot{m}/\rho$. Assuming isothermal flow at some absolute temperature θ, the density of the gas will be linearly related to the pressure by the ideal gas law, $p = \rho\mathscr{R}\theta$, where $\mathscr{R}$ is the gas constant (for the specific gas in question – not the "universal" gas constant), with units of J/kg°K. Hence, $Q_x = \dot{m}\mathscr{R}\theta/p$. Inserting this expression into eq. (4.18) yields

$$\dot{m}\mathscr{R}\theta = -\frac{wh^3}{12\mu}p\frac{\partial p}{\partial x}. \tag{4.19}$$

The left-hand side of this equation is a constant, and so this equation can easily be integrated from the inlet of the fracture, $x = 0$, where $p = p_i$, to the outlet, $x = L$, where $p = p_o$, to yield

$$\dot{m}\mathscr{R}\theta L = -\frac{wh^3}{12\mu}\frac{p^2}{2}\bigg]_{p_i}^{p_o} = \frac{wh^3}{12\mu}\frac{(p_i^2 - p_o^2)}{2},$$

$$\text{i.e.,} \quad \dot{m} = \frac{wh^3}{12\mu\mathscr{R}\theta}\frac{(p_i^2 - p_o^2)}{2L}. \tag{4.20}$$

This result shows that the flow of gas through a fracture, or through porous media in general, is in some sense governed by the gradient of "pressure-squared," rather than by the pressure gradient itself (*cf*., Collins, 1990, Section 3.11; Zimmerman, 2018, Chapter 9).

It is instructive to use the identity $a^2 - b^2 = (a + b)(a - b)$ and write eq. (4.20) as

$$\dot{m} = -\frac{wh^3}{12\mu\mathscr{R}\theta}\frac{(p_o + p_i)}{2}\frac{(p_o - p_i)}{L}, \tag{4.21}$$

which can be interpreted as a traditional Darcy-like flow law, with a driving force equal to the macroscopically imposed pressure gradient, $(p_o - p_i)/L$, but with a transmissivity that is proportional to the mean pressure, $(p_o + p_i)/2$. If the pressure drop is small, then $(p_o + p_i)/2 \approx p$, and $1/\mathscr{R}\theta = \rho/p$, and eq. (4.21) reduces to eq. (2.11), which governs the flow of an essentially incompressible fluid, *i.e.*, a liquid, through a fracture.

At moderate-to-high Reynolds numbers, the flow of gas through the fracture can be assumed to be governed "locally" (*i.e.*, averaged across each plane that is normal to the direction of the macroscopic pressure gradient) by the Forchheimer equation,

$$-\frac{dp}{dx} = \frac{\mu Q_x}{T_o} + bQ_x^2. \tag{4.22}$$

As the quadratic term has its basis in inertial effects, it is convenient to factor out the fluid density and write the coefficient b as $\rho/T_o\kappa$, where κ is a parameter that has units of [m] (Zhou *et al.*, 2019), so that eq. (4.22) takes the form

$$-\frac{dp}{dx} = \frac{\mu Q_x}{T_o} + \frac{\rho Q_x^2}{T_o\kappa}. \tag{4.23}$$

Assuming as usual a constant mass flowrate, conservation of mass gives $\rho Q_x = \dot{m}$, after which the ideal gas law again shows that $Q_x = \dot{m}\mathscr{R}\theta/p$, as was the case in the Darcy regime. Using this relation, along with the ideal gas law, to express the right-hand side of eq. (4.23) in terms of the pressure, leads, after multiplication through by p, to

$$-p\frac{dp}{dx} = \frac{\mu\dot{m}\mathscr{R}\theta}{T_o} + \frac{\dot{m}^2\mathscr{R}\theta}{T_o\kappa} = \frac{\mu\dot{m}\mathscr{R}\theta}{T_o}\left(1 + \frac{\dot{m}}{\mu\kappa}\right). \tag{4.24}$$

The right-hand side is a constant, and so this equation can be integrated from the inlet plane to the outlet plane of the fracture, to yield (Houpeurt, 1959)

$$\frac{p_i^2 - p_o^2}{2L} = \frac{\mu\dot{m}\mathscr{R}\theta}{T_o}\left(1 + \frac{\dot{m}}{\mu\kappa}\right). \tag{4.25}$$

Recalling that $T_o = wh^3/12$ for a smooth-walled fracture, it is seen that for "low" flowrates, when the parenthesized term reduces to 1, this expression agrees with eq. (4.20). The parenthesized term on the right therefore represents the fractional increase in the "pressure drop," due to inertial effects.

By fitting the measured data to eq. (4.25), laboratory data for gas flow through a standard rectangular fracture configuration of length L (in the direction of flow) and width w (transverse to the direction of flow) can be used to estimate the two Forchheimer parameters, T_o and κ.

Problems

4.1 By comparing eqs. (4.8) and (4.9), derive the relation between the dimension*al* weak inertia coefficient, a, and the dimension*less* weak inertia coefficient, α.

4.2 By comparing eqs. (4.11) and (4.12), derive the relation between the dimension*al* Forchheimer coefficient, b, and the dimension*less* Forchheimer coefficient, β.

4.3 Inspect Fig. 14a of Yin *et al.* (2019), and extract the flowrate and pressure gradient data for the case of 10 MPa normal stress, and a temperature exposure of 800 °C. Convert these data to transmissivities and Reynolds numbers, and plot the data in a Skjetne plot, as in Fig. 4.3b. Then fit the data to a Forchheimer curve, and compute the dimensionless Forchheimer coefficient β, and the critical Reynolds number.

References

Al-Yaarubi, A. 2003. *Numerical and Experimental Study of Fluid Flow in a Rough-Walled Rock Fracture*, Ph.D. dissertation, Imperial College, London.

Barree, R. D. and Conway, M. W. 2004. Beyond beta factors: a complete model for Darcy, Forchheimer, and trans-Forchheimer flow in porous media, in *Proceedings of the SPE Annual Technical Conference and Exhibition*, Houston, 26–29 Sept. 2004. Paper SPE 89325.

Batchelor, G. K. 1967. *An Introduction to Fluid Dynamics*, Cambridge University Press, Cambridge and New York.

Bear, J. 1988. *Dynamics of Fluids in Permeable Media*, Dover Publications, Mineola, N.Y.

Briggs, S., Karney, B. W., and Sleep, B. E. 2017. Numerical modeling of the effects of roughness on flow and eddy formation in fractures. *Journal of Rock Mechanics and Geotechnical Engineering*, 9, 105–15.

Brown, S. R. and Scholz, C. H. 1985. Broad bandwidth study of the topography of natural surfaces. *Journal of Geophysical Research*, 90(B14), 12575–82.

Collins, R. E. 1990. *Flow of Fluids through Porous Materials*, Research and Engineering Consultants, Englewood, Colo.

Cunningham, D., Auradou, H., Shojaei-Zadeh, S., and Drazer, G. 2020. The effect of fracture roughness on the onset of nonlinear flow. *Water Resources Research*, 56, e2020WR028049.

Dontsov, E. V. and Peirce, A. P. 2017. Modeling planar hydraulic fractures driven by laminar-to-turbulent fluid flow. *International Journal of Solids and Structures*, 128, 73–84.

Dupuit, J. 1863. *Etudes Théoretiques et Pratiques sur le Mouvement des eaux dans les Canaux Découverts et à Travers les Terrains Perméables*, 2nd ed. [Theoretical and Practical Studies of the Flow of Water in Open Channels and over Permeable Ground], Victor Dalmont, Paris.

Emerman, S., Turcotte, D. L., and Spence, D. A. 1986. Transport of magma and hydrothermal solutions by laminar and turbulent fluid fracture. *Physics of the Earth and Planetary Interiors*, 41(4), 249–59.

Forchheimer, P. 1901. Wasserbewegung durch Boden [Water movement through soil]. *Zeitschrift des Vereines Deutscher Ingenieure*, 45, 1781–88.

Geertsma, J. 1974. Estimating the coefficient of inertial resistance in fluid flow through porous media. *Society of Petroleum Engineers Journal*, 14(4), 445–50.

Hasegawa, E. and Izuchi, H. 1983. On the steady flow through a channel consisting of an uneven wall and a plane wall. *Bulletin of the Japanese Society of Mechanical Engineers*, 26(214), 514–20.

Houpeurt, A. 1959. On the flow of gases in porous media. *Revue de l'Institut Francais du Petrole*, 24(11), 1468–684.

Iwai, K. 1976. *Fundamental Studies in Fluid Flow through a Single Fracture*, Ph.D. dissertation, University of California, Berkeley.

Jin, Y., Han, L., Xu, C., Meng, Q., Liu, Z., *et al.* 2020. Cement grout nonlinear flow behavior through the rough-walled fractures: an experimental study. *Geofluids*, 2020, 9514691.

Kim, D. and Yeo, I. W. 2022. Flow visualization of transition from linear to nonlinear flow regimes in rock fractures. *Water Resources Research*, 58, e2022WR032088.

Lage, J. L. and Antohe, B. V. 2000. Darcy's experiments and the deviation to nonlinear flow regime. *Journal of Fluids Engineering*, 122, 619–25.

Lucas, Y., Panfilov, M., and Buès, M. 2007. High velocity flow through fractured and porous media: the role of flow non-periodicity. *European Journal of Mechanics B/Fluids*, 26(2), 295–303.

Mei, C. C. and Auriault, J. L. 1991. The effect of weak inertia on flow through a porous-medium. *Journal of Fluid Mechanics*, 222, 647–63.

Oron, A. P. and Berkowitz, B. 1998. Flow in rock fractures: the local cubic law assumption re-examined. *Water Resources Research*, 34(11), 2811–25.

Pain, C. C. 2000. *Brief Description and Capabilities of the General Purpose CFD Code: Fluidity.* Internal Report, Imperial College, London, UK.

Panfilov, M., Oltean, C., Panfilova, I., and Buès, M. 2003. Singular nature of nonlinear macroscale effects in high-rate flow through porous media. *Comptes Rendus Mecanique*, 331, 41–48.

Sano, M. and Tamai, K. 2016. A universal transition to turbulence in channel flow. *Nature Physics*, 12, 249–53.

Schlichting, H. 1968. *Boundary-Layer Theory*, 6th ed. McGraw-Hill, New York.

Skjetne, E., Hansen, A., and Gudmundsson, J. S. 1999. High-velocity flow in a rough fracture. *Journal of Fluid Mechanics*, 383, 1–28.

Wang, C., Jiang, Y., Liu, R., Wang, C., *et al.* 2020. Experimental study of the nonlinear flow characteristics of fluid in 3D rough-walled fractures during shear process. *Rock Mechanics and Rock Engineering*, 53, 2581–2604.

Wang, M., Chen, Y.-F., Ma, G.-W., Zhou, J.-Q., *et al.* 2016. Influence of surface roughness on nonlinear flow behaviors in 3D self-affine rough fractures: lattice Boltzmann simulations. *Advances in Water Resources*, 96, 373–88.

Yeo, I. W., de Freitas, M. H., and Zimmerman, R. W. 1998. Effect of shear displacement on the aperture and permeability of a rock fracture. *International Journal of Rock Mechanics and Mining Sciences*, 35(8), 1051–70.

Yin, Q., Liu, R., Jing, H., Su, H., *et al.* 2019. Experimental study of nonlinear flow behaviors through fractured rock samples after high-temperature exposure. *Rock Mechanics and Rock Engineering*, 52, 2963–83.

Yu, Y., Liu, R., and Jiang, Y. 2017. A review of critical conditions for the onset of nonlinear fluid flow in rock fractures. *Geofluids*, 2017, 2176932.

Zhang, W., Dai, B. B., Liu, Z., and Zhou, C. Y. 2017. A pore-scale numerical model for non-Darcy fluid flow through rough-walled fractures. *Computers and Geotechnics*, 87, 139–48.

Zhou, J.-Q., Chen, Y.-F., Wang, L., and Cardenas, M. B. 2019. Universal relationship between viscous and inertial permeability of geologic porous media. *Geophysical Research Letters*, 46, 1441–48.

Zimmerman, R. W. 2018. *Fluid Flow in Porous Media*, World Scientific, Singapore and London.

Zimmerman, R. W., Al-Yaarubi, A. H., Pain, C. C., and Grattoni, C. A. 2004. Nonlinear regimes of fluid flow in rock fractures. *International Journal of Rock Mechanics and Mining Sciences*, 41(Supp. 1), 163–169.

5

Thermo-Hydro-Chemical-Mechanical Effects on Fracture Transmissivity

5.1 Introduction

Chemical processes alter the mineral composition of rocks by selectively dissolving components, which then diffuse and precipitate onto free surfaces. Dissolution and precipitation are relevant both as geological-scale processes that affect subsurface fractures over time and as by-products of relatively short-lived engineering or natural processes, such as injection of a reactive fluid or an earthquake. Thermo–chemical–mechanical deformation is of special interest in relation to the disposal of nuclear waste in subsurface repositories, for geothermal energy studies where fluids continuously react with the rock, and in the study of seismicity and induced seismicity in the context of reactive fluid migration.

Mechanical compaction can lead to dissolution along the plane of the fracture, which in turn causes a local decrease in roughness and occlusion of pathways, leading to flow channeling. These changes to the internal geometry of the fracture network influence preferential flow by blocking and opening new paths within the fractures and the rock matrix, thereby affecting the transmissivity of the system as a function of the mechanical and chemical balancing of the system. Dissolution and precipitation may contribute either to an increase or decrease in the transmissivity by modifying the pore space and the geometry of fracture surface walls. Mineral solubility increases with pressure and temperature, leading to preferential dissolution regions, which in tensile stress regimes are the regions around fracture tips, and in the more ubiquitous compressive regime, occur on the contacting regions of the fracture surfaces.

Simplified reactive models usually consider only the dissolution and precipitation of major minerals of the selected mineralogy, which for granite will typically be quartz and for limestone will typically be calcite. When observing fractures in the field, evidence of precipitation of the dissolved minerals is usually found in the form of precipitates filling mineralized veins (Fig. 5.1). These mineralized veins record, to some extent, the *in situ* aperture distributions of the system at the moment of precipitation and also often trap fluids that can be instrumental in analyzing the conditions under which precipitation occurred. On a small scale, precipitated crystals can be used as chemical markers of growth, capturing the details of the conditions of the chemical interactions that took place when they formed and during their burial history. Minuscule fluid inclusions trapped in the precipitates can be studied to constrain the timing, duration, and conditions of fracture development, holding clues to how the patterns formed over time and how their

Fluid Flow in Fractured Rocks, First Edition. Robert W. Zimmerman and Adriana Paluszny.

Companion website: www.wiley.com/go/zimmerman/fluidflowinfracturedrocks

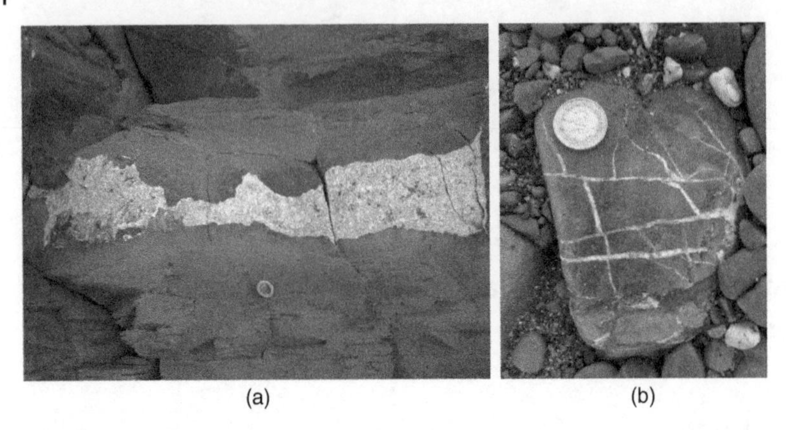

Figure 5.1 Mineralized veins in limestone rocks at the centimeter scale. (a) Plane view of a vein partially filled with precipitated calcite. (b) Cross-cut view of several intersecting fractures that have also been filled with minerals.

shape may have evolved, in some cases over millions of years. By affecting single fractures, dissolution and precipitation also affect the behavior of fracture networks buried deep in the subsurface, influencing their connectivity and their permeability (Laubach *et al.*, 2019).

Veins filled with mineral precipitates can extend into the scale of tens of meters, at which metal precipitates can also be found. Dissolution and precipitation also affect faults, as these processes serve to rapidly strengthen faults after earthquakes due to healing processes driven by dissolution of minerals along the fault planes and vicinal fractures during and after deformation. This behavior can be observed experimentally on a small scale and can also be observed in the field (Beeler and Hickman, 2004).

The present chapter will focus on analytical and numerical models for coupled thermal, hydraulic, mechanical, and chemical ("THMC") rock-water interactions. The classical Hertz model of contact between spherical asperities is briefly reviewed in Section 5.2. The thermodynamics and kinetics of pressure dissolution are discussed in Section 5.3. Models for the process of precipitation are introduced in Sections 5.4 and 5.5, based heavily on the work of Yasuhara and coworkers. The resultant changes in fracture aperture and the implications for fracture transmissivity are discussed in Sections 5.6 and 5.7. Section 5.8 discusses some basic concepts related to the numerical simulation of solution-precipitation in rock fractures. The models of Lehner–Leroy and Bernabé–Evans are presented in Sections 5.9 and 5.10, respectively. The chapter concludes with some comments on the differences between open and closed systems in Section 5.11.

5.2 Fracture Contact

The simplest analytical models of contact between fracture faces are based on the consideration of fracture contact regions and assume "Hertzian" contact between a fracture asperity and the opposing fracture surface. The stresses resulting from this interaction can be approximated by the classical Hertz solution for the contact between an elastic sphere and an infinite half-space (Fig. 5.2). The spherical asperity is assumed to be an elastic body

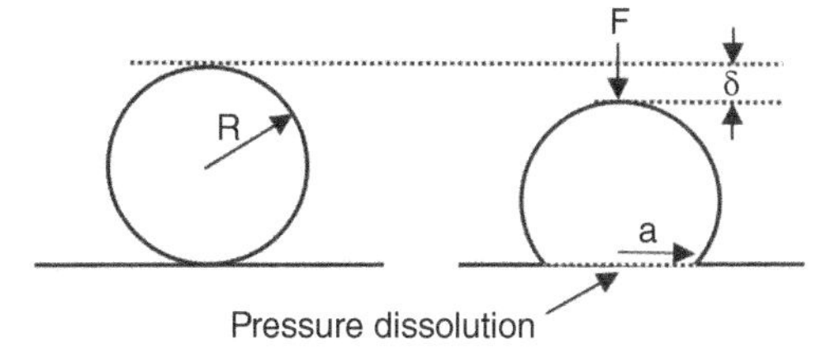

Figure 5.2 Contact of a spherical asperity of radius *R*, and a flat surface, showing (left) the geometry before any force is applied, and (right) after the application of a normal force, *F*. Pressure dissolution will occur along the dotted interface.

having elastic modulus, E [Pa], and Poisson's ratio, ν [–], whereas the half-space is assumed to be rigid. If the undeformed asperity is modeled as a hemispherical bump of radius R [m], and the total normal force acting between the asperity and the flat surface is F [N], the Hertzian contact model predicts that the asperity will deform until the region of contact between the asperity and the flat surface has the form of a circular region of radius $a < < R$, where a is given by (Timoshenko and Goodier, 1970)

$$a = \left[\frac{3FR(1-\nu^2)}{4E}\right]^{1/3}. \tag{5.1}$$

The deformed asperity has the shape of a truncated hemisphere whose tip has been sliced off to create a flat circular face of radius a (Fig. 5.2). Bearing in mind that $a \ll R$ (although the ratio a/R is exaggerated in the figure for illustrative purposes), simple geometrical considerations show that the normal displacement of the tip of the asperity is given by

$$\delta = \frac{a^2}{R} = \left[\frac{3F(1-\nu^2)}{4E\sqrt{R}}\right]^{2/3}. \tag{5.2}$$

The mean normal stress acting over the contact area is given by

$$\sigma_{mean} = \frac{F}{\pi a^2} = \left(\frac{16F}{9\pi^3 R^2}\right)^{1/3}\left(\frac{E}{1-\nu^2}\right)^{2/3}. \tag{5.3}$$

The *maximum* normal stress, which occurs at the center of the asperity, is 50% greater than the mean value, *i.e.*, $\sigma_{max} = 1.5\sigma_{mean}$.

The general form of the analytical solution of the Hertzian contact problem can also account for the case of two spherical bumps in contact, perhaps having different radii and different elastic properties (Timoshenko and Goodier, 1970). Although this basic Hertzian contact model neglects plastic and brittle deformation of the asperities, as well as interactions between nearby asperities, and other effects due to deviations from these highly idealized geometries, it nevertheless provides a starting point for understanding how fracture surface roughness and fracture stiffness change as a function of contact area.

If the macroscopic area of the nominal fracture plane is A_f [m^2], and the actual area of the nominal fracture plane over which the two rock faces are in contact is A_c [m^2], the contact area fraction, c (sometimes denoted by R_c), can be defined as

$$c = \frac{A_c}{A_f}. \tag{5.4}$$

Since $A_c \leq A_f$, it follows that $0 \leq c \leq 1$. The contact area fraction controls the relationship between the macroscopic normal stress and the actual contact stress, since if the "macroscopic" normal stress acting over the fracture plane is σ_{mac}, then a simple force balance shows that the mean normal stress acting over the actual contact regions is given by

$$\sigma_{mean} = \frac{\sigma_{mac}}{c}. \tag{5.5}$$

5.3 Pressure Dissolution

If the stress acting over the contact region exceeds some critical value, a process known as *pressure dissolution* will occur. In this process, minerals dissolve from the asperity, subsequently diffuse through an interfacial water film around the asperity, and eventually precipitate onto the fracture or pore walls (Fig. 5.3). This critical dissolution stress, which depends on temperature and varies from one mineral to another, can be expressed as (Revil, 1999)

$$\sigma_{crit} = \frac{E_m[1-(T/T_m)]}{4V_m}, \tag{5.6}$$

where E_m [J/mol] is the latent heat of fusion of the mineral, T [K] is the absolute temperature, T_m [K] is the melting point of the mineral, and V_m [m^3/mol] is the molar volume of the mineral. For quartz, for example, these values are $E_m = 8.57$ kJ/mol, $T_m = 1883$ K, and $V_m = 2.6 \times 10^{-5}$ m^3/mol, although values reported by different sources tend to differ slightly.

Rocks are often composed of multiple minerals, which have different critical dissolution stresses and dissolve at different rates. Usually, these poly-mineral systems contain a dominant mineral that contributes most of the volume reduction during dissolution. Consequently, and for simplicity, the models described below will focus only on capturing the behavior of the main interacting minerals of the rock. These main interacting minerals are often quartz, calcite, and silica, as found in granites, limestones, and calcium carbonate rocks (Fig. 5.4).

Dissolution is defined in terms of a mass flux, which is quantified as dM_{diss}/dt, [mol/s], which expresses the rate at which the dissolved mass leaves the asperity and enters the

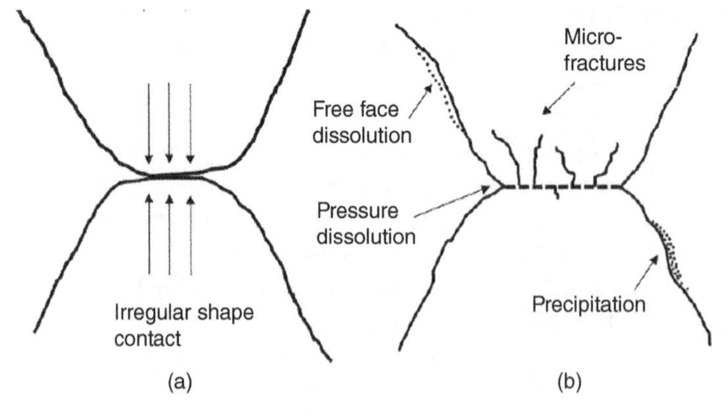

Figure 5.3 Schematic drawing of pressure dissolution at a contacting asperity. (a) Initial geometry of two contacting, irregular asperities. (b) Dissolution, precipitation, and micro-cracking due to normal stresses across the contact region.

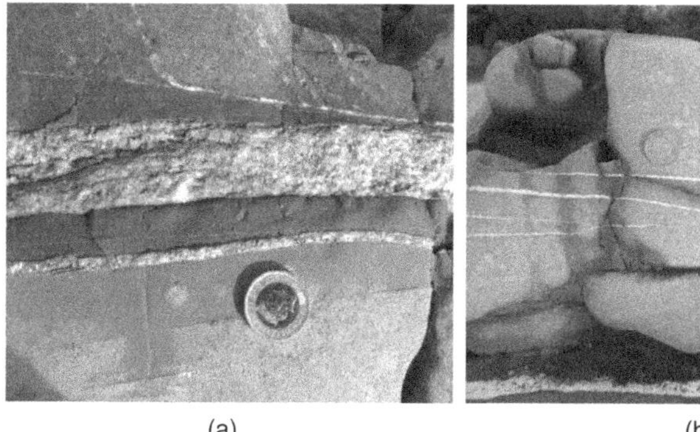
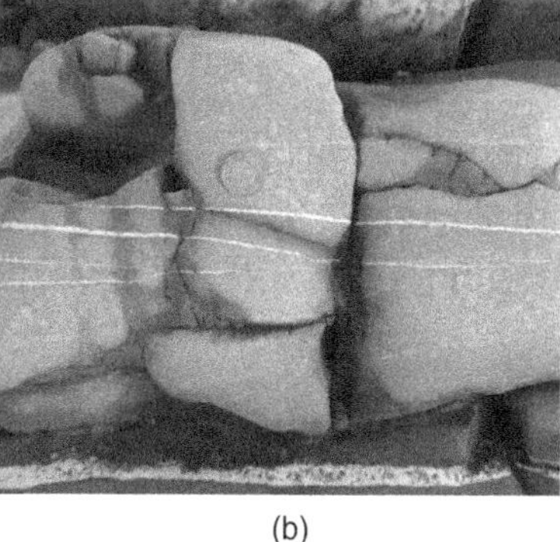

(a) (b)

Figure 5.4 Fractures filled with mineral precipitates can commonly be found in outcrops. (a) Three prominent vicinal fractures in limestone filled with mineral precipitates. (b) An array of fractures is highlighted by the precipitation in the same rock as shown in (a).

adjacent fluid film. This flux is driven by a combination of pressure and temperature and is ultimately governed by the difference between the chemical potential of the mineral when it is located in the stressed asperity and its chemical potential when it is located in an open, unstressed portion of the fracture surface. The difference is defined as the "final" chemical potential minus the initial chemical potential, $\Delta\mu = \mu_f - \mu_i$ [J/mol]. The chemical potential μ at the rock contact can be defined, considering non-hydrostatic and nonequilibrium thermodynamics, as (Heidug, 1995)

$$\mu = (\sigma_a + p)V_m + f - 2\kappa\gamma V_m, \tag{5.7}$$

where p [Pa] is the pore fluid pressure, f [J/mol] is the molar Helmholtz free energy, V_m [m^3/mol] is the molar volume, κ [1/m] is the mean local curvature of the solid-fluid interface, γ [J/m^2] is the specific interfacial energy at the contact, and σ_a [Pa], referred to by Yasuhara *et al.* (2003) as the "disjoining pressure," is the difference between the normal stress at the contact and the pore fluid pressure. For quartz, the molar volume, usually measured via X-ray crystallography, is 22.7×10^{-3} m^3/mol, and for calcite, the molar volume is 33.6×10^{-3} m^3/mol.

The pressure acting on the free face of the grains is the pore fluid pressure p, and so the chemical potential difference between the grain-to-grain contact area and the unstressed open pores is given by Yasuhara *et al.* (2003)

$$-\Delta\mu = \sigma_a V_m + \Delta f - 2\kappa\gamma V_m, \tag{5.8}$$

where Δf is the molar Helmholtz free energy difference between the stressed rock contact and the free surface. The first term on the right-hand side is an internal energy term due to stress, analogous to the pV term that appears in the thermodynamics of liquids, and the third term, $-2\kappa\gamma V_m$, is the surface energy at the contact, which is assumed to be the place where two rock grains meet and dissolve. The stress term usually dominates relative to the other two terms, in which case the second and third terms on the right-hand side can be ignored. This analysis assumes that the work done by the normal contact force is completely transformed into chemical potential. Thus, elastic strain, micro-fracturing, and plastic deformation of the asperity are not considered.

Once the system reaches thermodynamic equilibrium, it follows that $\Delta\mu = 0$, and therefore

$$\left(-\sigma_a = \frac{\Delta f}{V_m} - 2\kappa\gamma\right)_{eq}, \tag{5.9}$$

where the subscript *eq* denotes the equilibrium condition for each variable. At this state, the contact stress reaches the critical stress, $(\sigma_a)_{eq} = \sigma_{crit}$, resulting in

$$(\sigma_a)_{eq} = \sigma_{crit} = 2\gamma\kappa_{eq} - \frac{\Delta f_{eq}}{V_m}. \tag{5.10}$$

It follows that the chemical potential difference can be expressed as

$$-\Delta\mu \approx (\sigma_a - \sigma_{crit})V_m, \tag{5.11}$$

which defines the threshold stress value, σ_{crit}, that triggers pressure dissolution as a function of the chemical potential difference. The definition of the chemical potential difference can also be written as (Shimizu, 1995)

$$-\Delta\mu \approx \sigma_{eff}V_m. \tag{5.12}$$

The dissolution rate is proportional to the "activation energy coefficients" of the dissolution reactions taking place (Yasuhara *et al.*, 2003). Reaction rates in most materials, including rocks, follow an Arrhenius-type relationship, which implies that reactants must accrue a minimum amount of activation energy before reacting. In the context of dissolution and precipitation, the dissolution rate constant, k_+ [mol/m^{2}s], and the precipitation rate constant, k_-, follow relationships of the form

$$k_+ = k_+^0 e^{-E_{k_+}/\mathcal{R}T}, \tag{5.13}$$

$$k_- = k_-^0 e^{-E_{k_-}/\mathcal{R}T}, \tag{5.14}$$

where k_+^0 and k_-^0 are pre-exponential factors that are constant for each chemical reaction, $\mathcal{R} = 8.314$ J/mol K is the universal gas constant, T [K] is the absolute temperature, and E_{k_+} and E_{k_-} [J/mol] are the activation energy coefficients of the dissolution and precipitation reactions, respectively. These coefficients must be experimentally determined, and for quartz, they have been found to be $k_+^0 = 1.59$ mol/m^{2}s and $E_{k_+} = 71.3$ kJ/mol (Dove and Crerar, 1990) and $k_-^0 = 0.196$ mol/m^{2}s and $E_{k_-} = 49.8$ kJ/mol (Rimstidt and Barnes, 1980).

The rate of compaction of the rock is a function of the absolute rate of diffusion and the length of the diffusion path, which changes as dissolution progresses and the contact areas geometrically evolve. The strain rate can be expressed as (Yasuhara *et al.*, 2003)

$$\frac{d\varepsilon_{diss}}{dt} = \frac{3V_m^2 k_+}{2\mathcal{R}TR}(-\Delta\mu), \tag{5.15}$$

where ε_{diss} is the strain, normal to the fracture plane, due to dissolution, and R is the radius of the grain. Combining eqs. (5.11) and (5.15) yields

$$\frac{d\varepsilon_{diss}}{dt} = \frac{3V_m^2 k_+}{2\mathcal{R}TR}(\sigma_a - \sigma_{crit}) = \frac{3V_m^2 k_+}{2\mathcal{R}TR}\sigma_{eff}. \tag{5.16}$$

Following the discussion given by Yasuhara *et al.* (2003), who assumed two asperities of initial radius R in contact with each other, rather than a single asperity in contact with a rigid half-space, the strain rate in the direction normal to the nominal fracture surface can be expressed in terms of the local compaction as

$$\frac{d\varepsilon_{diss}}{dt} = \frac{1}{2R}\frac{d\delta}{dt}. \tag{5.17}$$

The surface area of the contact region is πa^2, and so the increment of dissolved volume is $2\pi a^2 d\delta$, and the increment of dissolved mass is $2\pi a^2 \rho_s d\delta$. It follows that the rate of change of the dissolved mass is $2\pi a^2 \rho_s(d\delta/dt)$. Consequently,

$$\frac{d\delta}{dt} = \frac{1}{2\pi a^2 \rho_s}\frac{dM_{diss}}{dt}, \tag{5.18}$$

where dM_{diss}/dt [kg/s] is the dissolution mass flux, and ρ_s [kg/m^3] is the density of the solid grains. Combining eqs. (5.16)–(5.18) yields the following expression for the dissolution mass flux:

$$\frac{dM_{diss}}{dt} = \frac{3\pi \rho_s V_m^2 a^2 k_+}{\mathscr{R}T}(\sigma_a - \sigma_{crit}). \tag{5.19}$$

The dissolution mass flux is inversely proportional to the temperature of the system and proportional to the compressive stress and the dissolution rate constant of the solid.

According to this model, the rate of aperture reduction increases proportionally to the increase in compressive stress. For systems dominated by dissolution, dependence of aperture reduction on stress is roughly linear, and doubling the stress will result in doubling the rate of reduction of aperture. However, the dependence on temperature is more pronounced, and the dissolution rate mainly depends on this variable. By substituting eq. (5.13) into eq. (5.19) and neglecting the effect of stresses and contact area, it is seen that (Yasuhara *et al.*, 2004)

$$\frac{dM_{diss}}{dt} \propto \frac{k_+^0}{\mathscr{R}T} e^{-E_{k+}/\mathscr{R}T}. \tag{5.20}$$

Therefore, changes in temperature can significantly accelerate the dissolution process. For example, for temperatures in the range of 80–200 °C, fracture healing (*i.e.*, the complete reduction of the fracture aperture) in granite can occur in a few years, whereas for a similar system at lower temperatures, healing would require hundreds of years. Compressive stresses of around 10 MPa would further accelerate fracture healing to a fraction of a year (Yasuhara *et al.*, 2004).

5.4 Diffusion Rates

The dissolved minerals flow into solution and are transported in the pore fluid via a diffusive process. According to Fick's first law of diffusion, the diffusion of the mineral solute under steady-state conditions can be expressed as (Ghez, 1988)

$$J = -DA\frac{dC}{dx}, \tag{5.21}$$

where C [kg/m^3] is the concentration of the mineral in the pore water, dC/dx is concentration gradient, D [m^2/s] is the diffusion coefficient of that mineral in water, A [m^2] is the cross-sectional area through which the mineral diffuses, and J [kg/s] is the mass flux rate. The concentration will be denoted by C to avoid confusion with the contact area fraction, which is denoted by c. Note also that in some equations in this chapter, C is quantified as a mass fraction, in which case it is dimensionless.

The diffusion coefficient varies with temperature according to an Arrhenius-type law,

$$D = D_o e^{-E_D/\mathscr{R}T}, \tag{5.22}$$

where D_o[m^2/s] is the pre-exponential factor, and E_D[J/mol] is an activation energy. For quartz, these parameters have the values $D_o = 5.2\times10^{-8}$ m^2/s and $E_D = 13.5$ kJ/mol (Revil, 1999).

If w [m] is the thickness of the pore fluid film at the interface of a circular contact region of radius a, then the radial flux through this contact region would be given by

$$J = \frac{dM_{diss}}{dt} = -2\pi rwD\frac{dC}{dr}, \tag{5.23}$$

where $2\pi rw$ represents the annular area through which that solution is diffusing and r is a radial coordinate lying parallel to the dashed contact area in Fig. 5.2. For quartz, the fluid film thickness, which can also be interpreted as the diffusion path width, was taken by Yasuhara *et al.* (2003) to be 4.0×10^{-9} m. Assuming that the minerals are uniformly dissolved over the stressed interface, this process is governed by a steady-state diffusion equation with a constant source term. Yasuhara *et al.* (2011) integrated this equation to find the following expression for the mass flux:

$$\frac{dM_{diss}}{dt} = 8\pi wD(C_i - C_p), \tag{5.24}$$

where C_i is the mean mineral concentration in the pore fluid within the interface region, and C_p is the mineral concentration in the fluid within the open fracture space, *i.e.*, at $r = a$.

5.5 Solute Precipitation

Once the minerals are dissolved, they will migrate through the fracture until the conditions for precipitation are met. In general, precipitation will reduce the amount of fracture void space, thereby changing both the mechanical and transport properties of the system. For pure quartz systems, precipitation may occur uniformly due to the homogeneous mineral content. In contrast, in rocks with multiple minerals, such as granite, precipitation can be more localized and can lead to the formation of crystals (Morrow *et al.*, 2001). The mass rate of precipitation of the solute on the walls of the fracture can be expressed as follows (Yasuhara *et al.*, 2003):

$$\frac{dM_{prec}}{dt} = V_p\frac{A}{M}k_{-}(C_p - C_{eq}), \tag{5.25}$$

where V_p [m^3] is the volume available for precipitation to occur, such as the volume of an empty fracture, k_{-} [kg/m^3 s] is the precipitation rate of the dissolved mineral expressed in terms of mass rather than moles, C_p [–] is the concentration of the dissolved mineral in the open pore space, and C_{eq} [–] is the equilibrium solubility of the dissolved mineral. As an example, the solubility of quartz is 150 ppm at 150 °C (Yasuhara *et al.*, 2003).

The dimensionless quantities A and M are the dimensionless relative fracture surface area and relative mass of the fluid, respectively (Rimstidt and Barnes, 1980).

For open and closed systems, a system of equations can be developed to model the relationship between changes in concentration, changes in temperature, and mass dissolution rates so as to predict the precipitation rate. It follows that the evolution of the mean concentrations of dissolved minerals in the interface between the stressed rock contacts and in the fracture "void" may be expressed as (Yasuhara *et al.*, 2003):

$$\begin{bmatrix} C_i \\ C_p \end{bmatrix}^{t+\Delta t} = \begin{bmatrix} D_1 + \frac{V_p}{4\Delta t} & -D_1 \\ -D_1 & D_1 + D_2 + \frac{V_p}{2\Delta t} \end{bmatrix}^{-1} \left\{ \begin{bmatrix} dM_{diss}/dt \\ D_2 C_{eq} \end{bmatrix}^{t+\Delta t} + \frac{V_p}{4\Delta t} \begin{bmatrix} 1 & 0 \\ 0 & 2 \end{bmatrix} \begin{bmatrix} C_i \\ C_p \end{bmatrix}^{t} \right\}, \tag{5.26}$$

where

$$D_1 = 8\pi w D, \quad D_2 = V_p \frac{A}{M} k_-, \tag{5.27}$$

and Δt denotes the time increment.

When modeling the three processes of dissolution, diffusion, and precipitation using these equations, in order to predict the reduction of fracture aperture and predict transport, the volumetric changes of the interface and free face region are not taken into account. This may result in the overestimation of the aperture reduction as the pressure dissolution and diffusion are not truly balanced. This can be accounted for by including an additional term to balance the pressure and diffusion mass flux. In numerical models, this effect is implicitly included as the regions of volume loss and accretion are explicitly tracked.

5.6 Aperture Changes

The rate of aperture reduction, or rate of fracture closure, dh/dt, can be expressed as a function of the flow rate, Q [m^3/s], the concentration of the mineral in the pore fluid, C_p [kg/m^3], and the contact area, A_c, as follows (Polak *et al.*, 2003):

$$\frac{dh}{dt} = \frac{QC_p}{\rho_s A_c}. \tag{5.28}$$

Based on a simple mass balance, the mineral concentration in the pore fluid can be expressed as a function of the mass of the dissolved mineral and the flow rate, as

$$C_p = \frac{1}{Q}\frac{dM_{diff}}{dt}. \tag{5.29}$$

Substituting eq. (5.29) into eq. (5.28), and assuming that total dissolution mass flux and diffusion mass flux balance each other, leads to

$$\frac{dh}{dt} = \frac{1}{\rho_s A_c}\frac{dM_{diff}}{dt}. \tag{5.30}$$

Next, by considering expression (5.19) for the dissolution mass flux, the previous equation can be written as

$$\frac{dh}{dt} = \frac{3\pi V_m^2 a^2 k_+}{\mathcal{R} T A_c}(\sigma_a - \sigma_c). \tag{5.31}$$

Assuming disk-shaped contact regions, and $A_c = \pi a^2$, the rate of aperture change can be expressed as

$$\frac{dh}{dt} = \frac{3V_m^2 k_+}{\mathcal{R}T}(\sigma_a - \sigma_c). \tag{5.32}$$

If the stresses acting over the contact regions are much larger than the critical stress, the term σ_c can be dropped, resulting in

$$\frac{dh}{dt} = \frac{3V_m^2 k_+ \sigma_a}{\mathcal{R}T}. \tag{5.33}$$

Noting that the stress σ_a acts only over the contact region, whereas the macroscopic stress σ_{mac} acts over the entire fracture plane, eq. (5.5) shows that the rate of compaction can also be expressed

$$\frac{dh}{dt} \approx \frac{3V_m^2 k_+ \sigma_{mac}}{c\mathcal{R}T}. \tag{5.34}$$

where c is the contact area fraction.

Unfortunately, for temperatures between 80 and 150 °C, the aperture closure rates predicted by this model are found to be 1–3 orders of magnitude smaller than those obtained by experimental measurements (Bond *et al.*, 2016, 2017). The underestimation of the fracture aperture closure is thought to be a reflection of the underestimation of the actual micro-scale surface area of dissolution. The measured macroscopic area of the fracture is a rough estimate of the microscopic fracture area, which is not only larger but also increases as a function of additional processes that occur during stress-driven dissolution – namely micro-fracture and fragmentation along the fracture surface.

A roughness factor, f_r, can be defined to account for this discrepancy, with the aim of accounting for the difference between the measured macroscopic surface area and the true surface area of the fracture, including its microscopic detail (Murphy and Helgeson, 1989). This factor effectively magnifies the dissolution and precipitation of the solid. It follows that the dissolution can be expressed as:

$$\frac{dM_{diss}}{dt} = \frac{3\pi \rho_s V_m^2 a^2 k_+ f_r}{\mathcal{R}T}(\sigma_a - \sigma_c). \tag{5.35}$$

This model can be evaluated in the long-time limit, where it is assumed that the pressure solution has dissolved away the surface roughness, without assuming a minimum critical stress. Likewise, the precipitation mass flux can be rewritten as

$$\frac{dM_{prec}}{dt} = f_r V_p \frac{A}{M} k_- (C_f - C_{eq}). \tag{5.36}$$

The roughness factor can be approximated by fitting the model to experimental data and results in relatively large values ranging from 40 to 3000, with higher values resulting from calibration with low-temperature experiments and lower values resulting from calibration with higher-temperature experiments (Yasuhara *et al.*, 2004). Additional factors that may contribute to the underprediction of dissolution rates may be related to unaccounted physical behaviors, such as free-face dissolution, and mechanical closure effects that are dominant at lower temperatures.

Once roughness factors are taken into account, predictions of aperture are in good agreement with the actual data. However, while the roughness factor leads to a match between experiments and models in predicting apertures, concentrations of the dissolved minerals obtained with these models still tend to underestimate experimental measurements by orders of magnitude. This discrepancy is partly due to the potentially dominating effects of free-face dissolution, which are not taken into account in these models.

5.7 Relationship Between Aperture, Contact Fraction, and Transmissivity

Yasuhara *et al.* (2004) approximated the mean fracture aperture, $\langle h \rangle \equiv h_m$, as a function of the evolving contact fraction, as follows:

$$\langle h \rangle = \gamma_1 + \gamma_2 e^{-(c/\gamma_3)}, \tag{5.37}$$

where $\{\gamma_i\}$ are fitting parameters. This model assumes that the rock is rigid, as opposed to behaving elastically, which corresponds to the state of the rock below the critical contact stress. It also assumes that there exists a minimum residual aperture that exists even if the fracture is "closed." This relationship suggests that there is no unique relationship between aperture and contact fraction, and that it depends on parameters derived from the specific conditions of the experiment.

Assuming as a zeroth-order approximation that $\langle h \rangle \equiv h_m \approx h_h$, mean mechanical apertures have been fit by eq. (5.37), independent of temperature and pressure conditions (Yasuhara *et al.*, 2004). Lang *et al.* (2015) showed that eq. (5.37) can be interpreted so that γ_1 [m] is the residual aperture, γ_2 [m] is the initial aperture, and γ_3 [-] is a dimensionless roughness amplitude of the composite surface, which is inversely proportional to the initial aperture of the fracture. For a damaged rock, γ_2 can be approximated by the fracture aperture of the damaged rock minus the residual fracture aperture. It follows that if residual apertures are not accounted for, then $\gamma_1 = 0$. For a three-dimensional siliciclastic fracture with a Hurst exponent (see Chapter 1) of $H = 0.8$, eq. (5.37) was found to have γ_3 values of 0.34 and 0.32, for fractures having $h_{rms} = 8$ and 16 μm, respectively (Lang *et al.*, 2015).

Walsh (1981) assumed that fracture contact areas were circular in form and derived the following relation between the hydraulic aperture and the mean aperture, as computed over the open regions of the fracture plane:

$$h_h^3 = \left(\frac{1-c}{1+c}\right) h_m^3, \tag{5.38}$$

where c is the dimensionless contact fraction, as defined in eq. (5.4). Zimmerman and Bodvarsson (1996) used effective medium theory and percolation theory to derive the following expression:

$$h_h^3 = h_m^3 \left(1 - 1.5\frac{\sigma_h^2}{h_m^2}\right)(1 - 2c), \tag{5.39}$$

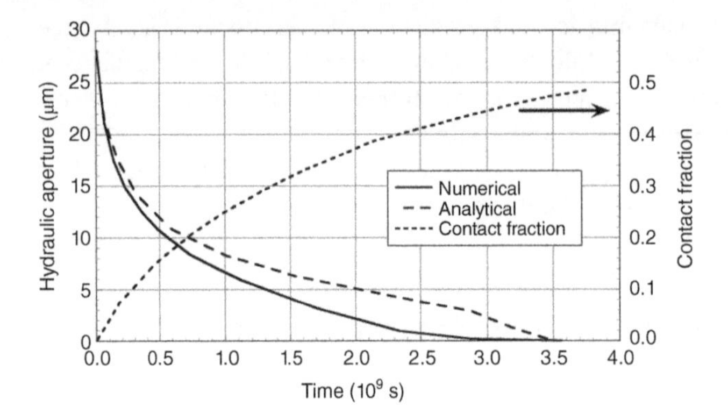

Figure 5.5 Hydraulic aperture and contact fraction predicted by the numerical simulations of Lang *et al.* (2015), for a siliciclastic rock fracture undergoing pressure dissolution and precipitation. The hydraulic aperture predictions obtained using the analytical model proposed by Zimmerman and Bodvarsson (1996), eq. (5.39), are shown for comparison. See text for details. Source: Adapted from Lang *et al.* (2015).

where h_m is the mean mechanical aperture, and σ_h is the standard deviation of the aperture distribution, both taken over the non-contacting regions of the fracture plane. As is the case for Walsh's model, this equation also assumes that the planform of the contact regions is circular. The percolation term, $1-2c$, implies that if the contact fraction reaches 50%, the hydraulic aperture, and therefore the transmissivity, will vanish.

Lang *et al.* (2015) used the Lehner–Leroy pressure dissolution model, described in Section 5.9, to compute the evolution of the mean mechanical aperture and contact fraction of several simulated siliciclastic rock fractures. Figure 5.5 shows the computed evolution of the hydraulic aperture, for a fracture having an initial mean aperture of 30 μm, $h_{rms} = 16$ μm, a Hurst exponent of $H = 0.8$, subjected to a macroscopic compressive stress of 10 MPa at a temperature of 473 K. Also shown is the evolution of the contact fraction. The hydraulic aperture predicted by the equation of Zimmerman and Bodvarsson (1996), eq. (5.39), is shown for comparison. Reasonably good agreement was found between the numerical values and the analytical model at early times, when the contact fractions were below about 0.2. After large times of pressure solution, as the contact area increases, the discrepancy between the analytical model and the simulated results increases. In this particular case, the percolation threshold, at which the fracture transmissivity went to zero, was reached at a contact fraction very close to 0.5, although for other cases the computed threshold varied between 0.41 and 0.52.

The models that have been discussed thus far in this chapter for pressure solution and precipitation are based on a number of simplifying assumptions. The rock is assumed to be rigid, and no elastic deformation is taken into account. In addition, contact shapes are simplified as being, for example, circular disks, and the rock material is assumed to dissolve homogeneously over the contact surface, whereas the contact stress is assumed to have an average and uniform value across the fracture. These models do not account for variations in the aperture distribution that often ensue due to hydro-mechanical chemically mediated processes occurring in the rock. Section 5.8 discusses some numerical models that are able to capture details of the deformation and geometric changes to the fracture surface, in order to better estimate aperture changes and their effect on fracture transmissivity.

5.8 Numerical Simulations of Pressure Solution

Variations in the stresses acting on the fracture often lead to normal compression and shearing deformation, which can create a mismatch between the two opposing fracture surfaces. This leads to a spatially heterogeneous distribution of the contacts, and a complex distribution of contact sizes and shapes, which in turn results in an uneven distribution of stress concentration, and consequently, uneven dissolution rates. Numerical simulations (Lang *et al.*, 2015) can capture local changes to the fracture surface that occur during contact. Instead of assuming average stresses and average aperture changes, the fracture surface is represented explicitly and is deformed and modified due to the mechanical loads and dissolution. This results in emerging fracture properties that directly affect fracture properties such as transmissivity and stiffness.

Simulations can be run in two or three dimensions, taking into account rigid, elastic, or plastic rock deformation. These simulations can consider the friction coefficient of the rock, a parameter that links the small-scale mechanical effects of the fracture rough surface to its behavior as a planar surface. They can also take into account the explicit geometric representation of the fracture. In small-scale simulations, a small region of the fracture is generated explicitly as a self-affine geometry, so as to follow the statistics of roughness measurements of real fractures of interest, which are made in the laboratory on fractured rock samples of sandstone or granite, for example, which are either fractured in the laboratory or collected in the field in a fractured state. Once the statistics of the fracture are quantified, a two-dimensional or three-dimensional model of the fracture is generated, as described in Chapter 1.

The steps of the thermo–hydro–mechanical–chemical numerical simulations can be summarized as follows. The first step is to generate the self-affine surface geometry. Considering that fractures have two walls, two different surfaces are created. These two surfaces can be assumed to be perfectly matched, such that if they were in contact, the resulting aperture would go to zero. However, in nature, and also experimentally, it is almost impossible for two fracture surfaces to perfectly match after they have been created. This is largely due to tangential displacement of the fracture walls or shearing, as well as fragmentation and fracturing of the surface, which prevent the surfaces from being truly "mated." Non-matedness results in an initial contact state that has a distribution of apertures across the surface. The height of the fracture at each location, $h(\mathbf{x})$, is initially composed of the heights of the fracture at each side of the fracture wall. This is illustrated for self-affine surfaces in Chapter 1, where Fig. 1.6 shows how the fracture surface differs on both sides of the fracture, due to minimal shearing in this case, resulting in a compound height surface that captures the variation of the lower and upper fracture surface profiles, recording differences in the roughness of the fracture on both sides. A single "composite" curve, at the top, captures the distance between the two, h, which represents the aperture variability of the system.

The distribution of apertures on the fracture surface can be captured numerically by superposing both surfaces and creating a composite surface, which represents the variation in the aperture field as the distance between a composite surface and a plane. Figure 5.6 (Lang, 2016) shows an example of a composite surface, resulting from the aperture variation of two impinging fracture walls, assuming that the deformation of the walls is negligible. This results in a surface that lacks large wavelengths when the two surfaces are strongly

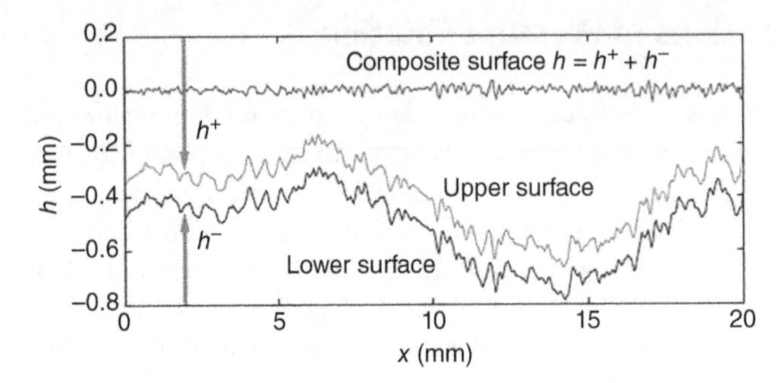

Figure 5.6 Upper and lower surfaces of a rock fracture, along with the "composite surface" that transforms the same aperture field into the space between a single composite surface and a flat plane (Lang, 2016). This transformation has the effect of removing long-wavelength features. Source: Adapted from Lang (2016).

correlated, which is often the case. Two random uncorrelated surfaces are found to form a composite surface that has the same Hurst exponent as its parents, and a lower h_{rms} than the parents (Hyun *et al.*, 2004).

Once the composite surface is created, the contact between that surface and a planar surface is modeled using linear elasticity theory. Hence, the stress at a specific contact is a function of the applied stresses, the local geometry, and the mechanical properties of the rock, symbolically of the form $\sigma(\mathbf{x}) = f(h, \sigma_{mac}, E, \nu, \mu)$, where h represents the surface geometry of the contact, σ_{mac} represents the macroscopic normal stress, and $\{E, \nu, \mu\}$ are the Young's modulus, Poisson's ratio, and friction coefficient of the intact rock. The stress and deformation can be computed assuming linear elastic deformation, using methods such as finite elements or finite differences. More complex numerical models can assume that the integrity of the rock reduces, according to a "damage model" (Ogata *et al.*, 2020), or can assume that the rock behaves plastically (Schuler *et al.*, 2020; Zou *et al.*, 2020).

The contact between the surfaces can be modeled numerically, for example, with the penalty method or Augmented Lagrangian method (Simo and Laursen, 1992; Wriggers and Zavarise, 1993), which take into account material properties and deformation of the rock during contact. Solving for the contact will lead to a local estimate of the contact radius and the local stress concentration across the fracture. Therefore, the stresses will be distributed across the fracture along the multiple contact regions that ensue. At each of these locations, the stresses and contact area then serve to predict the diffusion distance. This diffusion distance can then be used to determine where precipitation will take place.

5.9 Lehner–Leroy Model for Pressure Dissolution

Assuming that there is equilibrium of the concentration of the minerals at the hydrostatically stressed contact, the total dissolution rate can be expressed as a change of the asperity height (Lehner and Leroy, 2004), by expressing this change in terms of a constitutive relationship that governs the convergence rate between two spherical grains, as follows (Bernabé and Evans, 2007):

$$\frac{dh}{dt} \approx \frac{V_m^2 \langle \sigma_{eff} \rangle k_+}{\mathscr{R}T\left[1 + \frac{\rho_s a^2 V_m k_+}{8\rho_f w C_{eq} D^*}\right]}, \tag{5.40}$$

where dh/dt [m/s] is the spatial average of the total thickness of the layers of dissolved solid on both sides of the contact interface, a [m] is the contact radius between the two surfaces, V_m [m^3/mol] is the molar volume, $\mathscr{R}$ = 8.314 J/mol K is the universal gas constant, ρ_s [kg/m^3] is the solid density, ρ_f [kg/m^3] is the fluid density, k_+ [mol/m^2 s] is the dissolution rate constant, σ_{eff} [Pa] is the local effective contact stress, D^* [m^2/s] is the diffusion coefficient in the grain-to-grain or asperity-to-asperity boundary, C_{eq} [–] is the equilibrium concentration in the pore space expressed as a mass fraction, and w [m] is the thickness of the pore fluid film at the interface between the two asperities. Comparison of eqs. (5.40) and (5.34) indicates that the Lehner–Leroy model can be thought of as an extension of eq. (5.34) to account for diffusion within the contact region.

Lang *et al.* (2015) extended this relationship into a more general form, in order to account for the local variation of stresses and subsequent dissolution:

$$\frac{dh}{dt} \approx \frac{V_m^2 \sigma_{eff}(\mathbf{x}) k_+}{\mathscr{R}T\left[1 + \frac{\rho_s r_d^2(\mathbf{x}) V_m k_+}{8\rho_f w C_{eq} D^*}\right]}, \tag{5.41}$$

where $r_d(\mathbf{x})$ [m] is the distance to the nearest free pore space within the fracture, and $\sigma_{eff}(\mathbf{x})$ [Pa] is the effective normal stress at the contact.

The contact radius of an asperity under compression will change during a time interval Δt, and this change can be expressed as a change in the height of the asperity as a function of time, dh/dt. In the specific case of an idealized spherical grain of initial radius R, it follows that (Bernabé and Evans, 2007)

$$h(t + \Delta t) = \left[a^2(t) - \frac{d(h^2)}{dt}\Delta t + \sqrt{R^2 - a^2(t)}\frac{dh}{dt}\Delta t\right]^{1/2}, \tag{5.42}$$

where h is the height of the asperity. This relationship could also be expressed in terms of average stress change, by noting that, by virtue of a simple force balance,

$$\langle \sigma_{eff}(t) \rangle = \langle \sigma_{eff}(0) \rangle \frac{a^2(0)}{a^2(t)}, \tag{5.43}$$

where $\sigma_{eff}(t)$ is the effective stress at time t, $\sigma(0)$ is the initial effective stress, $a(t)$ is the contact radius at time t, $a(0)$ is the initial contact radius, and $\langle \sigma_{eff} \rangle$ denotes the spatial average of the stress.

Using this approach, after the elastic problem is solved yielding $\sigma_{eff}(\mathbf{x})$, the diffusion distance $r_d(\mathbf{x})$ is computed from eq. (5.41), and the local geometry is modified by adjusting the height of the asperity from $h(\mathbf{x}, t)$ to

$$h(\mathbf{x}, t + \Delta t) = h(\mathbf{x}, t) + [dh(\mathbf{x})/dt]\Delta t. \tag{5.44}$$

Once the amount of dissolution and the diffusion distance are computed, the next step is to compute the manner in which the surface changes, and modify its geometry. This corresponds to shifting the position of the surface, so as to capture its reduction in volume and the change in its shape. Next, the time of the simulation is increased, and the entire process is repeated in order to evaluate how the changes to the fracture geometry will affect the stress and dissolution rate, and so on. Precipitation is modeled by evaluating whether or not the conditions for precipitation are met, resulting in geometric modifications to account for the accretion of the solid mineral across the fracture surface. This may occur in the region immediately proximal to the dissolution, or further away along the fracture surface.

Figure 5.7 shows an example of a fracture surface dissolved under compression over a period of one hundred years, using the Lehner–Leroy method. The fracture is generated as a composite self-affine surface, as described in Chapter 1, and has a Hurst exponent of 0.8, an aperture field characterized by $h_{rms} = 8\,\mu m$, and a grid size of 256×256 cells. The fracture is confined under a normal stress of 10 MPa at a temperature of 423 K. The images show how the contact areas grow over time, and how the pressure distributes itself across the contacts, with the highest pressures occurring at the center of the largest contact regions, and also distributed across a number of smaller amorphous asperities in contact. Fluid circulates around the contact regions, creating conditions of channeling that become more and more pronounced as time goes by. Numerical experiments also show that pressure solution surfaces that lead to large flattened surfaces in contact also affect the stiffness of the fracture, whereby for low confining pressures, contact between the fracture walls occurs primarily over the flattened surfaces, leading to exponential increase in stiffness, while for high confining pressures, the contact between the walls occurs also between the rough portions of the fracture, resulting in a linear increase in stiffness with stress (Lang *et al.*, 2016).

5.10 Bernabé–Evans Model for Pressure Dissolution

Bernabé and Evans (2007) extended the model of Lehner and Leroy (2004) by considering the pressure dissolution at the contacting asperities in more microscopic detail. The first step, as before, is to compute the solution of the elastic contact problem numerically, yielding a distribution of stresses, $\sigma_{eff}(\mathbf{x})$, that corresponds to the surface profile $h(\mathbf{x}, t)$, and the macroscopic confining stress, σ_{mac}. The second step is to compute the concentration profile $C(\mathbf{x}, t)$, followed by computing the dissolution flux $J(\mathbf{x}, t)$, leading to the change of the surface at location $\mathbf{x}$, and finally to the computation of the new roughness profile, $h(\mathbf{x}, t + \Delta t)$, according to eq. (5.44).

The mass fraction of the concentration of the mineral at each location is bounded as follows:

$$C_{eq,s}(\sigma) \leq C(\mathbf{x}, t) \leq C_{eq}, \tag{5.45}$$

where C_{eq} is the concentration at equilibrium, and $C_{eq,s}$ is the stress-dependent equilibrium concentration due to the compression in the region where asperities are in contact. Consequently, the minimum value of the concentration scales with contact stress. The relationship between stress and concentration can be expressed as (Neretnieks, 2014)

$$C_{eq,s} = C_{eq} e^{V_m(\sigma_{eff,s} - \sigma_{eff})/\mathcal{R}T}, \tag{5.46}$$

where C_{eq} is the concentration of the mineral at the rock effective stress σ_{eff}, and $C_{eq,s}$ is the concentration at the dissolving surface under stress $\sigma_{eff,s}$. This relationship can also be interpreted as a relation for the solubility of the mineral at different stress states.

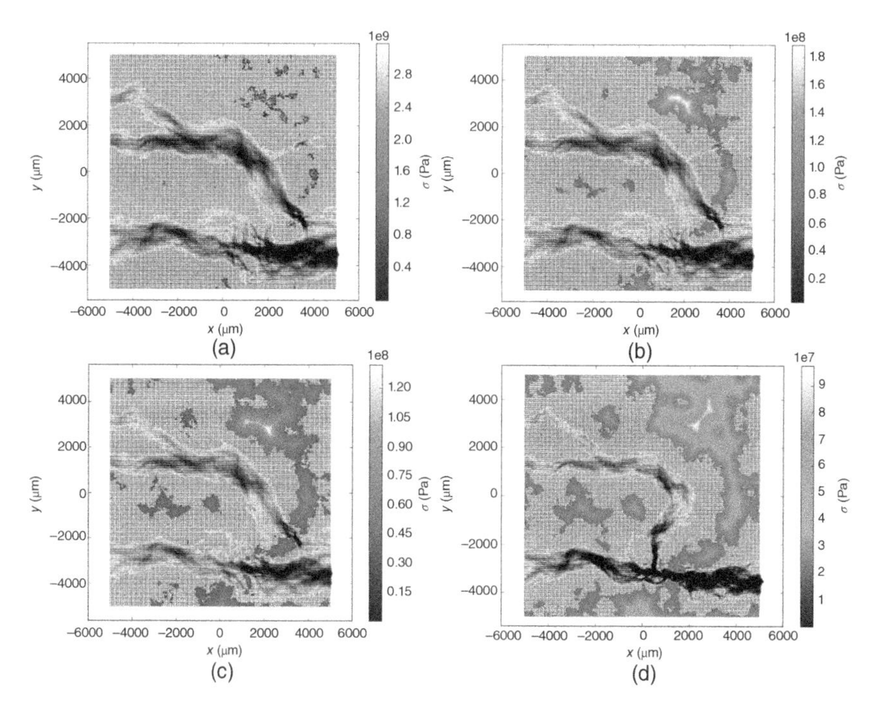

Figure 5.7 A fracture compacted due to pressure dissolution, simulated using the Lehner–Leroy method (Lang *et al.*, 2015). (a) initial state; (b) after 19 years; (c) after 48 years; (d) after 100 years. Fluid is flowing from left to right, flow vectors are represented by semi-transparent arrow glyphs, and the contact stresses are represented by the different gray levels. Source: Adapted from Lang *et al.* (2015).

The pressure dissolution mass flux at the contact zone can be expressed as follows (Bernabé and Evans, 2007), using the present notation:

$$J(\mathbf{x}, t) = \rho_s k_+ V_m \left[\frac{V_m \sigma_{eff}(\mathbf{x}, t)}{\mathcal{R} T} - \ln \frac{C(\mathbf{x}, t)}{C_{eq}} \right], \tag{5.47}$$

where σ_{eff} is the difference between the normal stress and the pore fluid pressure. If the logarithmic term in eq. (5.47) is omitted, the mass flux could be modeled as a linear function of the contact stress, similar to eq. (5.34), except for a different numerical factor due to the assumption of slightly different geometries. The dissolution rate can then be computed as follows:

$$\frac{dh(\mathbf{x}, t)}{dt} = \frac{J(\mathbf{x}, t)}{\rho_s}. \tag{5.48}$$

In the numerical formulation, the concentration is computed by solving the diffusion-reaction Poisson problem, with the mass flux rate serving as a source term:

$$\nabla^2 C(\mathbf{x}, t) = -\frac{J(\mathbf{x}, t)}{w \rho_f D(\mathbf{x})}, \tag{5.49}$$

where w is the interface layer thickness at the asperity contact, and ρ_f is the fluid density (Bernabé and Evans, 2007). It follows, after substituting eq. (5.47) into eq. (5.49) and rearranging, that

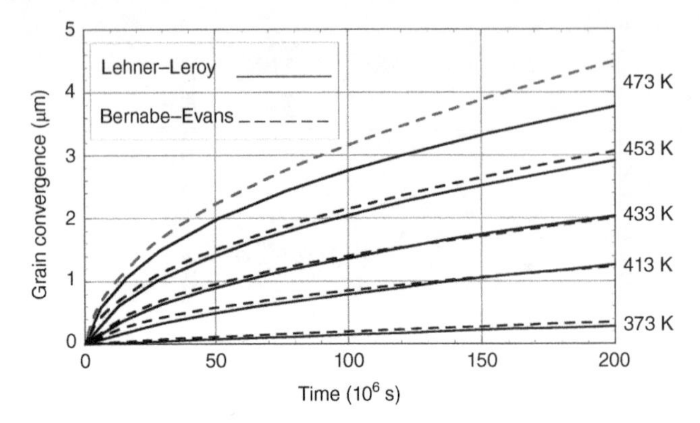

Figure 5.8 Convergence of a single asperity during pressure solution of a single three-dimensional grain, as a function of time over a six-year period, as computed by Lang *et al.* (2015) using the Lehner–Leroy and Bernabé–Evans models. Source: Adapted from Lang *et al.* (2015).

$$\nabla^2 C(\mathbf{x}, t) - \frac{\rho_s k_+ V_m}{w \rho_f D(\mathbf{x})} \left[\frac{V_m}{\mathcal{R} T} \sigma_{eff}(\mathbf{x}, t) - \ln \frac{C(\mathbf{x}, t)}{C_{eq}} \right] = 0. \tag{5.50}$$

This model implicitly assumes a steady-state aperture-averaged concentration distribution within the fracture contact zone and a sufficiently thin interface aperture, w. The diffusion coefficient $D(\mathbf{x})$ is defined over the surface of the fracture to equal the free-space diffusion coefficient D in open regions of the fracture and to equal fD in contact zones, where f is some factor that satisfies $0 < f < 1$, and accounts for the reduced diffusivity of the compressed contact zone, as compared to the free pore space (Revil, 2001). At the interface between the pore space and the contact, the harmonic average of D and fD can be used.

The nonlinear differential equation (5.50) can be solved using methods such as the finite volume scheme, leading to a distribution of concentrations along the fracture surface. The value of the time interval Δt must be chosen carefully; too large a value will lead to convergence issues, whereas too small a value causes the simulation to execute numerous inconsequential timesteps, in which no dissolution or precipitation takes place. More specifically, Δt must be a function of the maximum dissolution ratio and the geometry of the surface at the regions where those maxima are located. For simulations of dissolution and precipitation of granite under pressure, Δt must range from 10^6 to 10^7 seconds, for temperatures of 200 and 150 °C, respectively.

The pressure dissolution of a single grain can be modeled with the methods presented above. Figure 5.8 shows the convergence of a single asperity over a period of 2×10^8 seconds, which roughly corresponds to 6.3 years, as computed using the Lehner–Leroy and Bernabé–Evans models, for a range of temperatures. At lower temperatures, the two models agree well, whereas at higher temperatures the Bernabé–Evans model predicts higher grain convergence than does the Lehner–Leroy method. This discrepancy may be attributed to the transition toward diffusion-dominated dissolution at higher temperatures, leading to larger contact surface areas, which is perhaps better captured by the Bernabé–Evans model.

5.11 Dissolution and Precipitation in Open and Closed Systems

There are two types of thermo-hydro-mechanic systems with chemical mediation: "open" and "closed." In an open system, fluids can migrate in and out of the region of contact, thereby influencing the balance between the dissolved, diffused, and precipitation flux rates. In a closed system, the solute is assumed to migrate and precipitate locally, facilitating the quantification of these dissolution and precipitation rates. Depending on a combination of factors, including temperature, fluid pressure gradient, and the *in situ* stresses, the solute may precipitate close to the solution site, or further away in a smaller distant fracture.

Assuming that fractures can be represented by self-affine surfaces, where asperities appear at different scales, the process of dissolution can be found to influence the shape of these asperities both at the contact and away from the contact areas in systems where geometry is also modified through precipitation. In "closed" systems, where precipitation occurs relatively close to the locations where dissolution is occurring, the growth of contact surfaces occurs more rapidly, as it is a product of both dissolution and precipitation close to the dissolution sites. This leads to fewer, larger contact regions in closed systems, as opposed to more numerous, smaller contact regions in open systems.

The dissolution process is a function of the contact stress, and to a lesser degree also depends on the roughness of the asperities. For both open and closed systems, the contact fraction increases in a similar manner, and the distribution of $dh(\mathbf{x})/dt$ varies as a function of the diffusion distance and contact stress. In the closed system, precipitation occurs primarily in the close vicinity of the contacts. Dissolution within the contact is nonuniform, with more dissolution occurring in the center of the contact, where the stresses are highest.

Dissolution and precipitation are ubiquitous in the subsurface, with solutes often migrating over many kilometers, charging brines that then react with the rocks as they travel through them. Dissolution generally increases permeability, while precipitation generally decreases permeability. The combination of both, in turn, may serve to either increase or decrease the permeability of the system, depending on how the contact surfaces evolve and how the channels between them morph.

Problems

5.1 Consider the dissolution of a spherical quartz grain in contact with a flat surface. What pressure must be applied to dissolve half of the grain at 100, 200, and 300 °C? What are ideal and less favorable conditions for the dissolution?

5.2 Consider the dissolution of a spherical calcite grain in contact with a flat surface. What pressure must be applied to dissolve half of the grain at 100, 200, and 300 °C? What are ideal and less favorable conditions for the dissolution?

5.3 What are the ideal conditions for the precipitation of quartz? How do conditions vary with subsurface depth? What conditions would lead to fast precipitation?

5.4 Given that mechanical and hydraulic mean apertures decrease as a function of average contact area fraction, what is the expected decrease in the fracture permeability as a function of this reduction?

References

Beeler, N. M., and Hickman, S. H. 2004. Stress-induced, time-dependent fracture closure at hydrothermal conditions *Journal of Geophysical Research – Solid Earth*, 109, B02211.

Bernabé, Y., and Evans, B. 2007. Numerical modelling of pressure solution deformation at axisymmetric asperities under normal load, in *Geological Society Special Publication 284*, C. David, and M. Le Ravalec-Dupin, eds., Geological Society, London, pp. 185–205.

Bond, A. E., Brusky, I., Chittenden, N., Feng, X.-T., *et al.* 2016. Development of approaches for modelling coupled thermal–hydraulic–mechanical–chemical processes in single granite fracture experiments. *Environmental Earth Sciences*, 75, 1313.

Bond, A. E., Brusky, I., Cao, T. Q., Chittenden, N., *et al.* 2017. A synthesis of approaches for modelling coupled thermal-hydraulic-mechanical-chemical processes in a single novaculite fracture experiment. *Environmental Earth Sciences*, 76(1), 12.

Dove, P. M., and Crerar, D. A. 1990. Kinetics of quartz dissolution in electrolyte solutions using a hydrothermal mixed flow reactor. *Geochimica et Cosmochimica Acta*, 54, 955–69.

Ghez, R. 1988. *A Primer of Diffusion Problems*, Wiley, New York.

Heidug, W. K. 1995. Intergranular solid-fluid phase transformations under stress: the effect of surface forces. *Journal of Geophysical Research Solid Earth*, 100(B4), 5931–40.

Hyun, S., Pei, L., Molinari, J.-F., and Robbins, M. O. 2004. Finite-element analysis of contact between elastic self-affine surfaces. *Physical Review E*, 70(2), 026117.

Lang, P. S. 2016. *Multi-Scale Modelling of Thermo-Hydro-Mechanical-Chemical Processes in Fractured Rocks*, Ph.D. dissertation, Imperial College, London, UK.

Lang, P. S., Paluszny, A., and Zimmerman, R. W. 2015. Hydraulic sealing due to pressure solution contact zone growth in siliciclastic rock fractures *Journal of Geophysical Research – Solid Earth*, 120, 4080–4101.

Lang, P. S., Paluszny, A., and Zimmerman, R. W. 2016. Evolution of fracture normal stiffness due to pressure dissolution and precipitation. *International Journal of Rock Mechanics and Mining Sciences*, 88, 12–22.

Laubach, S. E., Lander, R. H., Criscenti L.J., and Anovitz, L. M., *et al.* 2019. The role of chemistry in fracture pattern development and opportunities to advance interpretations of geological materials. *Reviews of Geophysics*, 57, 1065–1111.

Lehner, F., and Leroy, Y. 2004. Sandstone compaction by intergranular pressure solution. In Guéguen, Y. and Boutéca, M., editors, *Mechanics of Fluid-Saturated Rocks*, Academic Press, San Diego, pp. 115–68.

Morrow, C. A., Moore, D. E., and Lockner, D. A. 2001. Permeability reduction in granite under hydrothermal conditions *Journal of Geophysical Research – Solid Earth*, 106(B12), 30551–60.

Murphy, W. M., and Helgeson, H. C. 1989. Thermodynamic and kinetic constraints on reaction rates among minerals and aqueous solutions. IV. Retrieval of rate constants and activation parameters for the hydrolysis of pyroxene, wollastonite, olivine, andalusite, quartz, and nepheline. *American Journal of Science*, 289(1), 17–101.

Ogata, S., Yasuhara, H., Kinoshita, N., and Kishida, K. 2020. Coupled thermal-hydraulic-mechanical-chemical modeling for permeability evolution of rocks through fracture generation and subsequent sealing. *Computational Geoscience*, 24(5), 1845–64.

Neretnieks, I. 2014. Stress-mediated closing of fractures: impact of matrix diffusion *Journal of Geophysical Research – Solid Earth*, 119(5), 4149–63.

Polak, A., Elsworth, D., Yasuhara, H., Grader, A. S., *et al.* 2003. Permeability reduction of a natural fracture under net dissolution by hydrothermal fluids. *Geophysical Research Letters*, 30(20), 2020.

Revil, A. 1999. Pervasive pressure-solution transfer: a poro-visco-plastic model. *Geophysical Research Letters*, 26(2), 255–58.

Revil, A. 2001. Pervasive pressure solution transfer in a quartz sand. *Journal of Geophysical Research*, 106(B5), 8665–86.

Rimstidt, J. D., and Barnes, H. L. 1980. The kinetics of silica-water reactions. *Geochimica et Cosmochimica Acta*, 44(11), 1683–99.

Schuler, L., Ilgen, I. G., and Newell, P. 2020. Chemo-mechanical phase-field modeling of dissolution-assisted fracture. *Computer Methods in Applied Mechanics and Engineering*, 362, 112838.

Shimizu, I. 1995. Kinetics of pressure solution creep in quartz: theoretical considerations. *Tectonophysics*, 245(3, 4), 121–34.

Simo, J., and Laursen, T. 1992. An augmented Lagrangian treatment of contact problems involving friction. *Computers and Structures*, 42(1), 97–116.

Timoshenko, S. P, and Goodier, J. N. 1970. *Theory of Elasticity*, 3rd ed. McGraw-Hill, New York.

Walsh, J. B. 1981. Effect of pore pressure and confining pressure on fracture permeability. *International Journal of Rock Mechanics and Mining Sciences*, 18(5), 429–35.

Wriggers, P., and Zavarise, G. 1993. Application of augmented Lagrangian techniques for non-linear constitutive laws in contact interfaces. *Communications in Numerical Methods in Engineering*, 9(10), 815–24.

Yasuhara, H., Elsworth, D., and Polak, A. 2003. A mechanistic model for compaction of granular aggregates moderated by pressure solution *Journal of Geophysical Research – Solid Earth*, 108(B11), 2530.

Yasuhara, H., Elsworth, D., and Polak, A. 2004. Evolution of permeability in a natural fracture: significant role of pressure solution *Journal of Geophysical Research – Solid Earth*, 109(B3), B03204.

Yasuhara, H., Kinoshita, N., Ohfuji, H., Lee, D. S., *et al.* 2011. Temporal alteration of fracture permeability in granite under hydrothermal conditions and its interpretation by coupled chemo-mechanical model. *Applied Geochemistry*, 26, 2074–88.

Zimmerman, R. W., and Bodvarsson, G. S. 1996. Hydraulic conductivity of rock fractures. *Transport in Porous Media*, 23(1), 1–30.

Zou, L. C., Li, B., Mo, Y. Y., and Cvetkovic, V. 2020. A high-resolution contact analysis of rough-walled crystalline rock fractures subject to normal stress. *Rock Mechanics and Rock Engineering* 53, 2141–55.

6

Solute Transport in a Single Fracture

6.1 Introduction

The first four chapters of this book have focused on the flow of a single-phase fluid through a rock fracture. In Chapter 5, the motion of dissolved minerals within the pore space of a fracture was considered, with an emphasis on how the dissolution and precipitation of these minerals might alter the transmissivity of the fracture. The process of the migration of these minerals within a fracture, as discussed in Chapter 5, is a special case of the more general process by which dissolved species may migrate through a fracture. These dissolved species may be minerals that have dissolved from the fracture walls, or they may be contaminants such as metals that have leached from landfills, chemicals that have been disposed of by subsurface injection, radionuclides that have escaped from an underground disposal facility for nuclear waste, *etc.* Moreover, although Chapter 5 considered the migration of the dissolved minerals to occur solely due to molecular *diffusion*, in the general case, the solute will also migrate as a result of being *advected* along with the flowing fluid.

This chapter will discuss the transport of solutes within a single fracture. This discussion will set the stage for the treatment of the topic of solute transport through *fractured rock masses*, in Chapter 11, which will necessarily need to follow after the discussion of models for fluid flow through fractured rock masses, as given in Chapters 7–9.

If the fluid in a fracture is stagnant, solutes will migrate through the fracture due to the process of molecular diffusion, which is governed by the phenomenological law known as Fick's law, which is defined quantitatively in Section 6.2. If, hypothetically, there was no molecular diffusion but fluid was flowing through the fracture, solutes would still migrate through a fracture due to being advected along with the flowing fluid. It is therefore evident that in the general case, both modes of solute migration, diffusion and advection, will occur simultaneously. Moreover, due to the spatially varying nature of the flow field, the solute will not generally advect in the form of "plug flow," but will also spread out due to a process known as dispersion, which, as will be seen in Section 6.3, can in many situations be described mathematically in a manner that is analogous to the mathematical description of diffusion. If this is the case, then on a local scale, but within the context of a continuum model, solute transport can be described by a partial differential equation known as the advection dispersion equation (ADE).

The ADE is derived in Section 6.2 in a general context that is applicable to porous, as well as fractured, media. The classical Taylor–Aris analysis of flow and transport through

Fluid Flow in Fractured Rocks, First Edition. Robert W. Zimmerman and Adriana Paluszny.

Companion website: www.wiley.com/go/zimmerman/fluidflowinfracturedrocks

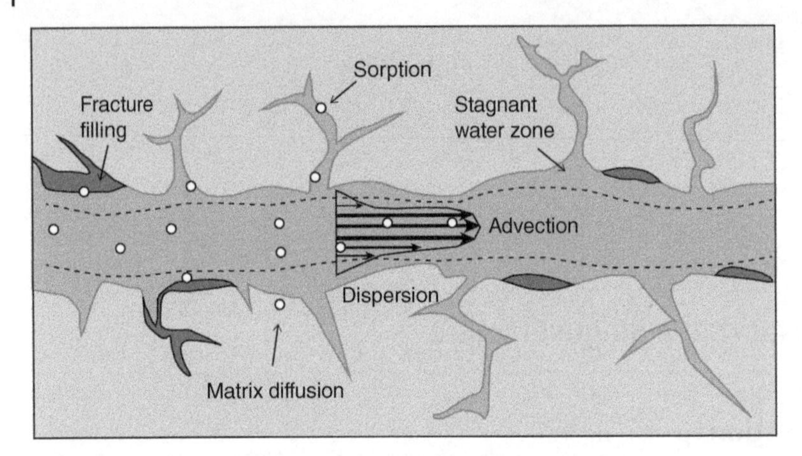

Figure 6.1 Schematic illustration of processes that influence solute transport in a single rock fracture. Adapted from Zhang *et al.* (2022). Solute molecules/particles are advected along with the flow, and also migrate due to molecular diffusion. Some of these particles may adsorb to the fracture walls or migrate into the adjacent rock matrix. Solute may also be temporarily "trapped" in a stagnant water zone within a local bulge of the fracture.

a uniform channel is presented in Section 6.3. This analysis provides, in a quantitative way, albeit for a simplified geometry, some justification for using an ADE-type equation to model solute transport in a rock fracture. Moreover, it explains why the effective dispersion coefficient for the solute is expected to be larger than the molecular diffusion coefficient and in fact increases with a dimensionless parameter known as the Peclet number, which quantifies the relative strengths of advection and molecular diffusion. Section 6.4 describes the influence of fracture morphology and Peclet number on transport within rock fractures. Section 6.5 discusses and contrasts Fickian and non-Fickian transport, where, roughly speaking, non-Fickian transport is that which cannot be governed by an ADE using a fixed value of the dispersion coefficient. Finally, Section 6.6 discusses the effects of adsorption of solute particles along the fracture walls and diffusion of solutes into the rock matrix. A schematic illustration of some of the factors that influence solute transport in a single rock fracture is shown in Fig. 6.1.

6.2 Advection–Diffusion Equation

The one-dimensional advection–diffusion/dispersion equation for solute transport can be derived by combining the mass flux due to diffusion and the mass flux due to advection and then applying the principle of conservation of mass. Molecular diffusion is the process by which solutes migrate from areas of high concentration into areas of low concentration due to the Brownian motion of individual solute particles. On a microscopic scale, this Brownian motion is essentially random, with each solute particle having the same probability of moving to the right as it has of moving to the left. However, consider a situation in which, at some time t, the solute concentration at location x is greater than the concentration at location $x + \Delta x$, *i.e.*, $C(x, t) > C(x + \Delta x, t)$. Due solely to the fact that there are more solute particles at location x than at location $x + \Delta x$, it will necessarily be the case that more particles move from x to $x + \Delta x$, than *vice versa*. Hence, there will be a net flux of particles from x to $x + \Delta x$, which is to say, there will be a flux of particles from regions where the concentration is higher to regions where the concentration is lower (Ghez, 1988).

In the usual "linear" model, known as Fick's first law, this diffusive flux is assumed to be linearly proportional to the concentration gradient. Using the notation introduced in Chapter 5, the flux of the solute is related to the gradient of solute concentration by (Bear, 1988)

$$\mathbf{j}_d = (1/A)\mathbf{J}_d = -D_m \nabla C, \tag{6.1}$$

where C is the concentration of the solute in the fluid [kg/m^3], $\mathbf{j}_d$ is the diffusive flux vector of the solute [kg/m^{2}s], D_m is the molecular diffusion coefficient [m^2/s], and ∇ [1/m] is the gradient operator. The molecular diffusion coefficient of most solutes in an unconfined ("free") fluid is typically on the order of 10^{-9} m^2/s (de Marsily, 1986). Similar to Darcy's law for fluid flow, the diffusive solute flux occurs in the direction opposite to the concentration gradient, which is to say that the solute diffuses from regions of higher concentration to regions of lower concentration. The one-dimensional form of eq. (6.1) is $j_d = -D_m(\partial C/\partial x)$.

In a porous medium, as opposed to an unconfined fluid, the effective molecular diffusion coefficient will, in general, be less than the free-fluid diffusion coefficient, due to the fact that the solute can only flow through the fluid but not through the mineral grains and also due to the tortuous path that the solute must travel in order to traverse the pore space (Tartakovsky and Dentz, 2019). Hence, the flux equation (6.1) in a porous medium is usually written in the form

$$\mathbf{j}_d = -D_{meff} \nabla C. \tag{6.2}$$

In some formulations, a porosity term ϕ is included on the right side of this equation. These two formulations are equivalent and merely correspond to different ways of defining the effective diffusivity.

The solute can also migrate through the porous medium by "advection," due to being transported through the pore space along with the flowing fluid. If the fluid is moving through the porous medium at a flowrate $\mathbf{q}$ [m/s], then the advective flux of the solute will be given by

$$\mathbf{j}_a = \mathbf{q}C, \tag{6.3}$$

the one-dimensional form of which is $j_a = qC$. The total flux of solute is given by the sum of the advective and diffusive fluxes, *i.e.*,

$$j = qC - D_{meff} \frac{\partial C}{\partial x}. \tag{6.4}$$

Consider now the mass of solute contained in a thin slab-like region between x and $x + \Delta x$, having cross-sectional area A in the plane normal to the flux. If $m(t)$ [kg] is the mass of solute stored in this region at time t, then applying the law of conservation of mass to this region over the time interval between t and $t + \Delta t$, leads to the following mass balance equation:

$$[Aj(x,t) - Aj(x + \Delta x, t)]\Delta t = m(t + \Delta t) - m(t), \tag{6.5}$$

where $Aj(x, t)\Delta t$ [kg] is the mass of solute entering the region from the left, $Aj(x+\Delta x, t)\Delta t$ is the mass exiting from the right, and $m(t+\Delta t) - m(t)$ is the change in the mass stored in the region over this time interval.

The mass of solute contained in this region is equal to the product of the pore volume and the solute concentration in the pore fluid, *i.e.*, $m(t) = \phi A \Delta x \overline{C}(x \to x + \Delta x, t)$, where $\overline{C}(x \to x + \Delta x, t)$ denotes the mean concentration in the region between x and $x+\Delta x$. Hence, eq. (6.5) can be written as

$$[Aj(x, t) - Aj(x + \Delta x, t)]\Delta t = \phi A \Delta x[\overline{C}(x \to x + \Delta x, t + \Delta t) - \overline{C}(x \to x + \Delta x, t)]. \tag{6.6}$$

Dividing both sides of this equation by $A\Delta x \Delta t$, taking the limit as $\Delta x \to 0$ and $\Delta t \to 0$, and noting that as $\Delta x \to 0$, $\overline{C}(x \to x + \Delta x, t)$ reduces to $C(x, t)$, leads to the following local mass balance equation:

$$\frac{\partial j}{\partial x} + \phi \frac{\partial C}{\partial t} = 0. \tag{6.7}$$

Insertion of the expression for the mass flux, as given by eq. (6.4), into the above equation and assuming that the fluid flux is constant in time, yields the one-dimensional advection–diffusion equation:

$$\phi \frac{\partial C}{\partial t} + q \frac{\partial C}{\partial x} = D_{meff} \frac{\partial^2 C}{\partial x^2}. \tag{6.8}$$

In the literature on solute transport, the fluid velocity magnitude q is usually denoted by u [m/s]. Equation (6.8) assumes steady-state, one-dimensional flow in a porous medium having spatially uniform properties. This formulation can be extended to three dimensions by starting with the vector expressions for the diffusive and advective fluxes, as given by eqs. (6.2) and (6.3), taking the mass balance over a cubical region, and making use of the divergence theorem. The 3D version of the advection–diffusion equation has the following form:

$$\phi \frac{\partial C}{\partial t} + \mathbf{q} \cdot \nabla C = D_{meff} \nabla^2 C, \tag{6.9}$$

where ∇ is the gradient operator, and ∇^2 is the Laplacian operator. For the specific case of a fracture, as opposed to a porous medium, the porosity term can be set to unity.

A basic and important one-dimensional solute transport problem is that of a porous medium in which the fluid flowrate is constant, and the solute concentration is initially zero everywhere, after which at $t = 0$, the concentration at the "source" (located at $x = 0$) is instantaneously changed to, and maintained at, some value C_o. The solution to this problem is (Ogata and Banks, 1961)

$$\frac{C(x,t)}{C_o} = \frac{1}{2}\left[\operatorname{erfc}\left(\frac{x - ut}{\sqrt{4Dt}}\right) + \exp\left(\frac{ux}{D}\right)\operatorname{erfc}\left(\frac{x + ut}{\sqrt{4Dt}}\right)\right], \tag{6.10}$$

where for notational simplicity, the diffusivity is denoted by D, and erfc, the *complementary error function*, is defined by (Ghez, 1988)

$$\operatorname{erfc}(x) = \frac{2}{\sqrt{\pi}} \int_x^{\infty} e^{-\eta^2} d\eta. \tag{6.11}$$

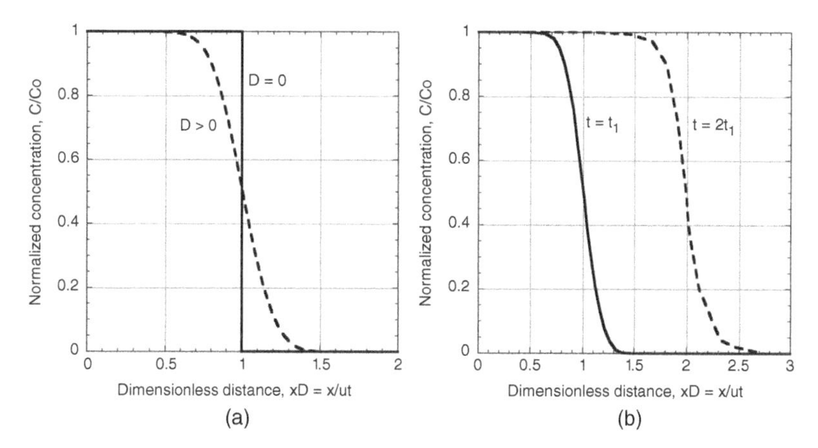

Figure 6.2 Schematic solute concentration profiles given by the Ogata–Banks equation (6.10). (a) Profile at a given time for two cases with the same mean flow velocity, with either zero diffusivity or a finite diffusivity. (b) Profiles for a system having a finite diffusivity, at two different values of the elapsed time.

The complementary error function has the properties that $\mathrm{erfc}(-\infty) = 2$, $\mathrm{erfc}(0) = 1$, and $\mathrm{erfc}(\infty) = 0$. The solute profiles described by eq. (6.10) are plotted in Fig. 6.2 to illustrate the spread of the solute plume with time, and the effect of the diffusion coefficient.

Although it is frequently asserted or assumed that the second term within the brackets can be neglected when using eq. (6.10) to analyze field or laboratory data, Cavalcante and de Farias (2013) have shown that this approximation often leads to poor results, and concluded that in general, both terms should be retained.

The solution displayed in eq. (6.10) illustrates a concept that is very useful when discussing transport, and which is based on quantifying the relative strengths of advection and diffusion. The time required for a solute particle to travel over a distance x by advection would be x/u, whereas the time required for a solute particle to travel over a distance x by diffusion would be roughly x^2/D (Ghez, 1988). The (dimensionless) ratio of these two characteristic times, $t_{diff}/t_{adv} = ux/D$, is known as the *Peclet number*, *Pe*. Processes having small Peclet numbers are referred to as being diffusion-dominated, whereas processes in which the Peclet number is large are referred to as being advection-dominated. As $Pe \rightarrow 0$, eq. (6.10) reduces to the purely diffusive profile, $C(x, t) = C_o \mathrm{erfc}(x/\sqrt{4Dt})$ (Crank, 1975, p. 21). As $Pe \rightarrow \infty$, eq. (6.10) reduces to a piston-type profile given by $C(x, t) = C_o$ for $x < ut$, and $C(x, t) = 0$ for $x > ut$ (see Problem 6.1).

Another important one-dimensional transport problem is that in which a finite amount of solute is introduced "instantaneously" into a uniform flow field, at $t = 0$ and $x = 0$. If the amount of introduced mass, per cross-sectional area, is m/A [kg/m^2], then the resulting concentration profile is given by (Kreft and Zuber, 1978; Tsang *et al.*, 1991)

$$C(x, t) = \frac{m/A}{\sqrt{4\pi Dt}} \exp\left[\frac{-(x - ut)^2}{4Dt}\right]. \tag{6.12}$$

According to this equation, the solute is advected along with the flow in the x-direction, while also diffusing in both the $+x$ and $-x$ directions. If the flowrate is zero, this solution reduces to the purely diffusive profile that occurs for the problem of a finite source of solute mass introduced instantaneously over a plane at $x = 0$ (Crank, 1975, p. 12). In the other extreme case of zero diffusivity, the profile given by eq. (6.12) reduces to a very narrow, highly peaked "Dirac delta function" centered at $x = ut$ that has the property that the spatial integral of the concentration profile equals m/A for all values of time.

A salient feature of the solution given by eq. (6.12) is that the center of mass of the solute plume grows linearly in time, according to $\langle x \rangle = ut$, whereas the standard deviation of the solute mass about its center of mass grows proportional to $\sqrt{t}$, specifically as $\sigma = \sqrt{4Dt}$; see Problem 6.2. Transport processes that follow this type of behavior are often referred to as "Fickian," since the diffusive component of the process is governed by Fick's law, and sometimes as "Gaussian," since the concentration profile for the problem of an instantaneous injection of solute follows a Gaussian-type curve. Behavior that differs from this scheme, such as is discussed in Section 6.5, is consequently referred to as non-Fickian, non-Gaussian, or "anomalous."

Numerous other closed-form solutions to the advection–diffusion equation, in one or more dimensions, have been compiled by Kreft and Zuber (1978), Javandel *et al.* (1984), de Marsily (1986), and Bodin (2015), among others.

6.3 Taylor–Aris Problem in a Uniform Channel

Based on the derivation of the advection–diffusion equation presented in Section 6.2, it might be thought that since an unfilled fracture essentially has a porosity of unity and generally does not follow a particularly tortuous path, solute transport in a fracture should be governed by eq. (6.8), with the effective diffusivity given by the free-fluid molecular diffusivity. However, even in the simple parallel-plate geometry, there is another physical process that gives rise to an additional mode of solute migration that, on a macroscopic scale, is in many cases mathematically analogous to molecular diffusion. This process stems from the fact that in a parallel-plate fracture, the local velocity varies parabolically from 0 at the fracture walls, to a maximum value of 1.5 times the mean velocity at the center line of the fracture (see Section 2.2). Hence, some portion of the solute mass will travel faster than the mean velocity of the solute particles, whereas some of the solute will travel more slowly than the mean velocity, resulting in an additional spreading of the solute plume, which will occur in addition to that due to molecular diffusion.

This process was first analyzed mathematically by Taylor (1953) and Aris (1956) for the problem of solute transport in a circular tube. Their analysis was later extended by several researchers to the case of a thin parallel-plate channel (Aris, 1959; Doshi *et al.*, 1978; Chatwin and Sullivan, 1982). In the Taylor–Aris analysis, the solute is assumed to exist in the form of individual molecules dissolved in the pore water, or small neutrally buoyant particles that will not agglomerate with one another upon contact. It is further assumed that the solute molecules/particles will not adsorb onto the fracture walls, or diffuse into the rock matrix.

Consider a rectangular channel, as shown in Fig. 6.3, with fluid flowing through it under an applied pressure gradient in the x-direction. The starting point of the Taylor–Aris

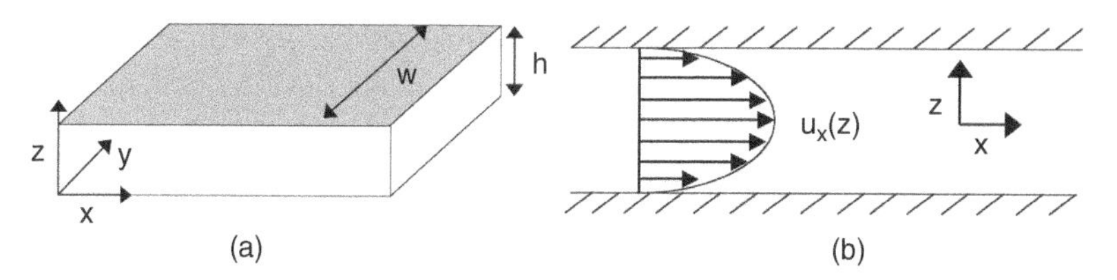

Figure 6.3 (a) Parallel-plate fracture having uniform aperture, h. The applied pressure gradient and the velocity vector are aligned with the x-direction. (b) Resultant parabolic velocity profile, with fluid traveling faster than the mean velocity in the center of the channel and slower than the mean velocity near the walls.

analysis is the local version of eq. (6.9), which applies at each point within the fluid-filled channel:

$$D_m\left(\frac{\partial^2 C}{\partial x^2}+\frac{\partial^2 C}{\partial y^2}+\frac{\partial^2 C}{\partial z^2}\right)=\frac{\partial C}{\partial t}+u_x(z)\frac{\partial C}{\partial x}, \tag{6.13}$$

where the porosity is set to 1, since the equation is only intended to apply within the open channel. For the case of an infinitely wide channel, with $w \to \infty$, the derivatives with respect to y will disappear, and the velocity is given by the parabolic expression given by eq. (2.10), which can be written as $u_x = 1.5\,u[1-(2z/h)^2]$, where u is the mean velocity in the x-direction. Equation (6.13) thereby takes the form

$$D_m\left(\frac{\partial^2 C}{\partial x^2}+\frac{\partial^2 C}{\partial z^2}\right)=\frac{\partial C}{\partial t}+1.5u\left[1-\left(\frac{2z}{h}\right)^2\right]\frac{\partial C}{\partial x}. \tag{6.14}$$

Following Taylor (1953), it is assumed that $\partial^2 C/\partial x^2 << \partial^2 C/\partial z^2$. This assumption can be justified by a simple order-of-magnitude analysis, which implies that $\partial^2 C/\partial x^2 \approx C_o/x^2$, whereas $\partial^2 C/\partial z^2 \approx C_o/h^2$. Under this assumption, eq. (6.14) simplifies to

$$D_m\frac{\partial^2 C}{\partial z^2}=\frac{\partial C}{\partial t}+1.5u\left[1-\left(\frac{2z}{h}\right)^2\right]\frac{\partial C}{\partial x}. \tag{6.15}$$

A convected coordinate is now defined, $\zeta = x - ut$, which moves along with the mean fluid velocity. This change of variables essentially has the effect of removing advection from the governing partial differential equation, which now takes the form

$$D_m\frac{\partial^2 C}{\partial z^2}=\frac{\partial C}{\partial t}+\frac{u}{2}\left[1-3\left(\frac{2z}{h}\right)^2\right]\frac{\partial C}{\partial \zeta}. \tag{6.16}$$

Taylor then assumes that at large times, the derivative $\partial C/\partial t$, computed with ζ and z held constant, will have essentially died out, in which case the concentration profile in the z-direction can be computed from the following equation:

$$\frac{\partial^2 C}{\partial z^2}=\frac{u}{2D_m}\left[1-3\left(\frac{2z}{h}\right)^2\right]\frac{\partial C}{\partial \zeta}, \tag{6.17}$$

along with the assumption that $\partial C/\partial \zeta$ does not vary with z. The concentration profile that satisfies this equation, and also satisfies the no-flux boundary condition for the solute at the fracture walls, which stipulates that $\partial C/\partial z = 0$ at $z = \pm h/2$, is

$$C(z,\zeta)=C(z=0,\zeta)+\frac{uh^2}{4D_m}\left[\left(\frac{z}{h}\right)^2-2\left(\frac{z}{h}\right)^4\right]\frac{\partial C}{\partial \zeta}. \tag{6.18}$$

The mass flux of solute through a plane of constant ζ is equal to the integral of uC over the entire cross-sectional area of flow. For a region having width w in the y-direction, this flux can be computed as follows, noting that the local velocity u should be expressed in terms of the (z,ζ) coordinate system, *i.e.*, using eq. (6.16) rather than eq. (6.15):

$$\begin{aligned}
J &= w\int_{-h/2}^{+h/2}\frac{u}{2}\left[1-3\left(\frac{2z}{h}\right)^2\right]\left\{C(z=0,\zeta)+\frac{uh^2}{4D_m}\left[\left(\frac{z}{h}\right)^2-2\left(\frac{z}{h}\right)^4\right]\frac{\partial C}{\partial \zeta}\right\}dz\\
&= \frac{uw}{2}C(z=0,\zeta)\int_{-h/2}^{+h/2}\left[1-12\left(\frac{z}{h}\right)^2\right]dz+\frac{u^2h^2w}{8D_m}\frac{\partial C}{\partial \zeta}\\
&\quad\times\int_{-h/2}^{+h/2}\left[1-12\left(\frac{z}{h}\right)^2\right]\left[\left(\frac{z}{h}\right)^2-2\left(\frac{z}{h}\right)^4\right]dz\\
&= \frac{u^2h^2w}{8D_m}\frac{\partial C}{\partial \zeta}\int_{-h/2}^{+h/2}\left[\left(\frac{z}{h}\right)^2-14\left(\frac{z}{h}\right)^4+24\left(\frac{z}{h}\right)^6\right]dz\\
&= \frac{u^2h^2w}{8D_m}\frac{\partial C}{\partial \zeta}\left[\frac{z^3}{3h^2}-\frac{14z^5}{5h^4}+\frac{24z^7}{7h^6}\right]_{-h/2}^{+h/2}=\frac{u^2h^2w}{4D_m}\frac{\partial C}{\partial \zeta}\left[\frac{h}{24}-\frac{14h}{160}+\frac{24h}{896}\right]\\
&= -\frac{u^2h^3w}{210D_m}\frac{\partial C}{\partial \zeta}=-\left(\frac{u^2h^2}{210D_m}\right)A\frac{\partial C}{\partial \zeta}.
\end{aligned}\tag{6.19}$$

The first integral on the right side of the second line vanishes because the *mean* velocity in the new advected (z,ζ) coordinate system is zero. It is apparent that the flux expression given by eq. (6.19) is mathematically equivalent to Fick's law, with an effective diffusion coefficient of (Aris, 1959; Wooding, 1960)

$$D_{eff}=\frac{u^2h^2}{210D_m}.\tag{6.20}$$

If the Peclet number is defined using the fracture aperture h as the characteristic length scale, *i.e.*, $Pe = uh/D_m$, then eq. (6.20) can be written as $D_{eff} = D_m Pe^2/210$.

The foregoing analysis assumed that the fracture has an infinite extent in the lateral direction, perpendicular to the direction of flow. Doshi *et al.* (1978) investigated the Taylor–Aris problem for a rectangular channel having a finite value of w (see Fig. 6.3), bounded by two side walls, and found that in the limit as $h/w \to 0$, the effective diffusivity does *not* approach the value obtained for an infinitely wide channel that lacks side walls. Instead, it approaches a value that is larger by a multiplicative factor of 7.9512. This unexpected result was re-derived by Chatwin and Sullivan (1982) using an independent method. This situation is in sharp contrast to the laminar flow problem in a rectangular channel, in which the transmissivity reduces, in the limit as $h/w \to 0$, to the value $h^3w/12$ that corresponds to a parallel-plate geometry without side walls, as expected (Sisavath *et al.*, 2001).

Aris (1956) found that the effective diffusivity of a thin slit-like channel having an elliptical cross-section scales as the square of the semi-*major* axis, whereas the effective diffusivity of a thin slit-like channel having a rectangular cross-section scales as the square of the semi-*minor* axis. More generally, several researchers (Guell *et al.*, 1987; Parks and Romero, 2007) have pointed out the extreme sensitivity of the effective Taylor–Aris diffusivity to small variations in channel geometry. Taken as a whole, these results severely limit the ability of the Taylor–Aris analysis to provide even a first-order approximation to the effective diffusivity in a real rock fracture. Rather, the usefulness of the Taylor–Aris analysis, for rock fractures, is that it provides a rational conceptual model that explains the phenomenon of an "effective diffusivity" that (a) may greatly exceed the molecular diffusivity, (b) increases with increasing flowrate, and (c) arises due to spatial variations in the local flow velocity.

The Taylor–Aris analysis hinges on the assumption that sufficient time has elapsed for the solute gradient in the z-direction to reach a quasi-steady state. As the characteristic length scale of the parallel-plate fracture is h, and the length scale over which molecular diffusion is appreciable grows as $\sqrt{D_m t}$ (Ghez, 1988), the time required for the concentration gradient in the z-direction to reach a quasi-steady-state is on the order of h^2/D_m. At times much greater than this, the solute plume evolves according to the advection–diffusion equation and follows the Fickian behavior discussed at the end of Section 6.2. At earlier times, however, the transport is "anomalous," and the mean and standard deviation of the solute plume each follow more complicated evolution laws that depend sensitively on the initial conditions (Camassa *et al.*, 2010).

6.4 Influence of Fracture Morphology on Solute Transport

The Taylor–Aris analysis implies that, at long times, solute transport during unidirectional flow in a rock fracture is governed by the one-dimensional advective diffusion equation, with an "effective" diffusivity that increases quadratically with flow velocity. Comparison both with experiments and detailed fine-scale numerical simulations indicates that this model has some validity, with the exception that the effective diffusivity seems to scale *linearly* with velocity. It is therefore common to write the effective diffusion/dispersion coefficient as $D_L = u\alpha_L$, where u is the mean velocity, and α_L is the *longitudinal dispersivity* (Gelhar and Axness, 1983). A similar expression is used for the effective diffusion coefficient in the transverse direction (the y-direction in Fig. 6.3), namely, $D_T = u\alpha_T$. In principle, the molecular diffusion coefficient should be added to these effective diffusivity terms, but in practice, the molecular diffusivity is usually much smaller and is often neglected.

However, the main contribution to the solute dispersion that occurs in actual rock fractures is now known to not be due to Taylor–Aris dispersion, which is caused by velocity variations in the z-direction, but rather to velocity variations from point to point within the fracture plane, *i.e.*, within the $x-y$ plane in Fig. 6.3. These velocity variations can be caused by spatial variations in the fracture aperture (Keller *et al.*, 1999), and/or by the tortuosity created by the fluid being forced to flow around contact areas (Beaudoin and Farhat, 2021).

Based on the assumption that the logarithm of the fracture aperture distribution is a statistically stationary, isotropic, two-dimensional Gaussian random field and using the Reynolds equation to model fluid flow through the fracture, Gelhar (1993) derived the following expression for the longitudinal dispersion coefficient:

$$D_L = u\alpha_L = u\left[3 + G\left(\sigma_{\ln h}^2\right)\right]\sigma_{\ln h}^2\lambda,$$
$$G(\sigma) = 1 + 0.205\sigma^2 + 0.16\sigma^4 + 0.045\sigma^6 + 0.0115\sigma^8, \quad (6.21)$$

in which $\sigma^2_{\ln h}$ is the log-variance of the aperture distribution, and λ is the correlation length of the aperture field. Gelhar's analysis assumes that $\sigma^2_{\ln h} < 5$, and that a sufficient distance has been traversed by the solute so that the dispersivity has converged to the "fully developed" value given by eq. (6.21).

Keller *et al.* (1999) conducted solute transport experiments in a naturally fractured granite core, using a 10% by weight solution of potassium iodide as a nonsorbing tracer. The fracture had a mean aperture of 0.825 mm and a standard deviation of 0.683 mm. The core had a total length of 166 mm and a diameter (corresponding to w in Fig. 6.3) of 52.5 mm. After establishing a steady-state flowrate under a fixed pressure gradient, tracer was injected along the inlet face of the fracture, and the solute concentration of the effluent fluid was measured to yield the "breakthrough curve," which is the solute concentration at the outlet, as a function of time. Measurements were made at five different flowrates, with three to five tests conducted at each flowrate.

The effective dispersion coefficient for each test was determined by fitting the breakthrough curve to the Ogata–Banks analytical solution, eq. (6.10). The mean dispersion coefficient, computed over the 3–5 tests at each flowrate, is plotted in Fig. 6.4a, against the mean flowrate. The results are highly consistent with the assumption that the dispersion coefficient is proportional to the velocity, *i.e.*, $D_L = u\alpha_L$, with a slope, reported by Keller *et al.* (1999), of $\alpha_L = 0.043$ m. The measured effective dispersion coefficients at each flowrate were at least two orders of magnitude larger than the molecular diffusion coefficient of potassium iodide in water, which is roughly 10^{-9} m^2/s, confirming that the observed solute dispersion was indeed caused by local variations in velocity, rather than by intrinsic molecular diffusion.

One of the measured breakthrough curves is shown in Fig. 6.4b, where it is compared with the "theoretical" breakthrough curve that can be obtained from the Ogata–Banks solution, eq. (6.10), using the dispersivity computed from Gelhar's model, eq. (6.21). The dimensionless time is defined such that $t_D = 1$ when one "pore volume" of fluid has been

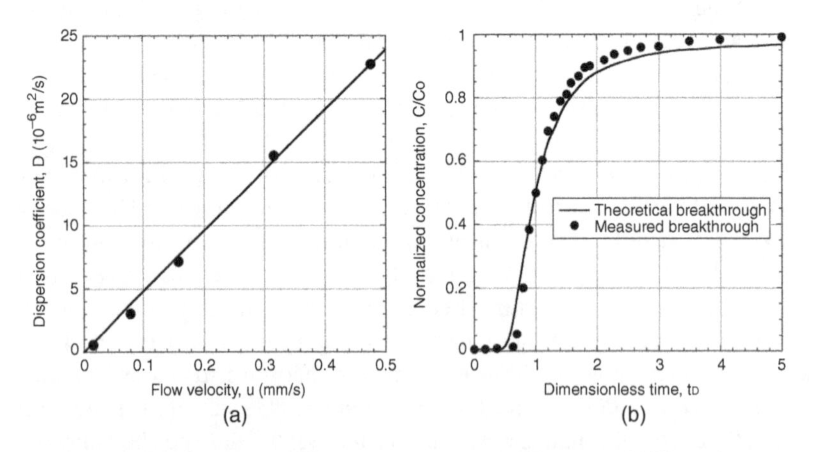

Figure 6.4 (a) Dispersion coefficients measured by Keller *et al.* (1999) during solute transport tests in a granite fracture, as a function of mean flow velocity. Each dot represents the mean value of three to five dispersion coefficients obtained from tests conducted at that flowrate. (b) Solute breakthrough curve for the one of these tracer tests, showing a comparison with "theoretical" curve. See text for details. Source: Adapted from Keller *et al.* (1999).

injected. The agreement is seen to be reasonably close. Details of the construction of the theoretical curve, and some possible explanations for the discrepancy between theory and measurements, are given by Keller *et al.* (1999).

6.5 Non-Fickian Transport in Rock Fractures

As originally pointed out by Taylor (1953) in the context of the circular pipe geometry, and mentioned at the end of Section 6.3, the Taylor–Aris analysis for a channel will not become valid until times that are greater than about h^2/D_m. Since molecular diffusivities in water are on the order of 10^{-9} m^2/s, this time scale will often be longer than the duration of laboratory transport experiments in rock fractures. For example, in the experiments of Keller *et al.* (1999), the mean aperture was 0.825 mm, and so this characteristic time is 680 s. In the case shown in the Figure 6.2, the normalized concentration at the fracture outlet had already reached about 0.9 by this time.

With regards to dispersion due to in-plane velocity heterogeneity, Gelhar and Axness (1983) assert that downstream displacements on the order of $10 - 100 \times \alpha_L$ will be required before expressions such as eq. (6.22) are strictly applicable. The dispersivity estimated by Keller *et al.* (1999) was $\alpha_L = 0.043$ m, and so a fracture whose length was at least 430 mm would have been required in order to safely utilize Gelhar's model, whereas the laboratory fracture had a length of only 166 mm. These points are consistent with the comments made at the end of Section 4.1 of Keller *et al.* (1999).

The numerical simulations of Beaudoin and Farhat (2021) confirmed that at any given flowrate, the dispersivity, and hence the effective diffusion coefficient, increases with time (by about one order of magnitude), before stabilizing to its asymptotic value. But this stabilization often required total mean displacements of roughly 10^3 dispersivity lengths, or roughly 10^3 correlation lengths, *i.e.*, $10^3\alpha_L$ or $10^3\lambda$. Such lengths will generally not be achievable in a laboratory.

Moreover, it is far from guaranteed that these lengths would be achieved by a fracture in the field, before an intersection with a neighboring fracture disrupts the flow and concentration profiles in that fracture. For example, the fracture utilized in the experiments of Keller *et al.* (1999) had a correlation length of 5.7 mm. Intersections with nearby fractures would need to be on the order of at least six meters apart for the dispersivity of such a fracture to have reached its asymptotic value. Although it might be thought that this issue could be handled by using "effective dispersivities" that varied with distance or time, the difficulty is more fundamental. Consider the basic problem discussed in Section 6.2 of a finite mass of solute introduced into a uniform flow field at $t = 0$ and $x = 0$, for which the ADE predicts a concentration profile described by eq. (6.12). The concentration profiles and breakthrough curves that actually obtain in the pre-asymptotic regime cannot be fit to equations such as (6.12), by *any* choice of D (Camassa *et al.*, 2010). Specifically, in this problem, some of the solute will typically arrive earlier than predicted by eq. (6.12), *i.e.*, "early breakthrough," which is due to the existence of localized fast paths within the fracture plane. At later times, the breakthrough curve will typically have a "longer tail" than is predicted by the ADE as some of the solute will have been held up in low-velocity zones within the fracture.

As pointed out by Zheng *et al.* (2019), two fundamentally different types of approaches have been used to model non-Fickian transport. One approach is to use a local version of the ADE, and allow the dispersion coefficient to vary in space (Stoll *et al.*, 2019; Zheng *et al.*, 2019). Depending on the degree of heterogeneity present, numerical solutions of this equation yield behavior that may be more complex than would be predicted by a "standard" ADE model that uses a spatially uniform dispersion coefficient.

Zheng *et al.* (2019) started with the following depth-averaged local ADE,

$$\frac{\partial(hC)}{\partial t} + \nabla \cdot (\mathbf{q}C) = \nabla \cdot (h\mathbf{D}\nabla C), \tag{6.22}$$

in which the components of the flux vector $\mathbf{q}$ [m^2/s] are equivalent to $h\overline{u}(x,y)$ in the notation of eqs. (2.36) and (2.37), where $\overline{u}(x,y)$ has been averaged in the z-direction of Fig. 6.3. The local aperture h and the dispersion tensor $\mathbf{D}$ were allowed to vary within the fracture plane. They aligned their x-axis with the macroscopic flow direction, and used the same boundary conditions that were used in the problem of eq. (6.10), *i.e.*, the solute concentration is initially zero everywhere, after which at $t = 0$, the concentration at $x = 0$ is instantaneously changed to, and maintained at, C_0. The local dispersion tensor was taken to have the following form:

$$\mathbf{D} = \begin{bmatrix} D_m + \dfrac{q_x^2}{210D_m} & 0 \\ 0 & D_m + \dfrac{q_y^2}{210D_m} \end{bmatrix}, \tag{6.23}$$

where the first term of each component is the intrinsic molecular diffusivity and the second term is the additional Taylor–Aris dispersion coefficient for channel flow, as given by eq. (6.20).

Using aperture fields obtained by scanning a fracture in a Santana tuff sample, Zheng *et al.* (2019) compared the breakthrough curves predicted by numerical solution of eq. (6.22) with those predicted by a presumably more accurate random-walk particle-tracking model. The local ADE was found to be quite accurate, for Peclet numbers as large as 450, where $Pe = u\langle h\rangle/D_m$ was defined based on the mean velocity and the mean aperture. Ignoring the Taylor–Aris terms in eq. (6.23) restricted the applicability of the ADE model to the range $Pe > 70$.

Non-Fickian solute profiles can also be caused by the presence of large amounts of contact area between the two opposing fracture walls, as was shown by Wang *et al.* (2021). They conducted numerical experiments within a fracture geometry obtained by scanning the surface of a tensile fracture in granite. Different normal displacements were numerically applied to the fracture plane, to generate fractures that had different contact fractions and different values of relative roughness. Simulations of flow and transport were conducted within each of the four resultant profiles, using the local ADE, and the "standard" conditions of a steady flow field with an initial solute concentration of $C(x, t = 0) = 0$, after which solute at some concentration C_0 was injected at $x = 0$, for all $t \geq 0$. The process was simulated using a local 3D version of the ADE, with a constant value of the intrinsic diffusivity, but a locally varying flow velocity. The diffusion coefficient was taken to be the intrinsic molecular diffusivity of lithium ions in water, *i.e.*, 10^{-9} m^2/s, and was not "corrected" to account for Taylor–Aris effects.

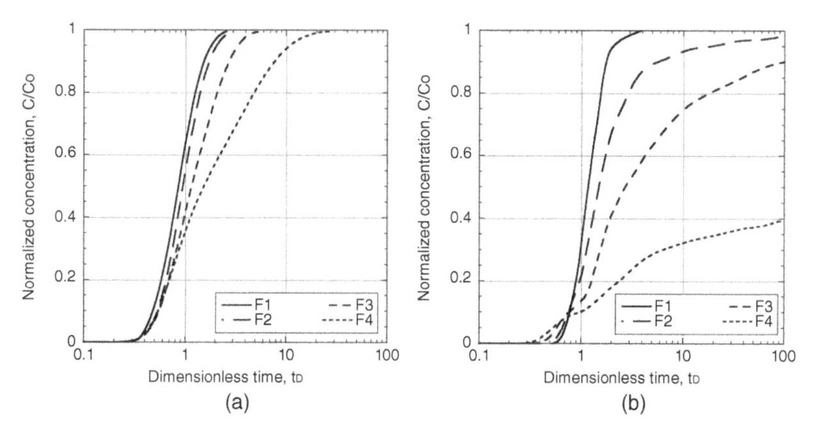

Figure 6.5 Breakthrough curves computed by Wang *et al.* (2021) for flow through a granite fracture that was "mathematically compressed" so as to yield four different geometries {F1, F2, F3, and F4} having increasing contact fractions: (a) $Pe = 0.1$, and (b) $Pe = 1000$. The dimensionless time is equivalent to the number of "pore volumes" of water that have been injected. Source: Adapted from Wang *et al.* (2021).

The four fracture profiles, labeled {F1, F2, F3, and F4}, had relative roughness values of {0.38, 0.47, 0.55, and 0.62}, and contact fractions of {0.013, 0.061, 0.206, and 0.448}, respectively. Simulations were conducted at a range of flowrates and hence a range of Peclet numbers. Wang *et al.* (2021) found that as *Pe* increases, a small amount of early breakthrough, and a large degree of "tailing," were observed in the breakthrough curves, with both effects increasing as the fraction of contact area increased, and as the Peclet number increased. The breakthrough curves that were computed for the two extreme values of Peclet number, 0.1 and 1000, are shown in Fig. 6.5.

Non-Fickian behavior is essentially due to inhomogeneity in the velocity field. An extreme type of velocity inhomogeneity would be the existence of recirculation zones within the fracture, within which the fluid is essentially (macroscopically) stagnant. Such zones are liable to form within small protruding "bumps" along the fracture wall (Fig. 6.1), and may occur at any value of the Reynolds number. Cardenas *et al.* (2007) performed numerical simulations of flow and transport in two-dimensional fracture profiles obtained from a rhyolitic tuff from Texas, at Reynolds numbers below 1, and found that the occurrence of stagnant recirculation zones led to long tails in the breakthrough curves. Similar simulations based on the same tuff fractures, as well as granite fractures from the Beishan area in northwestern China, were conducted in three-dimensional fracture profiles by Wang *et al.* (2020). Their simulations showed that recirculation zones, and the associated long tails in the breakthrough curves, were enhanced at Reynolds numbers that were large enough to cause non-Darcy flow.

Yoon and Kang (2021) conducted an extensive numerical investigation of the interplay between fracture roughness, Reynolds number, and Peclet number, and found that recirculation zones may either induce or suppress the existence of anomalous transport, depending on the Peclet number. In this regard, see also Dou *et al.* (2019), who investigated the effects of recirculation zones on non-Fickian transport behavior, particularly in the Forchheimer flow regime (see Chapter 4). The effect that the stress acting across the fracture plane may have on anomalous transport has been investigated by Kang *et al.* (2016, 2019) and Zou and Cvetkovic (2019). An extensive numerical investigation of the effect of shear displacement on solute transport has recently been conducted by Zou *et al.* (2022).

The above considerations have led to growing interest in developing mathematical models of non-Fickian transport that do *not* take the local ADE as their starting point. Detailed reviews and discussions of these non-Fickian models have been given by Berkowitz *et al.* (2006) and Neuman and Tartakovsky (2009), among others. A salient feature of such models is that ADE equation, which is a local partial-differential equation, is replaced by a nonlocal integro-differential equation. In particular, the models based on the Continuous Time Random Walk approach (CTRW; see Berkowitz *et al.* (2006), and references therein), have been found to provide a robust framework for modeling non-Fickian transport in both fractured and porous media.

6.6 Influence of Adsorption, Matrix Diffusion, and Radioactive Decay

There are several other factors that influence the transport of solutes through a fracture, many of which are related to the adjacent rock matrix. Solute particles may become *adsorbed* to the walls of the fracture, in which case they will no longer migrate through the fracture plane. Many different mathematical formulations have been proposed and used to model this process (see de Marsily, 1986), the simplest of which is to assume that the mass of solute adsorbed to the fracture walls, per unit area, is proportional to the local solute concentration.

With reference to the mass balance considerations presented in Section 6.2 and the geometric terminology of Fig. 6.3, a fracture region of length Δx will contain a fluid volume given by $A\Delta x = hw\Delta x$, and a total surface area of $S = 2w\Delta x$, where the factor of 2 accounts for the two opposing fracture surfaces. The mass of adsorbed solute per unit surface area is assumed to be given by $m_a = K_d SC$, where the parameter K_d [m] is known as the *distribution coefficient*, or *surface-sorption coefficient* (Bodin *et al.*, 2003). If this adsorbed mass is subtracted from the left-hand side of the mass balance equation (6.2), the following modified version of eq. (6.8) is eventually obtained (Tang *et al.*, 1981):

$$D\frac{\partial^2 C}{\partial x^2} = u\frac{\partial C}{\partial x} + \frac{\partial C}{\partial t} + \frac{2K_d}{h}\frac{\partial C}{\partial t} = u\frac{\partial C}{\partial x} + \left(1 + \frac{2K_d}{h}\right)\frac{\partial C}{\partial t}, \tag{6.24}$$

where the porosity term is taken to be unity within the fracture plane, and the fluid velocity term is written as u instead of q, as is standard in the solute transport literature.

The parenthesized term in eq. (6.24) is referred to as the (dimensionless) *retardation coefficient*, R:

$$R = 1 + \frac{2K_d}{h}. \tag{6.25}$$

According to this model, matrix adsorption essentially has the mathematical effect of dividing the time variable by R, *i.e.*,

$$D\frac{\partial^2 C}{\partial x^2} = u\frac{\partial C}{\partial x} + R\frac{\partial C}{\partial t} = u\frac{\partial C}{\partial x} + \frac{\partial C}{\partial t'}, \tag{6.26}$$

where $t' = t/R$. Consequently, mathematical solutions such as those given by eq. (6.10) or (6.12) continue to be valid, with t replaced by t/R (de Marsily, 1986). Since in the presence of adsorption, it is inherently the case that $R > 1$, solute migration is *retarded* (slowed down) relative to the manner in which they would proceed without adsorption.

Other adsorption models have been proposed, in which the relation between the adsorbed mass and the solute concentration in the flowing fluid is nonlinear, or in which different values of K_d are assumed to hold for adsorption and desorption (de Marsily, 1986; Bodin *et al.*, 2003).

Another process that has the effect of impeding the migration of solute along the fracture plane is diffusion of solute into the adjacent rock matrix. If the matrix has very low permeability, as is often the case in fractured hard rocks, solutes may still be able to diffuse into the matrix through the pore space, even if the permeability is so low as to essentially render the fluid in the matrix to be stagnant. In this situation, solute movement can be modeled by a pure diffusion equation in the matrix, and the loss of solute from the fracture into the matrix is included in the mass balance equation (6.26), by applying Fick's law at the interface (see Fig. 6.2), through a term such as $D_{matrix}(\partial C/dz)_{z=\pm h/2}$. This causes eq. (6.26) to take the form (Tang *et al.*, 1981; Carrera *et al.*, 1998)

$$R\frac{\partial C}{\partial t} + u\frac{\partial C}{\partial x} - D\frac{\partial^2 C}{\partial x^2} - \frac{2D_{matrix}}{h}\left.\frac{\partial C}{\partial z}\right]_{z=h/2} = 0, \tag{6.27}$$

where D is the dispersion coefficient for transport within the fracture, D_{matrix} is the diffusion coefficient in the matrix, and (by symmetry) matrix diffusion is assumed to be the same from each face of the fracture, with the two faces accounted for by the factor of 2. Equation (6.27) must be accompanied by the diffusion equation within the matrix, thereby leading to a set of two coupled partial differential equations. Numerical and analytical solutions to these equations, for various boundary and initial conditions, have been given by Grisak and Pickens (1980) and Tang *et al.* (1981), respectively, among others.

The effect of matrix diffusion is to a great extent similar to that due to surface adsorption, in that it removes solute from the flowing fluid in the fracture, and thereby retards the migration of the solute plume. This physical analogy is highlighted by the fact that eq. (6.27) is often simplified by modeling the fracture-to-matrix flux by a "first-order mass transfer model" (Haggerty and Gorelick, 1995; Carrera *et al.*, 1998), which in its simplest form has the effect of replacing the matrix diffusion term in eq. (6.27) with a term proportional to $\partial C/dt$, and thereby plays a role identical to the term $R(\partial C/dt)$ that arose due to surface adsorption.

If the solute is radioactive, solute migration within a fracture will also be retarded by the process of *radioactive decay*. On an atomic level, radioactive decay is a random process in which the probability of a given atom undergoing radioactive decay in a time interval Δt is proportional to Δt. Hence, the rate of change in concentration of that solute in the flowing fluid, due to radioactive decay, is given by a term proportional to its concentration. This introduces an additional term into eq. (6.27), leading to (Tang *et al.*, 1981)

$$\frac{\partial C}{\partial t} + \frac{u}{R}\frac{\partial C}{\partial x} - \frac{D}{R}\frac{\partial^2 C}{\partial x^2} - \frac{2D_{matrix}}{hR}\left.\frac{\partial C}{\partial z}\right]_{z=h/2} + \lambda C = 0, \tag{6.28}$$

where λ [1/s] is the decay constant. In the absence of advection, dispersion, or matrix diffusion, eq. (6.28) can easily be solved to show that an initial concentration C_o will decay according to $C(t) = C_o \exp(-\lambda t)$, thus indicating that $1/\lambda$ is the time constant for radioactive decay. Solutions to the full version of eq. (6.28) have been derived by Tang *et al.* (1981), Sudicky and Frind (1982), and Meng *et al.* (2018), among others.

Problems

6.1 Prove that as $Pe \to \infty$, the concentration profile given by eq. (6.10) reduces to the following piston-type profile: $C(x, t) = C_o$ for $x < ut$, and $C(x, t) = 0$ for $x > ut$. It will be helpful to make use of the approximation that, for large positive values of x, $\mathrm{erfc}(x) \approx e^{-x^2}/\sqrt{\pi}x$ (Crank, 1975, p. 37).

6.2 Prove that, for the concentration profile given by eq. (6.12), the center of mass of the solute plume grows according to $\langle x \rangle = ut$, whereas the standard deviation of the solute mass about its center of mass grows as $\sigma = \sqrt{4Dt}$. It will be helpful to make use of the fact that $\sigma^2 = \langle x^2 \rangle - \langle x \rangle^2$, where $\langle f \rangle$ is the concentration-weighted mean value of the quantity $\langle f \rangle$.

6.3 Consider the laboratory experiments of transport in a rock fracture conducted by Keller *et al.* (1999), with an effective dispersion coefficient of $D = 2.5 \times 10^{-5}\mathrm{m}^2/\mathrm{s}$ and a flowrate of $u = 5 \times 10^{-4}\mathrm{m/s}$. The length of the fracture in the direction of flow is 0.166 m. Fluid having solute concentration C_o is injected at $x = 0$. Use the solution of Ogata and Banks (1961), as given by eq. (6.10), to plot the normalized concentration profiles in the fracture, at times of 100, 200, 300, 400, and 500 s.

References

Aris, R. 1956. On the dispersion of a solute in a fluid flowing through a tube. *Proceedings of the Royal Society Series A*, 235(1200), 67–77.

Aris, R. 1959. On the dispersion of a solute by diffusion, convection and exchange between phases. *Proceedings of the Royal Society Series A* 252(1271), 538–50.

Bear, J. 1988. *Dynamics of Fluids in Permeable Media*, Dover Publications, Mineola, N.Y.

Beaudoin, A. and Farhat, M. 2021. Impact of the fracture contact area on macro-dispersion in single rough fractures. *Comptes Rendus Mécanique*, 349(2), 203–24.

Berkowitz, B., Cortis, A., Dentz, M., and Scher, H. 2006. Modeling non-Fickian transport in geological formations as a continuous time random walk. *Reviews of Geophysics*, 44, 2005RG000178.

Bodin, J. 2015. From analytical solutions of solute transport equations to multidimensional time-domain random walk (TDRW) algorithms. *Water Resources Research*, 51(3), 1860–71.

Bodin, J., Delay, F., and de Marsily, G. 2003. Solute transport in a single fracture with negligible matrix permeability: 2. Mathematical formalism. *Hydrogeology Journal*, 11, 434–54.

Camassa, R., Lin, Z., and Mclaughlin, R. M. 2010. The exact evolution of the scalar variance in pipe and channel flow. *Communications in Mathematical Sciences*, 8(2), 601–26.

Cardenas, M. B., Slottke, D. T., Ketcham, R. A., and Sharp, J. M. 2007. Navier-stokes flow and transport simulations using real fractures shows heavy tailing due to eddies. *Geophysical Research Letters*, 34(14), L14404.

Carrera, J., Sánchez-Vila, X., Benet, I., Medina, A., *et al.* 1998. On matrix diffusion: formulations, solution methods and qualitative effects. *Hydrogeology Journal*, 6, 178–90.

Cavalcante, A. L. B. and de Farias, M. M. 2013. Alternative solution for advective-dispersive flow of reagent solutes in clay liners. *International Journal of Geomechanics*, 13(1), 49–56.

Chatwin. P. C. and Sullivan, P. J. 1982. The effect of aspect ratio on longitudinal diffusivity in rectangular channels. *Journal of Fluid Mechanics*, 120, 347–58.

Crank, J. 1975. *The Mathematics of Diffusion*, 2nd ed., Clarendon Press, Oxford.

de Marsily, G. 1986. *Quantitative Hydrogeology*, Academic Press, San Diego.

Doshi, M. R., Daiya, M. D., and Gill, W. N. 1978. Three-dimensional laminar dispersion in open and closed rectangular conduits. *Chemical Engineering Science*, 33(7), 795–804.

Dou, Z., Sleep, B., Zhan, H., Zhou, Z., *et al.* 2019. Multiscale roughness influence on conservative solute transport in self-affine fractures. *International Journal of Heat and Mass Transfer*, 133, 606–18.

Gelhar, L. W. 1993. *Stochastic Subsurface Hydrology*, Prentice-Hall, Englewood Cliffs, N.J.

Gelhar, L. W. and Axness, C. L. 1983. Three-dimensional stochastic analysis of macrodispersion in aquifers. *Water Resources Research*, 19(1), 161–80.

Ghez, R. 1988. *A Primer of Diffusion Problems*, Wiley, New York.

Grisak, G. E. and Pickens, J. F. 1980. Solute transport through fractured media 1. The effect of matrix diffusion. *Water Resources Research*, 16(4), 719–30.

Guell, D. C., Cox, R. G., and Brenner, H. 1987. Taylor dispersion in conduits of large aspect ratio. *Chemical Engineering Communications*, 58(1–6), 231–44.

Haggerty, R. and Gorelick, S. M. 1995. Multiple-rate mass transfer for modeling diffusion and surface reactions in media with porescale heterogeneity. *Water Resources Research*, 31(10), 2383–2400.

Javandel, I., Doughty C., and Tsang, C. F. 1984. *Groundwater Transport: Handbook of Mathematical Models*, American Geophysical Union, Washington, D.C.

Kang, P. K., Brown, S., and Juanes, R. 2016. Emergence of anomalous transport in stressed rough fractures. *Earth and Planetary Science Letters*, 454, 46–54.

Kang, P. K., Lei, Q., Dentz, M., and Juanes, R. 2019. Stress-induced anomalous transport in natural fracture networks. *Water Resources Research*, 55(5), 4163–85.

Keller, A. A., Roberts, P. V., and Blunt, M. J. 1999. Effect of fracture aperture variations on the dispersion of contaminants. *Water Resources Research*, 35(1), 55–63.

Kreft, A. and Zuber, A. 1978. On the physical meaning of the dispersion equation and its solutions for different initial and boundary conditions. *Chemical Engineering Science*, 33, 1471–80.

Meng, S., Liu, L., Mahmoudzadeh, B., Neretnieks, I., *et al.* 2018. Solute transport along a single fracture with a finite extent of matrix: a new simple solution and temporal moment analysis. *Journal of Hydrology*, 562, 290–304.

Neuman, S. P. and Tartakovsky, D. M. 2009. Perspective on theories of anomalous transport in heterogeneous media. *Advances in Water Resources*, 32(5), 670–80.

Ogata, A., and Banks, R. B. 1961. A solution of the differential equation of longitudinal dispersion in porous media. *U.S. Geological Survey Professional Paper 411-A*, U. S. Geological Survey, Washington, D.C.

Parks, M. L. and Romero, L. A. 2007. Taylor–Aris dispersion in high aspect ratio columns of nearly rectangular cross section. *Mathematical and Computer Modelling*, 46(5–6), 699–717.

Sisavath, S., Jing, X. D., and Zimmerman, R. W. 2001. Laminar flow through irregularly-shaped pores in sedimentary rocks. *Transport in Porous Media*, 45, 41–62.

Stoll, M., Huber, F. M., Trumm, M., Enzmann, F., *et al.* 2019. Experimental and numerical investigations on the effect of fracture geometry and fracture aperture distribution on flow and solute transport in natural fractures. *Journal of Contaminant Hydrology*, 221, 82–97.

Sudicky, E. A. and Frind, E. O. 1982. Contaminant transport in fractured porous media: analytical solution for a system of parallel fractures. *Water Resources Research*, 18(6), 1634–42.

Tang, D. H., Frind, E. O., and Sudicky, E. A. 1981. Contaminant transport in fractured porous media: analytical solution for a single fracture. *Water Resources Research*, 17(3), 555–64.

Tartakovsky, D. M. and Dentz, M. 2019. Diffusion in porous media: phenomena and mechanisms. *Transport in Porous Media*, 130, 105–27.

Taylor, G. I. 1953. Dispersion of soluble matter in solvent flowing slowly through a tube. *Proceedings of the Royal Society Series A*, 219(1137), 186–203.

Tsang, C. F., Tsang, Y. W., and Hale, F. V. 1991. Tracer transport in fractures: analysis of data based on a variable-aperture channel model. *Water Resources Research*, 27(12), 3095–3106.

Wang, L., Cardenas, M. B., Zhou, J.-Q., and Ketcham, R. A. 2020. The complexity of nonlinear flow and non-Fickian transport in fractures driven by three-dimensional recirculation zones. *Journal of Geophysical Research: Solid Earth*, 125(9), e2020JB020028.

Wang, Z. H., Zhou, C. T., Wang, F., Li, C. B., *et al.* 2021. Channeling flow and anomalous transport due to the complex void structure of rock fractures. *Journal of Hydrology*, 601, 126624.

Wooding, R. A. 1960. Rayleigh instability of a thermal boundary layer in flow through a porous medium. *Journal of Fluid Mechanics*, 9(2), 183–92.

Yoon, S. and Kang, P. K. 2021. Roughness, inertia, and diffusion effects on anomalous transport in rough channel flows. *Physical Review Fluids*, 6, 014502.

Zhang X. Y., Ma, F. N., Dai, Z. X., Wang, J., *et al.* 2022. Radionuclide transport in multi-scale fractured rocks: a review. *Journal of Hazardous Materials*, 424, 127550.

Zheng, L., Wang, L., and James, S. C. 2019. When can the local advection–dispersion equation simulate non-Fickian transport through rough fractures? *Stochastic Environmental Research and Risk Assessment*, 33, 931–38.

Zou, L. and Cvetkovic, V. 2019. Impact of normal stress-induced closure on laboratory scale solute transport in a natural rock fracture. *Journal of Rock Mechanics and Geotechnical Engineering*, 12(4), 732–41.

Zou, L., Mas Ivars, D., Larsson, J., Selroos, J.-F., *et al.* (2022). Impact of shear displacement on advective transport in a laboratory-scale fracture. *Geomechanics for Energy and the Environment*, 31, 100278.

7

Analytical Models for the Permeability of a Fractured Rock Mass

7.1 Introduction

Previous chapters have focused exclusively on the hydraulic transmissivity of individual fractures and the response of the transmissivity of an individual fracture to mechanical and chemical influences. But the effect of rock fractures on the hydrological behavior of geothermal reservoirs, oil and gas reservoirs, radioactive waste repositories, and other underground systems and processes is, in most situations, dependent on fluid flow through a *fractured rock mass* rather than through individual fractures. In some cases, the matrix rock may have such a low permeability that the overall permeability of the fractured rock mass is controlled solely by the fracture network. In other cases, the overall permeability of the fractured rock mass is a function of both the fracture network properties and the matrix rock properties.

The effective macroscopic permeability of a fractured rock mass can be computed by numerically modeling the flow of fluid through a network of fractures, with or without accounting for the flow through the matrix (*e.g.*, Paluszny and Matthäi, 2010; Lang *et al.*, 2014; Hardebol *et al.*, 2015; Bisdom *et al.*, 2016; see also Chapter 8). However, this type of numerical modeling is computationally expensive. Moreover, given that the geometric information that is typically available on the lengths, apertures, and orientations of the fractures is stochastic in nature, many realizations are required in order to obtain representative values for the effective permeability. In addition, due to the high degree of variation and uncertainty inherent in fracture network properties, it is often of interest to quantify the effect of this uncertainty on the macroscopic effective permeability by probing a multi-dimensional parameter space, such as when assessing the suitability of sites for radioactive waste disposal, for example. Thus, the need is posed for simple, predictive analytical methods for estimating the macroscopic permeability of a fractured rock mass.

The present chapter will focus on analytical models for effective permeability. The starting point (Section 7.2) is the classical model developed by Snow (1969), which is based on the assumption of a set of infinitely long fractures embedded in an impermeable matrix. Next, for the more general case of fractures embedded in a *permeable* matrix, two sets of theoretical upper and lower bounds are presented (Section 7.3), due to Wiener (1912) and Hashin and Shtrikman (1962), which bound the possible values of the macroscopic permeability in terms of the permeabilities and volume fractions of the fractures and matrix rock.

Fluid Flow in Fractured Rocks, First Edition. Robert W. Zimmerman and Adriana Paluszny.

Companion website: www.wiley.com/go/zimmerman/fluidflowinfracturedrocks

Specific estimates of the macroscopic permeability are developed and investigated in Sections 7.4–7.6 by applying the formalism of "effective medium theories" to the geometrical model in which fractures are treated as thin and highly permeable spheroidal inclusions embedded in a homogeneous and less permeable rock matrix. Finally, the widely used heuristic equation proposed by Mourzenko *et al.* (2011), which was developed based on the analysis of a large number of numerical simulations and theoretical arguments, is discussed in Section 7.7.

7.2 Snow's Model of Planar Fractures of Infinite Extent in an Impermeable Matrix

The simplest conceptual model is that of a rock mass containing three orthogonal sets of infinite, planar fractures in an *impermeable* matrix rock. Let the fracture set that lies in the $y-z$ plane have spacing S_x, and aperture h_x, with a similar notation for the other two sets. This scenario is shown in Fig. 7.1a, looking only at the $x-y$ plane, for simplicity. Each fracture is assumed to be of infinite extent in its plane. A unit cell of this rock mass would have dimensions $\{S_x, S_y, S_z\}$ and would contain one fracture whose plane is normal to each of the three orthogonal directions. A pressure gradient in the z-direction would give rise to a fluid flux in the z-direction through each of the two fractures that lie in the $x-z$ and $y-z$ planes, according to eq. (2.11), so that the total flux in the z-direction is given by

$$Q_z = -\frac{S_x h_y^3}{12\mu}\frac{\partial p}{\partial z} - \frac{S_y h_x^3}{12\mu}\frac{\partial p}{\partial z}. \tag{7.1}$$

The cross-sectional area A of the unit cell in the direction perpendicular to the flux is $S_x S_y$, so the permeability in the z-direction is given by

$$k_z = \frac{-Q_z\mu}{A(\partial p/\partial z)} = \frac{h_y^3}{12S_y} + \frac{h_x^3}{12S_x}, \tag{7.2}$$

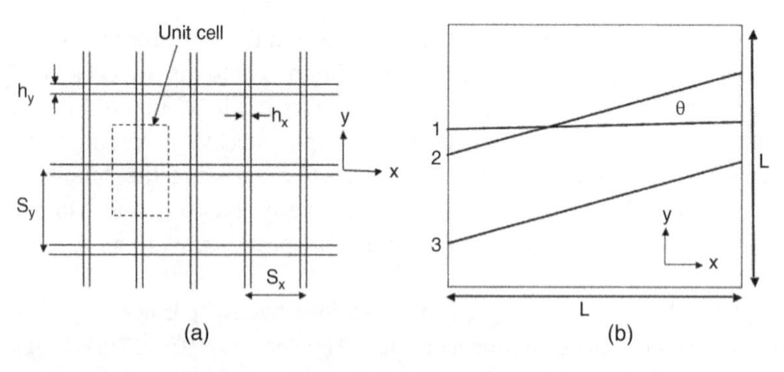

Figure 7.1 (a) Schematic diagram of a two-dimensional rock mass containing two sets of orthogonal fractures, showing a unit cell that is used to help in calculating the macroscopic permeability. The aperture of a fracture whose outward unit normal vector is in the x-direction is h_x, and the spacing of this fracture set is S_x. (b) Three fractures, extending infinitely far in the z-direction (into the page), connecting the left and right faces of a cubic block of side L; see Problem 7.2.

and similarly for the other two directions. This macro-scale permeability is proportional to the *cube* of the aperture and inversely proportional to the fracture spacing, *i.e.*, proportional to the spatial frequencies of the fracture sets. Similar equations for the permeabilities in the y-direction and z-direction can be obtained by permuting the subscripts in eq. (7.2); see Problem 7.1. Since the permeability is a second-order tensor, the components of the permeability tensor with respect to a coordinate system that is rotated by some arbitrary angle with respect to the $\{x, y, z\}$ axes can be obtained by the usual rules of tensor transformation (Bear, 1988; Jaeger *et al.*, 2007).

If the spacings of the three orthogonal fracture sets are equal and the apertures of the three orthogonal fracture sets are equal, this fractured rock mass will be hydraulically isotropic, with an effective macroscopic permeability given by

$$k_x = k_y = k_z = k_{eff} = \frac{2h^3}{12S}. \tag{7.3}$$

The porosity associated with the fracture set lying in the $y - z$ plane will be equal to h_x/S_x, and similarly for the other two fracture sets, and so in the isotropic case, the total fracture porosity will be $\phi_f = 3h/S$. The "permeability" of each *individual* fracture is, according to eq. (2.12), given by $k_f = h^2/12$. Hence, the macroscopic permeability of the fractured rock mass can be expressed as (Snow, 1969)

$$k_{eff} = \frac{2h^3}{12S} = 2 \cdot \frac{h^2}{12} \cdot \frac{h}{S} = \frac{2k_f\phi_f}{3}. \tag{7.4}$$

The above result for macroscopically isotropic rock masses, which will henceforth be written as $k_S = 2k_f\phi_f/3$, where the subscript S denotes "Snow," corresponds to a conceptual model of an otherwise impermeable rock mass containing three orthogonal sets of fractures, each of which extends infinitely far in the two in-plane directions. It is also true, although perhaps not intuitively obvious, that eq. (7.4) will hold for a medium containing fractures of infinite extent whose planes, as identified by their outward unit normal vectors, are *randomly* oriented in space with a uniform probability distribution (Adler *et al.*, 2013, p. 89).

Since real fractures will be of finite extent rather than infinite extent, it would be expected that, for a set of fractures having a given aperture and a given total fracture porosity, Snow's expression will *overestimate* the macroscopic permeability. This is indeed the case, as will be proven in Section 7.3.

Snow (1969) also developed expressions for a more general anisotropic case, in which the rock mass may contain several fracture sets having arbitrary orientations, not necessarily orthogonal to each other; in this regard, see also Parsons (1966) and Oda (1985). The basis of Snow's models is the fact that, if fractures are infinitely long in all directions, the amount of fluid that flows through an individual fracture due to an imposed far-field pressure gradient is *completely unaffected* by whether or not this fracture intersects other fractures (see Adler *et al.*, 2013, p. 89), and hence can be computed directly from equations such as eq. (2.11). This remarkable fact is proven in Problem 7.2, for the two-dimensional case, with the aid of Fig. 7.1b. However, if the fracture is of finite length, as are all real rock fractures, the flux through that fracture will indeed be influenced by intersections with other fractures. These effects are treated implicitly by the analytical methods discussed below and explicitly by the numerical simulations discussed in Chapter 8.

7.3 Upper and Lower Bounds on the Effective Permeability

As the actual geometry and topology of a fracture network will rarely be known in detail, it would be helpful to have recourse to theoretical upper and lower bounds that can constrain the possible range of numerical values of the effective permeability and which require knowledge of only a small number of parameters, such as k_m, k_f, and ϕ_f. The simplest bounds on the effective permeability of a two-component material, which are valid regardless of the specific microstructure or topology of the two components, are the "series" and "parallel" bounds of Wiener (1912); see also Torquato (2013, p. 556). In the context of a fractured rock mass, the two "components" are the matrix rock, which has permeability k_m and volume fraction $1-\phi_f$, and the fractures, which have permeability k_f and volume fraction ϕ_f. For this type of two-component medium, the Wiener bounds state that

$$\left(\frac{\phi_f}{k_f}+\frac{1-\phi_f}{k_m}\right)^{-1} \leq k_{eff} \leq \phi_f k_f + (1-\phi_f)k_m. \tag{7.5}$$

For fractured rock masses, it will necessarily be true that $k_m \ll k_f$ and $\phi_f \ll 1$, in which case the Wiener bounds essentially reduce to

$$k_m \leq k_{eff} \leq k_m + \phi_f k_f. \tag{7.6}$$

Hashin and Shtrikman (1962) derived a different pair of upper and lower bounds for the effective permeability of a two-component system. In the context of fractured rock masses, the Hashin–Shtrikman bounds can be written as

$$k_m + \frac{3k_m(k_f-k_m)\phi_f}{3k_m+(k_f-k_m)(1-\phi_f)} \leq k_{eff} \leq k_f - \frac{3k_f(k_f-k_m)(1-\phi_f)}{3k_f-(k_f-k_m)\phi_f}. \tag{7.7}$$

The Hashin–Shtrikman bounds are always at least as stringent as the Wiener bounds, and so for all practical purposes can be considered to supersede them. In the present context of fractured rock masses, for which $k_m \ll k_f$ and $\phi_f \ll 1$, the Hashin–Shtrikman bounds reduce to

$$k_m \leq k_{eff} \leq k_m + \frac{2k_f\phi_f}{3} = k_m + k_S. \tag{7.8}$$

The Hashin–Shtrikman lower bound states the obvious fact that the addition of conductive fractures into a rock mass can never cause the effective permeability to be less than the matrix permeability. The Hashin–Shtrikman upper bound also has a simple physical interpretation: the effective permeability can never exceed the sum of the permeability of the unfractured rock mass, k_m, plus the permeability due solely to an infinite set of fractures, which is given by Snow's equation, $k_S = 2k_f\phi_f/3$. Hence, in the case of an impermeable matrix, eq. (7.8) shows that the Snow equation provides an upper bound to the actual permeability.

For a fractured rock mass, the "fracture-related component" of the Hashin–Shtrikman upper bound, $2k_f\phi_f/3$, is less than the fracture-related component of the Wiener upper bound, $k_f\phi_f$, by a factor of one-third. This is essentially due to the fact that whereas the Wiener "parallel network" upper bound assumes that all of the fractures can transmit fluid in, say, the x-direction, a three-dimensional version of Fig. 7.1a, and the accompanying discussion, show that in an isotropic fracture network, only two-thirds of the fractures are able to transmit fluid in the x-direction. The Hashin–Shtrikman upper bound correctly accounts for this fact.

7.4 Spheroidal Inclusion Model of a Fractured Rock Mass

Snow's model is a "network model" that focuses on the geometry and topology of the fracture network. More sophisticated network-type models, that incorporate information about the lengths and interconnectivity of the fractures, have been developed by many researchers (Harris, 1990; de Dreuzy *et al.*, 2001; Mourzenko *et al.*, 2004; Leung and Zimmerman, 2012; Li and Li, 2015). These types of models have been reviewed in detail by Adler *et al.* (2013).

Another conceptual model of a fractured rock mass is that of a porous and permeable matrix, containing fractures that are represented by thin, oblate spheroidal inclusions (Fokker 2001; Pozdniakov and Tsang, 2004; Barthélémy, 2009; Sævik *et al.*, 2013, 2014). These inclusion-type models lend themselves in a natural way to the application of effective medium theories, which are well developed in many areas of mechanics (*cf*., Torquato, 2013). In the simplest version of this type of model, the porous matrix has permeability k_m, and each "inclusion" is treated as being composed of a porous material that obeys Darcy's law and has permeability k_f.

An objection may be raised that, although it is sensible to assign a "permeability" to flow *within* the fracture plane, it is not *a priori* clear that it is meaningful to assign that same permeability to flow within the fracture, but in the direction *normal* to the fracture plane. However, Vu *et al.* (2018) investigated this issue in detail, and verified the acceptability of this conceptual model, for the case of open (*i.e.*, unfilled) fractures without any surface "skin." Furthermore, Adler *et al.* (2013, p. 91) studied the consequences of assuming that the fractures have a *circular* shape in the nominal fracture plane, as opposed to being square or hexagonal, and found that planform shape has a very minor influence, as long as the shape is not too highly elongated in one direction. Hence, the oblate spheroidal inclusion model should not be dismissed as being unrealistic or overly simplistic.

A spheroidal inclusion has two equal semi-axis of length R, which can be referred to as the "fracture radius." The length of the third, shorter semi-axis is one-half of the maximum fracture aperture, *i.e.*, $h/2$. The aspect ratio α of the inclusion is then defined as the ratio of the minor semi-axis to the major semi-axis (Fig. 7.2a):

$$\alpha = h/2R \ll 1. \tag{7.9}$$

Several other conceptual models (*e.g.*, Adler *et al.*, 2013) treat the fractures as flat disks having a *uniform* aperture, b. These two different geometrical representations can be related to each other by equating the volumes of a single fracture, which yields the relation

$$b = \frac{4}{3}R\alpha. \tag{7.10}$$

A fracture set can be defined as a group of fractures that share common properties, such as radius and permeability. The number density, ρ_i, of fracture set i is the number of fracture centers, N_i, per volume, *i.e.*,

$$\rho_i = \frac{N_i}{V}, \tag{7.11}$$

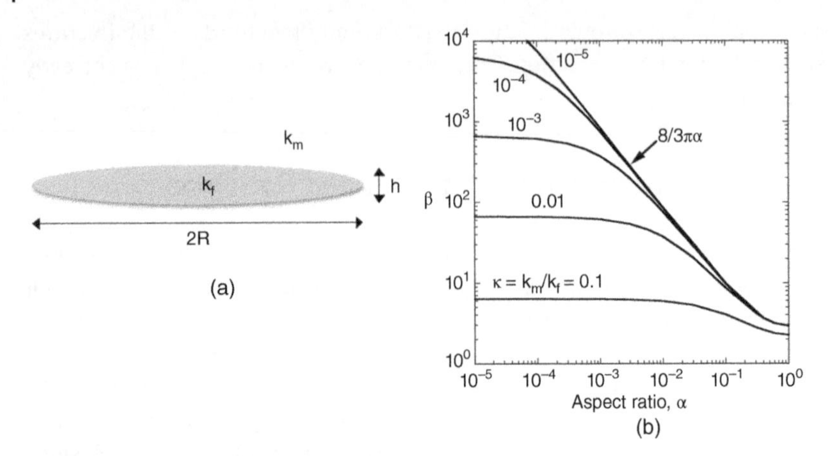

Figure 7.2 (a) Spheroidal inclusion of aspect ratio $\alpha = h/2R$, where h in this case is the maximum aperture, as a model for a thin fracture embedded in a porous matrix. The permeability ratio is defined as $\kappa = k_m/k_f$. (b) Factor β, as defined by eqs. (7.18) to (7.20), which quantifies the relative rate at which the effective permeability increases with increasing fracture porosity. This factor approaches the asymptotic line $\beta = 8/3\pi\alpha$ when $\alpha \gg \kappa$, and approaches the asymptotic value $\beta = 2/3\kappa$ when $\alpha \ll \kappa$.

where V is the volume of the domain for which the permeability is to be determined. The porosity of fracture set i is defined as the total volume occupied by these fractures, per unit volume of rock mass:

$$\phi_i = \frac{4}{3}\pi \rho_i R_i^3 \alpha_i. \tag{7.12}$$

If the rock mass contains n different fracture sets, the total fracture porosity would be found by summing up eq. (7.12) over all i, from $i = 1$ to n.

Another useful dimensionless parameter for assessing the influence of the fractures on rock mass permeability is the *fracture density parameter*, defined here by

$$\varepsilon_i = \rho_i R_i^3 = \frac{N_i R_i^3}{V}. \tag{7.13}$$

This definition agrees with the one given by eqs. (1.23) and (1.26), aside from the absence of the π^2 factor. Since the volume of an oblate spheroid of radius R and aspect ratio α is $4\pi R^3\alpha/3$, the fracture porosity is related to the fracture density parameter by

$$\phi_i = \frac{4}{3}\pi\, \alpha_i\, \varepsilon_i. \tag{7.14}$$

Several other definitions of fracture density have been used (Mourzenko *et al.*, 2011; Sævik *et al.*, 2014; Li and Li, 2015). The definition is given in eq. (7.13) is common, particularly in the context of inclusion-based effective medium theories.

Each fracture is assigned a permeability, k_f, which will depend on several factors, such as aperture, surface roughness, and filling material. Although various models, such as those discussed in Chapter 2, can be used for this purpose, in the following discussion it will occasionally be convenient to use the parallel plate model, eq. (2.12):

$$k_f = \frac{h^2}{12}, \tag{7.15}$$

in which the aperture h is assumed to be uniform throughout the plane of the fracture.

The effective macroscopic permeability of a fractured rock mass will depend strongly on the ratio of matrix to fracture permeability, defined as

$$\kappa = k_m / k_f. \tag{7.16}$$

For the case of unfilled fractures, which will be the main focus of this chapter, the fracture permeability is always several orders of magnitude greater than the matrix permeability, and so κ will typically be very small. For example, a rock fracture with an aperture on the order of 100 μm will have, according to eq. (7.15), a permeability on the order of 10^{-9} m^2, whereas the matrix rock will always have a permeability no greater than about 10^{-12} m^2. Hence, the permeability ratio κ will not be expected to exceed 10^{-3}. As will be seen in Sections 7.5 and 7.6, the permeability and flow behavior of a fractured rock mass will have qualitatively different characteristics depending on whether the aspect ratio is smaller than, or larger than, the permeability ratio, *i.e.*, whether $\alpha \ll \kappa$, or $\alpha \gg \kappa$.

Snow's model (Snow, 1969) is based on the assumption that each fracture extends infinitely far in its own plane. The fracture density parameter, as defined by eq. (7.13), cannot be directly computed for such scenarios, because the radius term becomes infinite. However, if eq. (7.14) is used to express the porosity in terms of fracture density and aspect ratio, Snow's equation (7.4) can be written in a form that explicitly contains the fracture density:

$$k_S = \frac{2k_f \phi_f}{3} = \frac{8\pi}{9} k_f \alpha\, \varepsilon. \tag{7.17}$$

For a rock mass containing a "low" concentration of fractures, exact expressions can be derived for the effective permeability. If the fractures are modeled as randomly oriented spheroidal inclusions, it follows from the analysis given by Fricke (1924) that, to first-order in fracture porosity (Zimmerman, 1989),

$$k_{eff} = k_m(1 + \beta\phi_f), \tag{7.18}$$

where

$$\beta = \frac{(r-1)}{3}\left[\frac{4}{2+(r-1)M} + \frac{1}{1+(r-1)(1-M)}\right], \tag{7.19}$$

$$M = \frac{2\theta - \sin 2\theta}{2\tan\theta \sin^2\theta}, \tag{7.20}$$

$\theta = \arccos\alpha$, and $r = 1/\kappa$. For thin spheroids, $\alpha \ll 1$, and so $M = \alpha\pi/2 \ll 1$ (Sævik *et al.*, 2013), which implies $1 - M \approx 1$. Combining these approximations with $r \gg 1$, which is the case of interest for fractured rocks, eq. (7.19) simplifies to

$$\beta = \frac{r}{3}\left[\frac{4}{2+(\alpha\pi r/2)}\right] = \frac{8}{12\kappa + 3\pi\alpha}, \tag{7.21}$$

in which case eq. (7.18) takes the form

$$k_{eff} = k_m\left[1 + \left(\frac{8}{12\kappa + 3\pi\alpha}\right)\phi_f\right]. \tag{7.22}$$

The variation of β with the aspect ratio α, for various values of the permeability ratio κ, is shown in Fig. 7.2b. This figure, and eq. (7.21), both clearly indicate the existence of two different asymptotic regimes:

$$\beta = \frac{2}{3\kappa} \quad \text{when} \quad \alpha \ll \kappa, \tag{7.23}$$

$$\beta = \frac{8}{3\pi\alpha} \quad \text{when} \quad \alpha \gg \kappa \tag{7.24}$$

The "dilute concentration" expression for the effective permeability that is given by eq. (7.22) is not expected to be applicable for dense and highly connected fracture networks, since it does not account for flow interactions between nearby or intersecting fractures. Dilute concentration results such as this are typically extended to higher concentrations by using an "effective medium" scheme, which attempts to account, in some simple and approximate manner, for the complex flow interactions between neighboring and intersecting fractures. Numerous such schemes for estimating the "effective conductivity" of heterogeneous media have been proposed since as far back as the late nineteenth century (Maxwell, 1873; Rayleigh, 1892); see also Zimmerman (1996) and Torquato (2013). Sævik *et al.* (2013) and Ebigbo *et al.* (2016) have reviewed many of these schemes in the specific context of the permeability of fractured rock masses.

Effective medium schemes are typically accurate for moderate values of the inclusion concentration (in this case, ϕ_f), but often have difficulty above the percolation threshold. However, the so-called "self-consistent" effective medium scheme (Sævik *et al.*, 2013) is reasonably accurate over a wide range of the parameter space and shows qualitatively correct behavior in all limiting cases.

Whereas eq. (7.22) was derived by considering the effect of a single inclusion placed in an infinite matrix having permeability k_m, in the self-consistent approach (sometimes referred to as the asymmetric self-consistent approach), the inclusion is placed within an effective medium whose permeability is already equal to k_{eff}. This has the effect of replacing k_m with k_{eff} in the second term on the right side of eq. (7.22), including where it appears in the term $\kappa = k_m/k_f$. The result, after being expressed as a function of the fracture density parameter ε, is

$$\frac{k_{eff}}{k_m} = 1 + \frac{32}{9}\left\{\frac{(k_{eff}/k_m)}{[4(k_{eff}/k_m)/\pi(\alpha/\kappa)] + 1}\right\}\varepsilon. \tag{7.25}$$

Since the ratio k_{eff}/k_m appears on both sides of this equation, the variation of k_{eff}/k_m with the parameters α/κ and ε is not easy to discern. The behavior of this equation, in the regimes of small and large values of the ratio α/κ, will be examined in detail in the following two sections.

7.5 Effective Permeability in the Regime $\alpha/\kappa \ll 1$

According to the self-consistent effective medium theory, eq. (7.25), the normalized effective permeability depends only on the fracture density parameter ε, and the dimensionless parameter α/κ. Assuming a typical aperture of 100 μm, the permeability ratio κ may

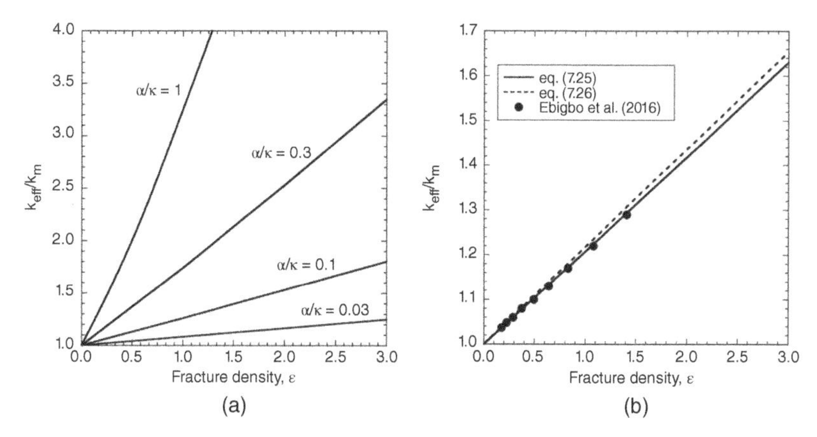

Figure 7.3 (a) Effective permeability of a fractured rock mass, as a function of fracture density, in the regime $\alpha/\kappa \leq 1$, as predicted by the self-consistent effective medium theory, eq. (7.25). (b) Effective permeability of a fractured rock mass, as a function of fracture density, for the case $\alpha/\kappa = 0.078$. Curves are from the self-consistent effective medium theory, eq. (7.25), and its linear approximation, eq. (7.26). Data points are from the finite element simulations of Ebigbo *et al.* Source: Adapted from Ebigbo *et al.* (2016).

range from 10^{-3} for fractures in permeable matrix rocks, down to 10^{-10} for fractures in low-permeability crystalline rocks. Assuming in-plane fracture radii that vary between 10 cm and 10 m, the fracture aspect ratio α may vary from 10^{-3} to 10^{-5}. The parameter α/κ may therefore vary from about 10^7 down to about 10^{-2}. For fractured rock masses having $\alpha/\kappa \ll 1$, such as might occur in fairly permeable sedimentary matrix rocks, and/or for fractures that contain some infill material, it is straightforward to show (see Problem 7.3) that eq. (7.25) reduces to

$$k_{eff} = k_m + \frac{8\pi}{9} k_f \alpha \varepsilon = k_m + k_S. \tag{7.26}$$

This result coincides exactly with the Hashin–Shtrikman upper bound, eq. (7.8).

According to eq. (7.26), the effective permeability in the regime $\alpha/\kappa \ll 1$ increases linearly with fracture density, with a slope that is proportional to the fracture permeability. Figure 7.3a shows the normalized effective permeability, k_{eff}/k_m, as a function of fracture density, for several values of $\alpha \leq \kappa$, as calculated from eq. (7.25). For values of α/κ as high as 1, the curves do not deviate very far from the linear approximation given by eq. (7.26), which was derived by assuming $\alpha/\kappa \ll 1$.

In the regime $\alpha/\kappa \ll 1$, the self-consistent effective medium theory is quite accurate. Effective permeability values obtained from the finite element simulations of Ebigbo *et al.* (2016), for a fractured rock mass having $\alpha/\kappa = 0.177$, are compared to the predictions of eq. (7.25), and its linear approximation, eq. (7.26), in Fig. 7.3b. The agreement between the data and the effective medium model is very good. The variability between the values obtained from simulations performed on different randomly generated fracture networks was found to lie roughly within the size of the data markers used in the figure.

An interesting feature of eq. (7.26) is that, since the effective permeability increases linearly with k_f, no abrupt increase in k_{eff} ever occurs, at any fracture density. In other words, fractured rocks for which $\alpha/\kappa \ll 1$ do *not* exhibit percolation-type behavior, which would

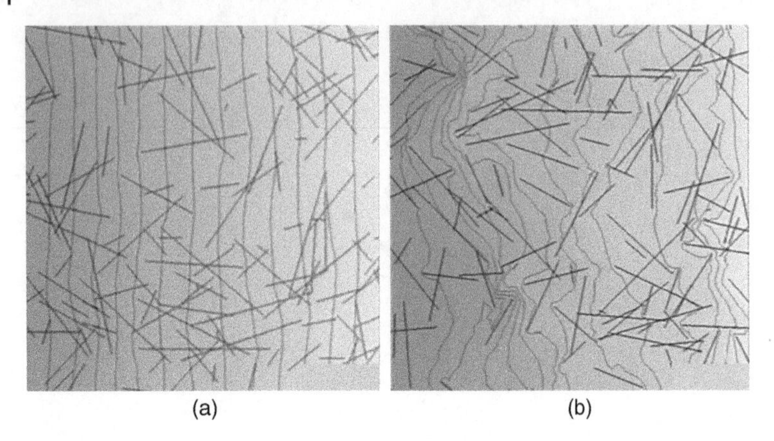

Figure 7.4 Flow characteristics of a fractured-rock mass, for (a) $\alpha/\kappa = 0.177$, and (b) $\alpha/\kappa = 177$. The fracture density is $\varepsilon = 1$ in both cases. Images depict slices through the middle of a 3D cube-shaped domain, with fluid flowing from left to right. The somewhat irregular lines passing from the top to the bottom of the flow region are contours of constant pressure. The fracture traces that intersect this plane appear as thick, straight-line segments. Source: Adapted from Ebigbo *et al.* (2016).

appear as a somewhat abrupt transition from an effective permeability dominated by the matrix, to an effective permeability dominated by the fractures, at some specific value of ε. Since the fracture aspect ratio α that appears in eq. (7.26) is necessarily very small, the effective permeability k_{eff} will never approach k_f. This prediction by the self-consistent effective medium theory, of a lack of percolative behavior in the regime $\alpha/\kappa \ll 1$, is consistent with the numerical simulations of Bogdanov *et al.* (2007), who found that percolative behavior tends to disappear when $\alpha/\kappa < 7.5$.

The lack of percolative behavior in this regime can also be understood by examining the pressure contours from a numerical simulation of flow through a fractured medium. Figure 7.4a shows the pressure contours for flow, from left to right, through a fractured rock mass with $\alpha/\kappa = 0.177$, and a fracture density of $\varepsilon = 1$, as simulated by Ebigbo *et al.* (2016) using the finite element method. When α/κ is "small," very little preferential flow occurs through the fractures, and the contours of constant pressure are only mildly perturbed from the shapes that they would have if no fractures were present. In this regime, it can be said that the flow is "matrix dominated" – *despite* the fact that the fracture density is above the percolation threshold. This latter fact can be seen by visually noting that a connected fracture pathway exists through the simulation volume, from left to right.

7.6 Effective Permeability in the Regime $\alpha/\kappa \gg 1$

The regime of values of α/κ that is probably of more relevance and practical importance to subsurface flow processes in fractured rocks is that of $\alpha/\kappa \gg 1$. Following the discussion given in the previous section, typical values of the fracture aspect ratio α are in the range of 10^{-3} to 10^{-5}. Again, assuming a typical aperture of 100 μm, the permeability of an individual fracture would be, by eq. (7.15), about 10^{-9} m^2. Hence, if the matrix permeability is less than about 10^{-16} m^2, the ratio α/κ will be greater than 100. Most hard, crystalline rocks would fall into this regime, as would some carbonate formations.

The behavior of eq. (7.25) in the regime $\alpha/\kappa \gg 1$ is more complex than in the regime $\alpha/\kappa \ll 1$. At sufficiently small values of the fracture density, it will still be true that $k_{eff}/k_m \ll \alpha/\kappa$ in the denominator on the right-hand side, in which case eq. (7.25) effectively reduces to (Ebigbo *et al.*, 2016; see also Problem 7.4)

$$\frac{k_{eff}}{k_m} = \left(1 - \frac{32}{9}\varepsilon\right)^{-1}. \tag{7.27}$$

Although k_{eff} begins to grow rapidly as ε approaches 9/32, its numerical value in this regime is still essentially controlled by the matrix permeability, rather than by the fracture permeability – which in fact does not appear in eq. (7.27) at all.

At larger values of the fracture density, when k_{eff}/k_m becomes roughly of the same magnitude as α/κ, eq. (7.25) approaches the following asymptotic straight line when plotted as a function of fracture density (Ebigbo *et al.*, 2016):

$$k_{eff} = k_m + \frac{8\pi}{9}k_f\alpha\varepsilon - \frac{\pi\alpha}{4}k_f = k_m + k_{Snow} - \frac{\pi\alpha}{4}k_f. \tag{7.28}$$

According to this model, at high fracture densities, the effective permeability increases linearly with ε but remains below the value predicted by Snow's model by a constant, finite amount. This asymptotic expression can be written in normalized form as

$$\frac{k_{eff}}{k_m} = \frac{8\pi}{9}\left(\frac{\alpha}{\kappa}\right)\varepsilon - \left[\frac{\pi}{4}\left(\frac{\alpha}{\kappa}\right) - 1\right] \approx \frac{8\pi}{9}\left(\frac{\alpha}{\kappa}\right)\varepsilon - \frac{\pi}{4}\left(\frac{\alpha}{\kappa}\right). \tag{7.29}$$

The predictions of eq. (7.25), for two different values of α/κ in the regime $\alpha/\kappa \gg 1$, are shown in Fig. 7.5a. Also shown is the small-ε approximation, eq. (7.27). The curves for all values of α/κ coalesce to eq. (7.27) for small values of ε. For large fracture densities, the curves approach the asymptotic lines given by eq. (7.29). The approximate expression (7.27) clearly shows that the effective medium theory correctly predicts the existence of a percolation threshold, at some critical fracture density $\varepsilon = \varepsilon_c$. However, the critical fracture density predicted by the theory, which is $9/32 = 0.28125$, differs slightly from the actual known value (Mourzenko *et al.*, 2011) of $\varepsilon_c = 0.244$.

Effective permeability values computed by Ebigbo *et al.* (2016) using the finite element method, for the case $\alpha/\kappa = 78$, are shown in Fig. 7.5b. Also shown are the predictions of the self-consistent effective medium theory, as given by eq. (7.25), and the predictions of Snow's equation, eq. (7.17). The self-consistent model provides a very good fit below the percolation threshold and correctly predicts that the graph of $k_{eff}(\varepsilon)$ eventually becomes linear, lying parallel to, but below, the prediction of Snow's equation. However, the offset between the actual effective permeabilities, and Snow's equation, is underestimated. Ebigbo *et al.* (2016) found qualitatively similar results for networks containing polydisperse fracture sets, in which each set has a different value of α/κ. In these cases, eq. (7.25) was applied by computing an effective value of α/κ, weighted by the porosity of each fracture set. A detailed prescription for applying the effective medium models to polydisperse fracture sets has been given by Wong *et al.* (2020). The error in the vertical offset of the effective permeability can be mitigated by using the semi-empirical equation developed by Mourzenko *et al.* (2011), which is also shown in Fig. 7.5b, and discussed in more detail in Section 7.7.

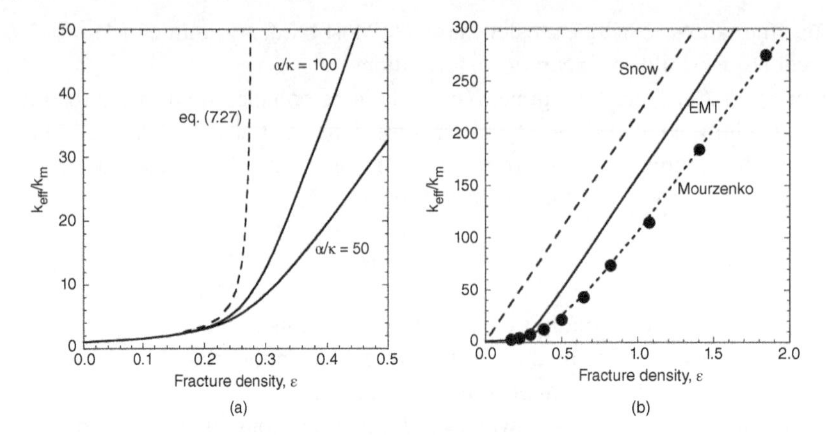

Figure 7.5 (a) Normalized permeabilities predicted by the self-consistent effective medium theory, eq. (7.25), for two different values of $\alpha/\kappa \gg 1$. Dashed curve shows the small-ε approximation, eq. (7.27), which also corresponds to the case of $\alpha/\kappa \to \infty$. (b) Effective permeability values computed by Ebigbo *et al.* (2016), for networks having $\alpha/\kappa = 78$, compared with the predictions of the Snow equation, the self-consistent effective medium theory, and the Mourzenko equation. Source: Adapted from Ebigbo *et al.* (2016).

The existence of a percolation threshold, at which the effective permeability transitions between being essentially controlled by k_m, to being controlled by k_f, is clearly seen in Fig. 7.5a. The occurrence of percolative behavior in the regime $\alpha/\kappa \gg 1$ is also consistent with the plots of the pressure contours that can be obtained from numerical simulations using the finite element method. Figure 7.4b shows the pressure contours computed by Ebigbo *et al.* (2016) for flow through a fractured rock mass having $\alpha/\kappa = 177$, and a fracture density of $\varepsilon = 1$, which is well above the geometrical percolation threshold. When α/κ is "large," the fluid flows preferentially through the fracture network, and the contours of constant pressure are strongly perturbed from the shapes that they would have if no fractures were present. In this regime, it can be said that the flow is "fracture dominated."

7.7 Semi-empirical Model of Mourzenko *et al.*

The inclusion-based effective medium model described in the previous sections qualitatively captures most of the trends exhibited by the macroscopic permeability as a function of the various parameters such as the aspect ratio α, the matrix-fracture permeability ratio κ, and the fracture density, ε. In the regime $\alpha/\kappa \ll 1$, the predictions of this model are very accurate. Nevertheless, when $\alpha/\kappa \gg 1$, the model fails to quantitatively match some aspects of the behavior, such as the asymptotic value of the amount by which, at high fracture densities, the effective permeability falls below the value predicted by Snow's model. A semi-empirical equation that matches the behavior of the effective permeability reasonably well, when $\alpha/\kappa \gg 1$ and the fracture density is above the percolation threshold, has been developed by Adler and his collaborators (Bogdanov *et al.*, 2003; Mourzenko *et al.*, 2004; Bogdanov *et al.*, 2007; Mourzenko *et al.*, 2011). The development of this equation was guided by various theoretical considerations, but some of its parameters were chosen so as to allow the equation to fit effective permeability values that were computed by explicit numerical simulations. For simplicity, this equation will be referred to hereinafter as the "Mourzenko equation."

Assuming that the fractures have a circular shape in the fracture plane, and using the present notation, the Mourzenko equation can be written as (Mourzenko *et al.*, 2011)

$$k_{eff} = k_m + k_{Snow}\left\{1 - \frac{1}{1+1.91(\alpha/\kappa)^{-0.7}}\left[1 - \frac{0.18\pi^2(\varepsilon-\varepsilon_c)^2}{\varepsilon[1+0.18\,\pi^2(\varepsilon-\varepsilon_c)]}\right]\right\}, \tag{7.30}$$

in which it is assumed that $\varepsilon > \varepsilon_c$, where $\varepsilon_c = 0.244$ is the critical fracture density that defines the percolation threshold. A more general form of this equation, which can account for different planform shapes of the fractures as well as for multiple families of fractures having different sizes, has been presented by Mourzenko *et al.* (2011) and Adler *et al.* (2013).

An example of the accuracy of the Mourzenko equation when $\varepsilon > \varepsilon_c$ is shown in Fig. 7.5b, where its predictions are compared to the permeabilities computed by Ebigbo *et al.* (2016), for fracture networks having $\alpha/\kappa = 78$. Whereas the self-consistent effective medium theory overestimates the effective permeability, by an amount that eventually stabilizes to some finite value, the Mourzenko equation matches the data quite well. For polydisperse fracture networks, Ebigbo *et al.* (2016) found that the Mourzenko equation worked well in some cases, but in other cases, it overestimated the effective permeability. Nevertheless, it generally performed better than any of the four different effective medium theories that were tested: the asymmetric effective medium theory, the symmetric effective medium theory, Maxwell's effective medium theory, and the differential effective medium theory.

An asymptotic expansion of eq. (7.30), for large values of α/κ, and large values of ε, shows that Mourzenko's equation eventually approaches the following straight line, when plotted as normalized permeability *vs.* fracture density:

$$\frac{k_{eff}}{k_m} = \frac{8\pi}{9}\left(\frac{\alpha}{\kappa}\right)\varepsilon - \left[\frac{8\pi}{9}\left(\frac{\alpha}{\kappa}\right) - 1\right] \approx \frac{8\pi}{9}\left(\frac{\alpha}{\kappa}\right)\varepsilon - \frac{8\pi}{9}\left(\frac{\alpha}{\kappa}\right). \tag{7.31}$$

Hence, the Mourzenko equation predicts that, at high fracture densities, the normalized effective permeability will fall *below* the value predicted by Snow's equation, by an amount equal to $8\pi\alpha/9\kappa \approx 2.79\alpha/\kappa$. This offset is more than three times larger than the offset predicted by the effective medium theory, which according to eq. (7.29) is $\pi\alpha/4\kappa \approx 0.785\alpha/\kappa$. It is also worth noting that the effective medium prediction reaches its asymptotic line for values of ε on the order of unity, whereas the Mourzenko equation does not reach its asymptotic behavior until ε is on the order of about 10.

Problems

7.1 Consider an otherwise impermeable rock mass that contains three mutually orthogonal sets of fractures. One set of fractures lies parallel to the $y-z$ plane and has an aperture of 100 μm and a spacing of 1 m. The second set of fractures lies parallel to the $x-z$ plane and has an aperture of 200 μm and a spacing of 2 m. The third set of fractures lies parallel to the $x-y$ plane and has an aperture of 300 μm and a spacing of 1 m. Calculate the macroscopic permeabilities k_x, k_y, and k_z.

7.2 Consider a cubic block, of side L, of an otherwise impermeable rock mass that contains three fractures, as shown in Fig. 7.1b. A pressure drop of ΔP is imposed across the two (left and right) faces of the block that are normal to the x-axis. Let the transmissivity of fracture 1 be T_1, and the transmissivity of fracture 2 be T_2. Fracture 3 is identical to, and parallel to, fracture 2, but does not intersect any other fractures. Apply Darcy's law to each fracture, and prove the (nonobvious) fact that the flowrates through fractures 2 and 3 are *exactly* the same. In other words, if the fractures pass fully through the medium, as is assumed in Snow's model, the flowrate through a given fracture is *unaffected* by intersections with other fractures. Hint: at the point of intersection of fractures 1 and 2, the pressure in each fracture must be the same.

7.3 Show that, in the regime $\alpha/\kappa \ll 1$, the general expression for the effective permeability that is given by eq. (7.25) reduces to the expression given by eq. (7.26).

7.4 Show that, in the regime $\alpha/\kappa \gg 1$, for fracture densities that are sufficiently low that $k_{eff}/k_m \ll \alpha/\kappa$, the effective permeability given by eq. (7.25) reduces to the expression given by eq. (7.27).

7.5 Consider a fractured rock mass having matrix permeability $k_m = 10^{-15}$ m^2 and fracture permeability $k_f = 10^{-9}$ m^2. Assume that the fractures are thin oblate spheroids of aspect ratio $\alpha = 10^{-3}$. Construct graphs of the effective permeability as a function of fracture density ε, according to the self-consistent effective medium theory, and the Mourzenko equation. Compare these two curves to the "small ε" approximation, eq. (7.27), to the "large ε" approximation, eq. (7.28), of the effective medium theory, and to the "large ε" approximation, eq. (7.31), of the Mourzenko equation.

References

Adler, P. M., Thovert, J.-F., and Mourzenko, V. V. 2013. *Fractured Porous Media*, Oxford University Press, Oxford.

Barthélémy, J. F. 2009. Effective permeability of media with a dense network of long and micro fractures. *Transport in Porous Media*, 76(1), 153–78.

Bear, J. 1988. *Dynamics of Fluids in Permeable Media*, Dover Publications, Mineola, N.Y.

Bisdom, K., Bertotti, G., and Nick, H. M. 2016. The impact of in-situ stress and outcrop-based fracture geometry on hydraulic aperture and upscaled permeability in fractured reservoirs. *Tectonophysics*, 690, 63–75.

Bogdanov, I. I., Mourzenko, V. V., Thovert, J. F., and Adler, P. M. 2003. Effective permeability of fractured porous media in steady state flow. *Water Resources Research*, 39(1), 1023.

Bogdanov, I. I., Mourzenko, V. V., Thovert, J. F., and Adler, P. M. 2007. Effective permeability of fractured porous media with power-law distribution of fracture sizes. *Physical Review E*, 76, 036309.

de Dreuzy, J., Davy, P., and Bour, O. 2001. Hydraulic properties of two-dimensional random fracture networks following a power law length distribution 1. Effective connectivity. *Water Resources Research*, 37(8), 2065–78.

Ebigbo, A., Lang, P. S., Paluszny, A., and Zimmerman, R. W. 2016. Inclusion-based effective medium models for the permeability of a 3D fractured rock mass. *Transport in Porous Media*, 113, 137–58.

Fokker, P. 2001. General anisotropic effective medium theory for the effective permeability of heterogeneous reservoirs. *Transport in Porous Media*, 44(2), 205–18.

Fricke, H. 1924. A mathematical treatment of the electric conductivity and capacity of disperse systems. *Physical Review*, 24, 575–87.

Hardebol, N. J., Maier, C., Nick, H., Geiger, S., *et al.*, 2015. Multiscale fracture network characterization and impact on flow: a case study on the Latemar carbonate platform. *Journal of Geophysical Research Solid Earth*, 120, 8197–8222.

Harris, C. K. 1990. Application of generalised effective-medium theory to transport in porous media. *Transport in Porous Media*, 5, 517–42.

Hashin, Z., and Shtrikman, H. 1962. A variational approach to the theory of the effective magnetic permeability of multiphase materials. *Journal of Applied Physics*, 33(10), 3125–31.

Jaeger, J. C., Cook, N. G. W., and Zimmerman, R. W. 2007. *Fundamentals of Rock Mechanics*, 4th ed., Wiley-Blackwell, Oxford.

Lang, P. S., Paluszny, A., and Zimmerman, R. W. 2014. Permeability tensor of three-dimensional fractured porous rock and a comparison to trace map predictions. *Journal of Geophysical Research Solid Earth*, 119(8), 6288–6307.

Leung, C. T. O., and Zimmerman, R. W. 2012. Estimating the hydraulic conductivity of two-dimensional fracture networks using network geometric properties. *Transport in Porous Media*, 93(3), 777–97.

Li, L. and Li, K. 2015. Permeability of microcracked solids with random crack networks: role of connectivity and opening aperture. *Transport in Porous Media*, 109(1), 217–37.

Maxwell, J. C. 1873. *Treatise on Electricity and Magnetism, Vol.* 1, Clarendon Press, Oxford.

Mourzenko, V. V., Thovert, J. F., and Adler, P. M. 2004. Macroscopic permeability of three-dimensional fracture networks with power-law size distribution. *Physical Review E*, 69, 066307.

Mourzenko, V. V., Thovert, J. F., and Adler, P. M. 2011. Permeability of isotropic and anisotropic fracture networks, from the percolation threshold to very large densities. *Physical Review E*, 84, 036307.

Oda, M. 1985. Permeability tensor for discontinuous rock masses. *Géotechnique*, 35(4), 483–95.

Paluszny, A. and Matthäi, S. K. 2010. Impact of fracture development on the effective permeability of porous rocks as determined by 2-D discrete fracture growth modeling. *Journal of Geophysical Research Solid Earth*, 115(B02), 203.

Parsons, R. W. 1966. Permeability of idealized fractured rock. *Society of Petroleum Engineers Journal*, 6(2), 126–36.

Pozdniakov, S. and Tsang, C. F. 2004. A self-consistent approach for calculating the effective hydraulic conductivity of a binary, heterogeneous medium. *Water Resources Research*, 40, W05-105.

Rayleigh, L. 1892. On the influence of obstacles arranged in rectangular order on the properties of a medium. *Philosophical Magazine E*, 34, 481–502.

Sævik, P. N., Berre, I., Jakobsen, M., and Lien, M. 2013. A 3D computational study of effective medium methods applied to fractured media. *Transport in Porous Media*, 100(1), 115–42.

Sævik, P. N., Jakobsen, M., Lien, M., and Berre, I. 2014. Anisotropic effective conductivity in fractured rocks by explicit effective medium methods. *Geophysical Prospecting*, 62(6), 1297–1314.

Snow, D. T. 1969. Anisotropic permeability of fractured media. *Water Resources Research*, 5(6), 1273–89.

Torquato, S. 2013. *Random Heterogeneous Materials: Microstructure and Macroscopic Properties*, Springer, Berlin.

Vu, M. N., Pouya, A., and Seyedi, D. M. 2018. Effective permeability of three-dimensional porous media containing anisotropic distributions of oriented elliptical disc-shaped fractures with uniform aperture. *Advances in Water Resources*, 118, 1–11.

Wiener, O. 1912. Die Theorie des Mischkorpers fur das Feld der stationaren Stromung [Field theory for a heterogeneous body under stationary flow]. *Der Abhandlungen der Mathematisch-Physischen Klasse der Konigl Sachsischen Gesellschaft der Wissenschaften*, 32(6), 507–604.

Wong, D. L. Y., Doster F., Geiger, S., and Kamp, A. 2020. Partitioning thresholds in hybrid implicit-explicit representations of fractured reservoirs. *Water Resources Research*, 56, 2019WR025774.

Zimmerman, R. W. 1989. Thermal conductivity of fluid-saturated rocks. *Journal of Petroleum Science and Engineering*, 3(3), 219–27.

Zimmerman, R. W. 1996. Effective conductivity of a two-dimensional medium containing elliptical inhomogeneities. *Proceedings of the Royal Society A: Mathematics, Physics and Engineering Science*, 452(1950), 1713–27.

8

Fluid Flow in Geologically Realistic Fracture Networks

8.1 Introduction

Modeling fluid flow in fractured rock masses accurately is a challenge, in particular because of the complex geometry of the fracture networks and the interplay between fluid flow and mechanical deformation that affects the permeability. In natural systems, individual fractures may arise from the intersection and coalescence of multiple smaller-scale fractures and are often the by-product of a deformation history having mechanical, fluid, and chemical footprints. The difference between the sizes of fractures, faults, flaws, and other heterogeneities present in the rocks makes modeling these systems numerically challenging, with an inherent difficulty directly related to the choice of which geometric features to represent explicitly in a distributed model, and which features to represent by means of parameters in constitutive relationships that describe a continuum.

The state of stress at any point in the Earth's crust is almost always anisotropic, with a significant deviatoric component. This is to say, the three principal stresses are generally not equal to each other. Fracture networks that form under these conditions will be geometrically anisotropic (Zoback *et al.*, 1989) and will display highly directional patterns that often record the deformation history of the rock. Depending on their individual orientations, fractures will experience different magnitudes of compression and shear displacement, both of which have a strong influence on hydraulic transmissivity, as discussed in Chapter 3.

Fracture networks are characterized using various concepts. Fractures that are formed during the same geological process usually have coherent orientations. For example, fractures formed during uplifting or folding will be mostly parallel, whereas fractures forming due to impact or intrusion will be concentric. Fractures that are systematically organized are usually defined to be part of a "fracture set." A subsurface network is, in general, composed of many superimposed and interacting fracture sets, which together constitute a fracture network. This chapter describes the techniques used to study the effects of spatial organization of fractures on permeability using Discrete Fracture Networks (DFNs). These are datasets that are created to directly capture the geometric description of the fractures, using specific "discrete" objects, such as curves and surfaces, to represent the fractures within the rock matrix. In an effort to generate realistic networks, three methodologies can be distinguished: (a) mapping of fracture networks from the field, (b) generation of discrete networks using stochastic methods, informed by distributions of properties and in some cases

Fluid Flow in Fractured Rocks, First Edition. Robert W. Zimmerman and Adriana Paluszny.

Companion website: www.wiley.com/go/zimmerman/fluidflowinfracturedrocks

informed by geomechanically inspired empirical rules, and (c) growth of geomechanically realistic fracture networks based on the principles of continuum mechanics.

Methods of generating fracture networks using stochastic and statistical methods are discussed in Section 8.2. The generation of fracture networks using a numerical geomechanics simulator is presented in Section 8.3. Fracture intersections and apertures are discussed in Sections 8.4 and 8.5, respectively. Numerical computation of the macro-scale permeability of fractured rock masses is treated in Section 8.6, and the influences of fracture density and stress on the macroscopic permeability are discussed in Sections 8.7 and 8.8. The occurrence of "flow channels" within a fractured rock mass is discussed in Section 8.9.

8.2 Stochastically Generated Fracture Networks

The difficulty of observing three-dimensional fracture networks at depth has led to a reliance on two-dimensional characterization of rocks exposed on the surface and to the creation of stochastically generated patterns by integrating geological data obtained from boreholes, reservoir analogs, and outcrops. Consequently, the most common type of DFN is created by a so-called Poisson process in which the *centers* of the fractures are randomly placed within a finite-sized matrix volume (Long *et al.*, 1982).

Stochastic networks that follow a lognormal or power-law fracture size distribution were considered for many years to be the most geologically realistic fracture models for fluid flow modeling. Fractures in two dimensions are represented by straight lines, whereas in three dimensions their shapes are approximated by disks, hexagons, or polygons. Their apertures and lengths can be assumed to take a fixed value for the entire network. Alternatively, they can be randomly generated to follow a uniform distribution within a specific set of bounds or can be defined to follow a correlated distribution, which assumes a fixed relationship between properties such as length and aperture (Bogdanov *et al.*, 2003). It follows from the analysis of field observations that the distribution of fracture radii are expected to follow a power-law distribution, as follows (Odling, 1997):

$$P(r) = Cr^{-n}, \quad r_{\min} \leq r \leq r_{\max}, \tag{8.1}$$

where $P(r)dr$ is the probability of observing fractures having radii lying between r and $r + dr$, the fracture radii all lie in the range $r_{\min} \leq r \leq r_{\max}$, and C is a constant whose value, which depends on $\{r_{\min}, r_{\max,} n\}$, is found through normalization (Mourzenko *et al.*, 2005).

Apart from a distribution of radii, which can be numerically generated by a uniform random distribution generator such as the Mersenne twister pseudorandom number generator (Matsumoto and Nishimura, 1998), the orientation of the fractures must also be defined. This can be achieved by defining the normal vector of the fractures, either randomly or following a distribution of orientations that mimic observations in the rocks. The variation of these properties can generate high-density fracture networks rapidly and in a consistent manner. For example, for a 3D domain of size $L \times L \times L$, an exponent of $n = 1.5$, and $\{r_{\min} = L/10;\ r_{\max} = L/2\}$, the generated fracture models have between 25 and 900 disk fractures. In terms of the fracture network parameters defined in Section 1.4, these generated networks range in dimensionless density ρ'_{3D} from 0.513 to 21.256, and in terms of P_{32} range from 0.013 to 0.448 m^{-1}; see Table 8.1.

Table 8.1 Number of fractures, n_f, fracture intensity, P_{32} [m^{-1}], and dimensionless density, ρ'_{3D}, for several 100 m × 100 m × 100 m cubic fracture network models.

n_f	25	50	100	200	300	500	700	800	900
P_{32}	0.013	0.023	0.047	0.1	0.148	0.241	0.335	0.389	0.448
ρ'_{3D}	0.513	0.943	2.113	4.666	6.781	11.035	15.716	18.058	21.256

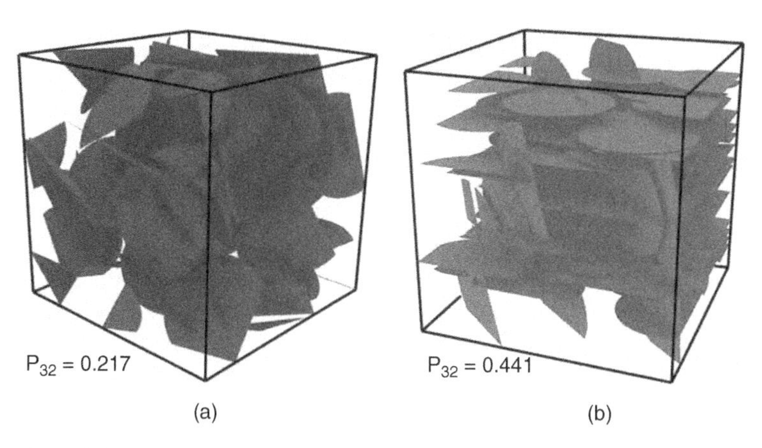

Figure 8.1 Three-dimensional stochastically generated fracture networks. In both cases, the locations of the centers of the disk-like fractures are random. In (a) the fracture orientations are random, with the objective of generating an isotropic network. In (b) there are two preferential directions, aiming to capture two main orientations observed in an outcrop.

Figure 8.1 shows two examples of stochastically generated fracture networks, taken from Thomas *et al.* (2020a). The process of stochastic fracture generation neglects the effects of mechanical interactions between fractures, resulting in networks with unrealistic spatial correlations, thereby often producing networks with fracture locations and orientations that are geomechanically improbable and geologically inconsistent. To remediate these shortcomings, *kinematic* fracture models have been devised that add additional mechanically informed rules to the stochastic process of fracture network generation. These kinematic, or rule-based, models can be extended to incorporate algorithms that simplistically model fracture nucleation, growth, and arrest (Maillot *et al.*, 2016; Davy *et al.*, 2010). Such models insert fractures following rules that account for first-order mechanical interaction effects between fractures. These rules govern the insertion of fixed-sized fractures into the models, aiming to avoid unrealistic scenarios that can be predicted based on the geometric constraints imposed by preexisting fractures. The initial flaws are then grown following mechanically inspired rules, such as the avoidance of shadow zones of other fractures, the artificial limitation of connectivity, and the imposition of maximum fracture sizes with respect to the model size. One such rule-based approach is the parent-daughter approach proposed by Bonneau *et al.* (2016), which is a stochastic, sequential, and nonstationary nucleation process to simulate hierarchical and spatially correlated DFNs that avoids the insertion of new fractures within the stress shadow zones of preexisting fractures.

Another rule-based model, proposed by Davy *et al.* (2010), mimics the growth of fractures with simplified kinematic rules of nucleation, growth, and arrest. This model goes beyond considering the simple geometry of the network and considers the first-order effect of elastic strain energy concentrations on nucleation and growth, introducing stress-dependent size-position correlations that organize the character of the intersections of fractures to account for size effects (Lavoine *et al.*, 2020). With regard to the "backbone" of the fracture network, *i.e.*, all fracture surfaces that contribute to flow during the permeability computation, in Poisson models the backbone represents 73–93% of the total fracture surface, whereas for kinematic models the backbone of the network ranges between 30–95% (Maillot *et al.*, 2016). This variation is less apparent when considering the fractal dimension, which for both types of models ranges between 2.8 and 3.0.

8.3 Geomechanically Generated Fracture Networks

The geometry and topology of fracture networks in subsurface rocks depart from those of idealized stochastic fracture networks in a number of ways. Consequently, effort has recently been expended to generate fracture networks based on the underlying physics of elastic deformation and the theory of linear elastic fracture mechanics (Renshaw and Pollard, 1994; Thomas *et al.*, 2020a). During fracture growth, geomechanically realistic fracture networks obey complex underlying processes that control the distribution of new surface area created during a growth event (see Fig. 8.2). Thus, the increase in density in the network may lead to very localized deformation that subsequently results in significant connectivity and permeability changes in the rock mass. The patterns subsequently shown in this chapter were generated using a finite element-based method that relies on the geometric growth of fractures based on stress intensity factors around the fracture tips, which control the extension, direction, and geometric change of the pattern in two dimensions (Paluszny and Matthäi, 2010) and three dimensions (Paluszny and Zimmerman, 2011).

As it is based on accepted principles of continuum mechanics, geomechanical fracture network generation is an attractive alternative to stochastic generation of fracture sets. Its main disadvantage with respect to the latter is that it is much more computationally expensive. However, the simplicity of randomly generated datasets does not capture the complex mechanical interactions between fractures that occur during growth, such as those shown in Figs. 8.2 and 8.3. Therefore, stochastically generated networks often contain patterns that do not exhibit fracture self-organization and are therefore capable only of partially reproducing some simple statistics of the original measured data, such as spacing and aperture distributions, without capturing the complex topology that is created by the mechanical interactions of the growing fractures. These patterns depend on the rock's deformation history and its material properties and have a strong influence on the flow behavior through the fractured rock mass. One example is fracture density: in randomly generated networks, density is increased by adding more or larger fractures, whereas in "geomechanical" networks, density is a by-product of fracture growth and coalescence, as well as nucleation. Furthermore, randomly generated networks cannot lead to fracture curving, as is observed in Figs. 8.2 and 8.3. These fractures are always planar, whereas in mechanically

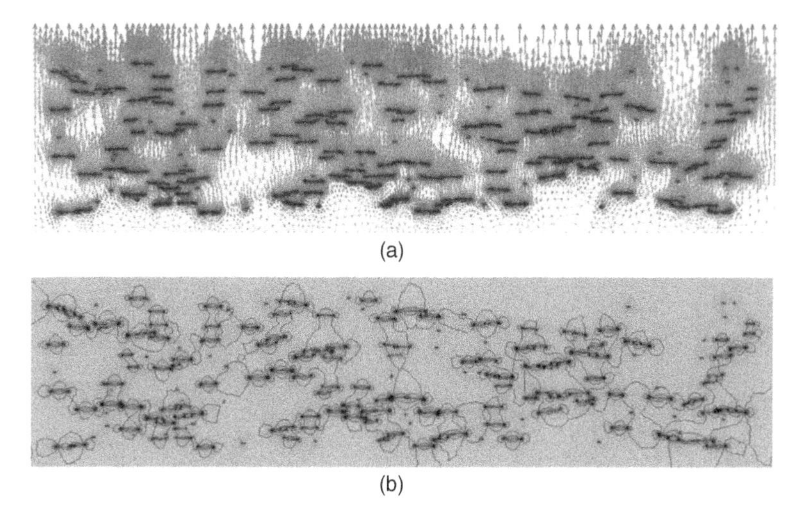

Figure 8.2 Two-dimensional growth of two hundred fractures due to a uniform extensional deformation at the top of the model. (a) The displacement field after growth, showing how fractures induce heterogeneities in the deformation and that fracture clusters may form even at low fracture densities. (b) Mean stress contours, indicating regions of equal mean stress; closely spaced contours indicate a spatially rapid variation in mean stress. Due to assumption of linear elasticity, units are not relevant.

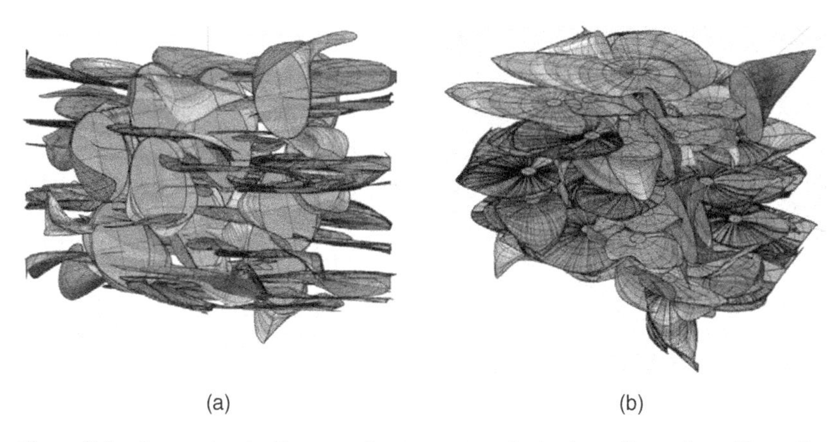

Figure 8.3 Geomechanically grown fracture networks in three dimensions. The self-organization in three-dimensional networks affects both fracture density and the spatial arrangement of the ensuing patterns.

informed simulations this is not the case, and fractures can grow in any shape and pattern, and thus, can greatly enhance connectivity without necessarily increasing fracture density significantly.

Early work on multiple crack propagation invariably disregarded fracture curving. In some cases, this assumption is permissible. For example, when fractures occur on a brittle coating layer of an otherwise flexible bending plate, the plate constrains the deformation, and the maximum principal stress is governed by the macroscopic bending and not by local fracture interactions (Wu and Pollard, 1992). Another case is the generation of fractures

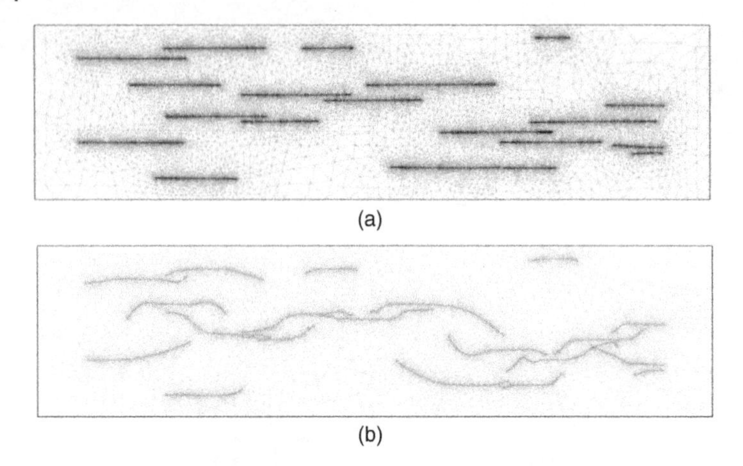

Figure 8.4 Straight versus curving paths under tensile stress. Comparison between two networks generated from the same initial set of flaws subjected to tensile stress. In (a), fractures are allowed to propagate in a straight path only. In (b), the fracture propagation angle is updated at every growth step. Fractures that are far from any neighbors will exhibit similar shapes in both cases.

in an inherently anisotropic medium that exhibits weakness in a preferred direction. Figure 8.4 shows two simulations performed using the same initial flaw distribution, in which fractures grow due to tensile displacements in the y-direction. In Fig. 8.4a, fractures were propagated on a straight path only. In Fig. 8.4b, the local stress field controls fracture curving, and the resulting patterns have distinct geometrical qualities. If sufficiently far from other fractures, fractures will propagate in nearly linear paths in both cases. But if fractures are allowed to grow in response to local stress field perturbations, curved patterns will develop, leading to a much higher degree of connectivity and consequently to higher permeabilities.

Curving is a manifestation of fracture interaction and is a strong indicator of the local stress conditions at the time of crack formation (Olson and Pollard, 1989). Curving significantly influences the final characteristics of a simulated pattern, affecting the relationship between density and connectivity effectively transforming planar surfaces into complex curved domains. An example that highlights the impact of modeling curvature is fracturing due to shrinkage. Shrinkage cracks can be modeled numerically by defining an isotropic contraction factor while fixing the displacement at the boundaries. At the onset of the simulation, the model contracts, and cracks propagate to accommodate the loss of volume. Figure 8.5 shows two simulations performed on the same initial set of flaws. Figure 8.5a shows the initial flaw distribution, whereas Figs. 8.5b and 8.5c show advanced stages of growth. Fractures grow as a function of stress concentration at their tips. In Fig. 8.5b, fractures are forced to propagate in a straight path. In Fig. 8.5c, crack paths are updated after every computational step, resulting in fracture curving. Cracks approach each other orthogonally. The simulation shown in Fig. 8.5b does not capture this emergent effect. It is apparent that curving has a significant influence on the final pattern: in Fig. 8.5b, fractures grow blindly across each other, whereas the network in Fig. 8.5c captures the self-organization of the fractures that creates internal polygonal regions (Goehring, 2013).

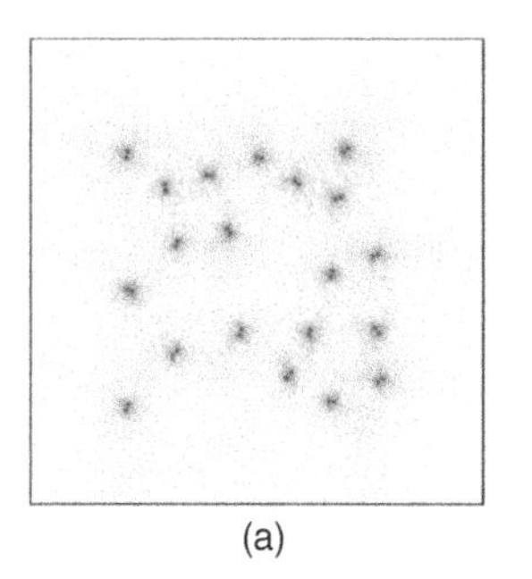

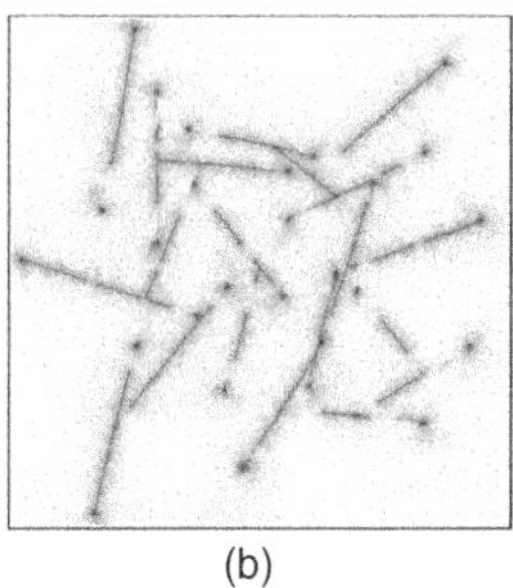

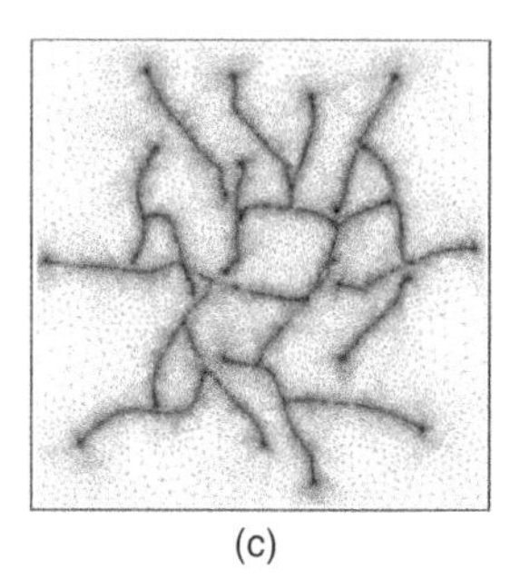

Figure 8.5 Shrinkage cracks modeled by applying a uniform isotropic shrinkage of 0.1% while fixing the displacement at the boundaries: (a) the initial flaw distribution; (b) fracture pattern created by allowing straight fracture growth only; (c) fracture pattern generated by updating the propagation angle at each iteration.

8.4 Intersections and Connectivity in Fracture Networks

Fracture intersections are expected to substantially increase the permeability of a fracture network (*cf*., Adler *et al.*, 2013). In well-developed stochastically generated networks, higher density leads to higher connectivity. In geomechanically generated networks, higher connectivity may be the result of interaction between fractures, and high connectivity may develop even in low-density fracture networks. In two dimensions, fracture intersections are points at which two fractures intersect. In three dimensions, fracture intersections are lines or polylines that demarcate the shared regions between the fractures. One way to characterize the connectivity of a fractured network is to measure the number of intersections. Thus, as with connectivity, intersections are both a measure of development and density and one of the parameters used to approximate the permeability of a network when detailed geometry is unknown (Leung and Zimmerman, 2012).

In uncorrelated Poisson stochastic models, intersections follow a power-law model that depends on fracture length. In contrast, the parent-daughter fracture nucleation model proposed by Bonneau *et al.* (2016) results in a distribution of intersections that depends on the proportion of simulated fractures per step, acting to increase the intersections for small fractures and reduce intersections for large fractures. These two effects result in the intensification of the spatial correlations between fractures, increasing connectivity, and producing the clusters that are often observed in nature. Figure 8.6 shows the increase in intersections for stochastic Poisson models, as compared to geomechanically generated fracture networks. For both cases, measurements are in three dimensions, and intersections, which increase linearly, are quantified as the sum of the lengths of the segments marking the intersection between two fractures.

In a numerical study conducted by Thomas and Paluszny (2020), the intersections of stochastically generated discrete fracture networks (SDFNs) were compared to geomechanically consistent networks of increasing density. These results are shown in Fig. 8.6. Two types of SDFNs were generated: the first (SDFN-1) having random fracture orientations, and the second (SDFN-3) that roughly followed the anisotropy that was observed in the geomechanically generated networks. It can be seen that intersections increase at a much slower rate for geomechanical networks, as compared to the stochastic DFN, SDFN-1,

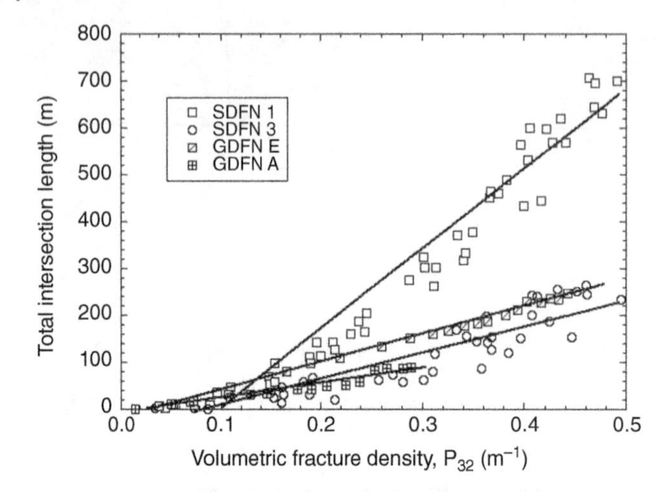

Figure 8.6 Intersections of fracture networks as a function of increasing density, comparing two sets of stochastically generated networks, SDFN-1 and SDFN-3, and two sets of geomechanically generated discrete fracture networks, GDFN-A and GDFN-E. Source: Adapted from Thomas and Paluszny (2020).

which is isotropic in nature. When the stochastic DFN is anisotropic, as in SDFN-3, the progression of intersections is slower, indicating that larger numbers of fractures can coexist within the region without drastically increasing the connectivity of the network. In stochastically generated networks, fracture density increases by adding more fractures, causing paths to link up and increasing connectivity and density simultaneously. Therefore, fracture density plays an important role in the increase or decrease of the overall conductivity of the network. In contrast, in geomechanically generated datasets, density increases primarily through the growth of existing fractures, and so connectivity is a by-product of coalescing fractures and not only a function of the number of fractures.

8.5 Fracture Apertures in Discrete Fracture Networks

Fractures and faults in three-dimensional models may be represented either by surfaces (*e.g.*, Reichenberger *et al.*, 2006) or by volumes (*e.g.*, Matthäi *et al.*, 2005). Using a lower-dimensional representation increases computational efficiency and is flexible enough to implicitly represent the characteristics of the fractures (Kim and Deo, 2000; Monteagudo and Firoozabadi, 2004). However, this approach is permissible only when the fractures are more permeable than the rock, thereby excluding sealed fractures and sealing faults. Moreover, numerical simulation of capillary pressure-driven fluid exchange between fractures and rock matrix (see Chapter 12) requires a fine computational grid orthogonal to the fracture plane, in order to represent sharp gradational or discontinuous saturation variations. In these cases, volumetric meshes are often preferred, as they allow for volumetric fluid accumulation to take place within the fracture.

Fracture apertures within a fracture network containing one or more fracture sets can be assumed to be uniformly or geomechanically distributed. In the first case, a single value of aperture is defined for each fracture, whereas in the second case, the aperture varies as a function of stress. The permeability of fractured rocks with uniform apertures and simplified geometries can be approximated analytically, as has been discussed in Chapter 7. For a fractured rock with uniform apertures, the higher the fracture density, the higher the equivalent permeability of the medium. As conductivity continues to increase beyond the percolation threshold, permeability increases further. More fractures result in more high-permeability flow pathways, more intersections, and larger macroscopic permeabilities. Comparatively, when mechanical effects are considered, the equivalent permeability of the system is much lower. In two dimensions, variations in the apertures and fracture permeability translate into variations in the effective connectivity of the network, as low-permeability regions having a reduced number of fractures can significantly reduce the overall fluid flow in the system (Paluszny and Matthäi, 2010). In three dimensions, mechanical apertures translate into spatial variations of the fracture permeability over its nonplanar surface, which not only affects permeability but also results in the formation of fluid flow channels. Figures 8.7 and 8.8 show examples of aperture distributions for a network of geomechanically generated fractures in two and three dimensions, respectively.

The local fracture aperture is a key ingredient needed in any numerical simulation of fluid flow through a fractured rock mass, as it controls the permeability of the fracture. As discussed in Chapter 2, the local permeability of an individual fracture is given by

$$k_f = \frac{h^2}{12}, \tag{8.2}$$

where h is the local fracture aperture, measured normal to the nominal fracture plane. In computational models, aperture is usually defined at the fracture nodes and must be interpolated to find values at the center of the fracture elements. When fractures are represented

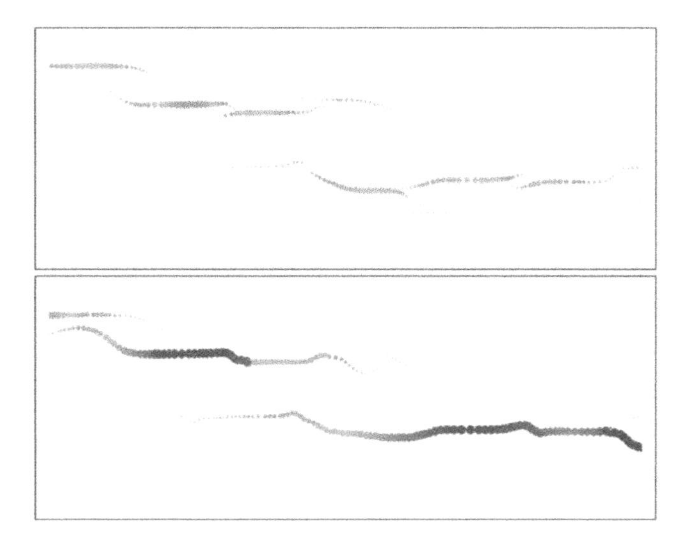

Figure 8.7 Aperture distribution at two stages of fracture growth, under extension in the vertical direction. The size and darkness of each glyph are proportional to the aperture of the fracture at that location. Glyph sizes are exaggerated by a factor of 1000. When fractures connect, they behave as one fracture, and the aperture distribution along the centerline changes.

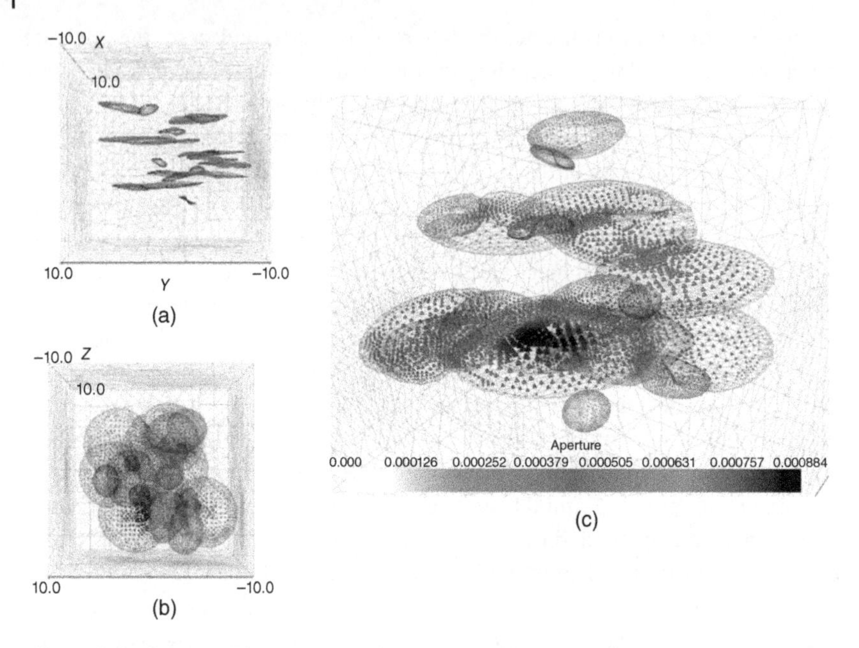

Figure 8.8 Groups of fractures growing under tension in the "vertical" direction. (a) Side view, (b) top view, (c) angled view, showing local apertures. Aperture units are arbitrary, with relative significance only.

in a "sub-dimensional manner," *i.e.*, discretized by lines in two-dimensional domains and by triangles and quadrilaterals in three-dimensional domains, properties such as permeability and porosity must be weighted by the local aperture in order to capture the actual thickness of the reduced element. It follows that in systems where fractures are represented by sub-dimensional elements, fracture permeability must be defined as

$$k_f = \frac{h^2}{12} \cdot h = \frac{h^3}{12}. \tag{8.3}$$

By using sub-dimensional elements, the discretization of fracture domains requires significantly fewer elements to capture the volumetric inner domain of the fracture, as compared to using volumetric elements, without compromising accuracy (Juanes *et al.*, 2002).

For apertures that are much larger than the roughness of the fracture, the cubic law yields good predictions of fracture permeability. For cases where the fracture roughness is of the same magnitude as the mean aperture, the variation of the aperture becomes more important in predicting the permeability (see, *e.g.*, Chapter 2 and Sisavath *et al.*, 2003). When defining apertures numerically, the parallel plate law can be applied in a stepped, local manner, assuming that the local cubic law is obeyed in each region, as opposed to attempting to identify a single hydraulic aperture representative of the flow capacity of the entire fracture. This is useful when modeling the mechanical deformation of the matrix in response to *in situ* stresses. Various methods for assigning apertures, which may vary within a single fracture, have been critically reviewed by Bisdom *et al.* (2016).

8.6 Numerical Computation of Fractured Rock Mass Permeability

Often, the flow analysis of DFNs assumes that the rock matrix is impermeable. If the analysis of the flow properties of the rock mass also incorporates the effects of the matrix, the models will also include a volumetric mesh composed of tetrahedra, hexahedra, prisms, and pyramids and are referred to as Discrete Fracture and Matrix (DFM) models. As discussed within an analytical framework in Chapter 7, when fractures are embedded in a rock matrix, they modify the macro-scale permeability of the rock mass. In the most common case, fractures are many times more permeable than the matrix, and so a fracture network will enhance the permeability of the rock mass. In a less common but also possible case, such as mineralized veins in an otherwise permeable matrix, the permeability of the fractures may be much lower than that of the matrix. In geologically realistic cases, there is a variation in the geometry and permeability of the fractures.

The macro-scale permeability of a fractured rock mass is often referred to as the "equivalent permeability" or the "effective permeability." In the context of analytical models, in which the rock mass is generally assumed to be of infinite extent, the term effective permeability is typically used (*cf*., Chapter 7). In the context of the numerical estimation of the macro-scale permeability, these two concepts must be distinguished from each other. The distinction is related to the size of the domain over which the permeability is evaluated, L_Ω, relative to the maximum size of the fractures in the region, L_f. When fracture lengths are well below the domain length, *i.e.*, $L_\Omega \gg L_f$, it can be assumed that the computed permeability is representative of the entire system, and in this case, the computed permeability can be said to represent the "effective permeability." It is usually assumed that the computational domain size L_Ω must be at least five times larger than the maximum fracture size L_f in order that the computed permeability can be assumed to represent the effective permeability that would be computed on a statistically equivalent infinite domain. However, in rock masses in which fracture lengths continue to increase as a function of scale, such an assumption becomes problematic, and the computed permeability is referred to as the "equivalent permeability."

The equivalent macroscopic permeabilities of the fracture network can be defined both for DFN and DFM models (Bogdanov *et al.*, 2003). In the former case, the macro-scale permeability will pertain to the fracture network only and can be denoted by k_{fn}. In the latter case, the equivalent permeability reflects both the contributions of the fractures and the matrix rock and can be denoted by k_{eq}.

A general framework for computing the macro-scale permeability of a fractured rock mass was developed by Durlofsky (1991) and extended by, among others, Wen *et al.* (2003) and Lang *et al.* (2014). Before outlining this method, first note that in the most general case, the rock mass may be anisotropic, in which case Darcy's law must be written in tensorial form, $\mathbf{q} = -(1/\mu)\mathbf{k}\nabla p$, where $\mathbf{k}$ is the second-order permeability tensor, which has nine components. Explicitly, this equation takes the form (Bear, 1988; Jaeger *et al.*, 2007)

$$\begin{bmatrix} q_x \\ q_y \\ q_z \end{bmatrix} = -\frac{1}{\mu}\begin{bmatrix} k_{xx} & k_{xy} & k_{xz} \\ k_{yx} & k_{yy} & k_{yz} \\ k_{zx} & k_{zy} & k_{zz} \end{bmatrix}\begin{bmatrix} \partial p/\partial x \\ \partial p/\partial y \\ \partial p/\partial z \end{bmatrix}. \tag{8.4}$$

In general, some or all of the off-diagonal components (the components with nonmatching subscripts) of the permeability tensor may be nonzero. In such cases, a pressure gradient in the x-direction may lead to a flux component in the y-direction, for example.

Although the symmetry of the permeability tensor has been subject to some disagreement and controversy, it is generally thought that the effective permeability tensor must be symmetric, which is to say that $k_{xy} = k_{yx}$, $k_{xz} = k_{zx}$, and $k_{yz} = k_{zy}$. If the permeability tensor is symmetric, it will necessarily possess three mutually orthogonal "principal" directions in which the permeability takes on a local maximum or minimum value. The permeabilities in these three directions are the three principal permeabilities, generally denoted by $\{k_1 \geq k_2 \geq k_3\}$. In mathematical terms $\{k_1, k_2, k_3\}$ are the three eigenvalues of the permeability tensor/matrix that appear in eq. (8.4). As explained by Bear (1988), the permeability tensor can be visualized as an ellipsoid in three dimensions, or an ellipse in two dimensions, in which the distance from the origin to a point on the ellipse/ellipsoid is proportional to the permeability in that direction. The three orthogonal semi-major axes of this ellipsoid will be $\{k_1, k_2, k_3\}$.

The first step in computing the macro-scale permeability tensor is to perform a three-dimensional "fine-scale" version of the calculation illustrated in Fig. 2.5. The computational region can be taken to be a cube of side L, although other domain geometries could be used as well. In a very general sense, the fractured rock mass can be thought of as a heterogeneous medium in which the local permeability $\mathbf{k}(\mathbf{x})$ [m^2] varies spatially, where $\mathbf{x} = (x, y, z)$ is the local position vector. The local volumetric flux vector $\mathbf{q}(\mathbf{x})$ [m/s] is given by Darcy's law, $\mathbf{q}(\mathbf{x}) = -(1/\mu)\mathbf{k}(\mathbf{x})\nabla p$, in which μ [Pa s] is the viscosity, and ∇p [Pa/m] is the pressure gradient. If the pore fluid is assumed to be incompressible, then conservation of mass is equivalent to conservation of volume, and so in the steady state, the divergence of the volumetric flux must vanish, *i.e.*, $\nabla \cdot \mathbf{q}(\mathbf{x}) = 0$. Consequently, fluid flow through this fractured rock mass is governed, after factoring out the term $-1/\mu$, by the following differential equation:

$$\nabla \cdot [\mathbf{k}(\mathbf{x})\nabla p] = 0. \tag{8.5}$$

To compute all of the components of the permeability tensor, several flow simulations are performed, based on eq. (8.5), each having the macroscopic pressure gradient aligned with one of the coordinate axes. Specifically, a uniform pressure p_i imposed over one face, a uniform pressure p_o is imposed over the opposing face, and the other four faces are taken to be impermeable, no-flow boundaries. In the two-dimensional case, the two simulations are denoted as I and II, for solutions aligned with the x and y axes, respectively. In three dimensions, the additional simulation III corresponds to a pressure gradient aligned with the z-axis. During the solution procedure, the pressure p is approximated at the nodes of an unstructured finite element mesh in two or three dimensions, employing the standard Galerkin method. When solving for the fluid pressure field, more complex characterizations of the flow field can be modeled using this approach, for example, by assuming hydro–mechanical coupling and assuming that the fractured mass is under compression with friction acting between the fracture walls.

As pointed out above, in an anisotropic medium, a pressure gradient in the z-direction may cause a flux in the x-direction, for example. But the no-flow boundary condition imposed on the faces of the cube that lie in the $\{y, z\}$ plane will not allow fluid to flow out of the cube in the x-direction and instead will lead to a distorted flow field near the boundary. To avoid these boundary effects, the permeability is computed by considering the flux and pressure fields only over a sub-domain of the cubical region. This approach is referred to as "undersampling," since the pressure field is computed over the entire model, but the fluxes and pressure gradients used in the computation of the permeability tensor are only sampled within a sub-domain that is sufficiently far away from the boundaries. One simple way to achieve this is to sample only those elements whose distance to any outer boundary exceeds some prescribed value, thereby effectively ignoring the outer margins of the simulated region. This technique, also referred to as "flow jacketing," or sometimes as "oversampling," reduces boundary artifacts while better reflecting the geometric connectivity of the sampled domain (Holden and Lia, 1992; Wu *et al.*, 2003).

After solving eq. (8.5) to find the pressure field, the flux vector in each finite element can be found by applying Darcy's law:

$$\mathbf{q}^e = -\frac{1}{\mu}\mathbf{k}^e \nabla p^e, \tag{8.6}$$

which results in a vector field of element piecewise constant fluxes $\mathbf{q}^e$, corresponding to the local element pressure gradient ∇p^e, and permeability, $\mathbf{k}^e$. For matrix elements, $\mathbf{k}^e$ takes the value of the matrix permeability, whereas for fracture elements, it takes the value of the fracture permeability.

Next, volume-averaged fluxes and pressure gradients are defined as (Durlofsky, 1991)

$$\langle q_i \rangle = \frac{1}{V_S}\sum_e \int_{V^e} q_i \, dV^e, \tag{8.7}$$

$$\left\langle \frac{\partial p}{\partial x_i} \right\rangle = \frac{1}{V_S}\sum_e \int_{V^e} \frac{\partial p^e}{\partial x_i} dV^e, \tag{8.8}$$

where the sums are taken over all elements e within the sampled sub-domain, V_S is the volume of the sampled sub-domain, and the generic coordinate x_i can represent x, y, or z. A macro-scale version of Darcy's law can then be written, relating the volume-averaged fluxes to the volume-averaged pressure gradients, for each of the two or three simulations, depending on whether the model is two-dimensional or three-dimensional.

In the two-dimensional case, these equations take the form (Azizmohammadi and Matthäi, 2017)

$$\langle q_x \rangle^{\mathrm{I}} = \frac{-1}{\mu}\left[k_{xx} \left\langle \frac{\partial p}{\partial x} \right\rangle^{\mathrm{I}} + k_{xy} \left\langle \frac{\partial p}{\partial y} \right\rangle^{\mathrm{I}} \right], \tag{8.9}$$

$$\langle q_y \rangle^{\mathrm{I}} = \frac{-1}{\mu}\left[k_{yx} \left\langle \frac{\partial p}{\partial x} \right\rangle^{\mathrm{I}} + k_{yy} \left\langle \frac{\partial p}{\partial y} \right\rangle^{\mathrm{I}} \right], \tag{8.10}$$

$$\langle q_x \rangle^{\mathrm{II}} = \frac{-1}{\mu}\left[k_{xx} \left\langle \frac{\partial p}{\partial x} \right\rangle^{\mathrm{II}} + k_{xy} \left\langle \frac{\partial p}{\partial y} \right\rangle^{\mathrm{II}} \right], \tag{8.11}$$

$$\langle q_y \rangle^{\mathrm{II}} = \frac{-1}{\mu}\left[k_{yx} \left\langle \frac{\partial p}{\partial x} \right\rangle^{\mathrm{II}} + k_{yy} \left\langle \frac{\partial p}{\partial y} \right\rangle^{\mathrm{II}} \right]. \tag{8.12}$$

These equations can be written in matrix form as follows:

$$\begin{bmatrix} \langle \partial p/\partial x\rangle^{\mathrm{I}} & \langle \partial p/\partial y\rangle^{\mathrm{I}} & 0 & 0 \\ 0 & 0 & \langle \partial p/\partial x\rangle^{\mathrm{I}} & \langle \partial p/\partial y\rangle^{\mathrm{I}} \\ \langle \partial p/\partial x\rangle^{\mathrm{II}} & \langle \partial p/\partial y\rangle^{\mathrm{II}} & 0 & 0 \\ 0 & 0 & \langle \partial p/\partial x\rangle^{\mathrm{II}} & \langle \partial p/\partial y\rangle^{\mathrm{II}} \end{bmatrix} \begin{bmatrix} k_{xx} \\ k_{xy} \\ k_{yx} \\ k_{yy} \end{bmatrix} = -\mu \begin{bmatrix} \langle q_x\rangle^{\mathrm{I}} \\ \langle q_y\rangle^{\mathrm{I}} \\ \langle q_x\rangle^{\mathrm{II}} \\ \langle q_y\rangle^{\mathrm{II}} \end{bmatrix}, \tag{8.13}$$

which can symbolically be written as $\mathbf{Ax} = \mathbf{b}$, where the components of the unknown 4×1 vector $\mathbf{x}$ are the four components of the two-dimensional permeability tensor. These components can then be found by solving this equation, formally as $\mathbf{x} = \mathbf{A}^{-1}\mathbf{b}$.

The equivalent permeability computed over a *finite region* is not guaranteed to be symmetric. However, it is generally thought that, on a larger scale, the computed permeability tensor will converge to an effective permeability tensor that *is* symmetric (Lesueur *et al.*, 2022). Symmetry can be imposed *a posteriori* by defining the permeability tensor to be the symmetric part of the tensor that was computed from the calculation $\mathbf{x} = \mathbf{A}^{-1}\mathbf{b}$ (Wen *et al.*, 2003; Lang *et al.*, 2014), *i.e.*,

$$k_{xy} = k_{yx} = \frac{1}{2}(k_{xy} + k_{yx})_{\mathrm{computed}}. \tag{8.14}$$

The three-dimensional version of this procedure follows in an obvious way. The three-dimensional version of eq. (8.13) comprises nine equations in nine unknowns, and the post-processing calculation to impose symmetry then requires two additional equations, analogous to eq. (8.14), for the pairs $\{k_{xz}, k_{zx}\}$ and $\{k_{yz}, k_{zy}\}$. The explicit form of these equations can be found in Lang *et al.* (2014).

In a modified version of this procedure, the symmetry conditions on the permeability tensor can be imposed *during* the calculation. In the two-dimensional context, this implies adding a fifth equation, of the form $k_{xy} - k_{yx} = 0$. The matrix eq. (8.13) then takes the form

$$\begin{bmatrix} \langle \partial p/\partial x\rangle^{\mathrm{I}} & \langle \partial p/\partial y\rangle^{\mathrm{I}} & 0 & 0 \\ 0 & 0 & \langle \partial p/\partial x\rangle^{\mathrm{I}} & \langle \partial p/\partial y\rangle^{\mathrm{I}} \\ \langle \partial p/\partial x\rangle^{\mathrm{II}} & \langle \partial p/\partial y\rangle^{\mathrm{II}} & 0 & 0 \\ 0 & 0 & \langle \partial p/\partial x\rangle^{\mathrm{II}} & \langle \partial p/\partial y\rangle^{\mathrm{II}} \\ 0 & 1 & -1 & 0 \end{bmatrix} \begin{bmatrix} k_{xx} \\ k_{xy} \\ k_{yx} \\ k_{yy} \end{bmatrix} = -\mu \begin{bmatrix} \langle q_x\rangle^{\mathrm{I}} \\ \langle q_y\rangle^{\mathrm{I}} \\ \langle q_x\rangle^{\mathrm{II}} \\ \langle q_y\rangle^{\mathrm{II}} \\ 0 \end{bmatrix}, \tag{8.15}$$

which again can be formally written as $\mathbf{Ax} = \mathbf{b}$. However, as this is an overdetermined system of *five* equations with *four* unknowns, it cannot be solved uniquely by matrix inversion techniques. Instead, both sides of this equation can be pre-multiplied by the transpose of $\mathbf{A}$, to yield $\mathbf{A}^{\mathrm{T}}\mathbf{Ax} = \mathbf{A}^{\mathrm{T}}\mathbf{b}$. This new system can be subsequently "solved" with the least squares minimization technique (Trefethen and Bau, 1997) to obtain the four permeability components. In the three-dimensional case, the analog of eq. (8.15) will be an overdetermined system of twelve equations in nine unknowns (Lang *et al.*, 2014).

Although symmetry has been "imposed" by including the equation $k_{xy} - k_{yx} = 0$, since the least-squares solution of an overdetermined system of equations will generally yield only an *approximate* solution, it is still not guaranteed that the resulting permeability tensor will be symmetric. Hence, the post-processing step given by eq. (8.14) can again be applied in order to produce a symmetric permeability tensor.

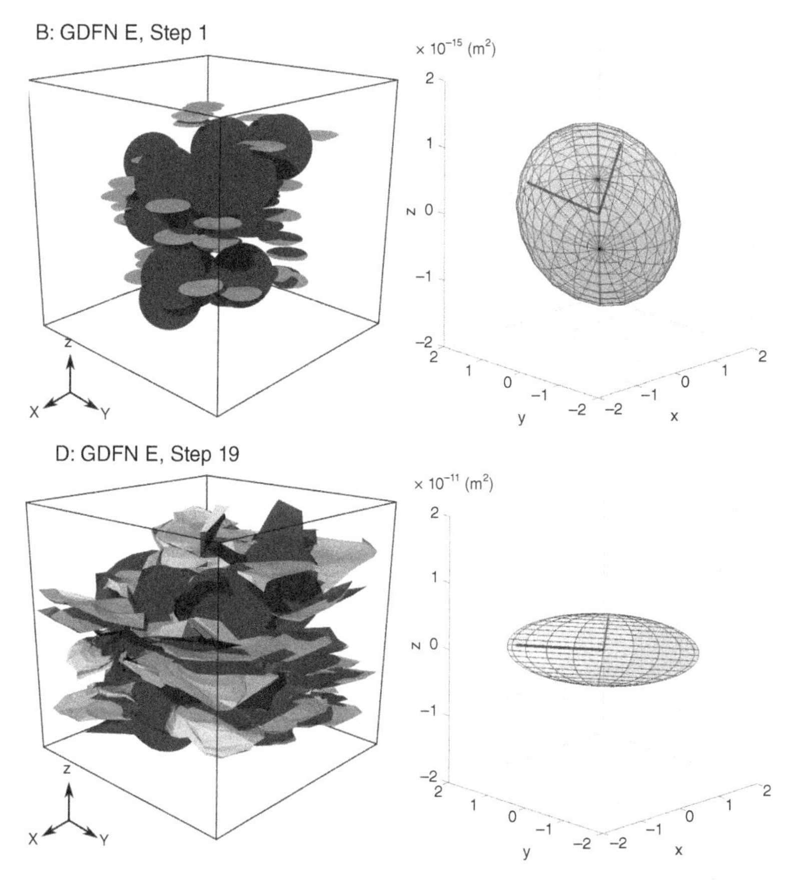

Figure 8.9 Two examples of fracture network geometry: (top) a less well-developed fracture network and (bottom) a more well-developed fracture network, along with their corresponding permeability tensors. The axes of the permeability tensor indicate the magnitudes and directions of the three principal permeabilities.

Figure 8.9 shows examples of a stochastically and a geomechanically generated fracture network and their computed permeability tensors (Lang *et al.*, 2014). These examples indicate the important fact that the principal directions of the permeability tensor are not always obvious *a priori*, as there may be local variations that occur due to the geometry of the network and the effects of the stresses applied to the network. These influences may affect the network geometry and modify its topology and, consequently, its connectivity. In contrast to many previously proposed methods, the method described above for computing the permeability tensor does not require that the principal directions be "guessed" or identified *a priori*.

8.7 Effect of Fracture Density on Equivalent Permeability

Results obtained from percolation theory have highlighted the lack of geometrically realistic and sufficiently resolved simulations of fluid flow in fractured media (Bogdanov *et al.*, 2007). These simulations often rely on stochastically generated fracture datasets that attempt to mimic reality but do not reproduce the internal structure arising due to mechanical interaction during fracture growth. Huseby *et al.* (1997) studied the effect of fracture density, as well as topological and geometric interconnectivity, on the flow properties of the fractured rock mass. Initial experiments included measuring the effective permeability of fracture sets with different connectivity levels. The experiments were conducted on randomly generated datasets consisting of equal-length, three-dimensional polygonal fractures. Subsequently, Bogdanov *et al.* (2007) conducted similar numerical experiments, but with fracture sizes that followed a power-law length distribution. In both cases, it was found that the effective permeability increased linearly as a function of fracture density.

The trend can be approximated by two linear segments: for fracture densities below the percolation threshold, permeability increases linearly with density, and above the percolation threshold the permeability values are several orders of magnitude higher, but again increase linearly with density. This behavior is roughly consistent with some of the analytical models discussed in Chapter 7; *cf*., Figs. 7.3 and 7.5. There is also a strong dependence on the fracture-to-matrix permeability ratio; the larger this ratio, the larger the jump in the effective permeability that occurs at the percolation threshold.

Apart from the uncertainty in the fracture geometry, fluid flow models for fractured porous media have often involved simplistic assumptions about the relationship between fracture length and aperture, or in some cases, collapsed the range of values into a single value. Matthäi and Belayneh (2004) have shown that such models may yield highly unrealistic results.

Bogdanov *et al.* (2007) studied the overall effect of density and percolation on the permeability of a porous rock mass containing a set of randomly generated fractures having a power-law size distribution, but with apertures that did not vary with fracture length/radius. In these simulations, the size of the smallest fracture controls the power-law length distribution. Permeability in all cases increased up to the percolation threshold. However, due to the finite size of the computational domain, at this point, some of the networks percolate, and others do not, leading to a transition range of the fracture density within which networks may or may not percolate. In the simulations of Bogdanov *et al.* (2007), an abrupt increase in equivalent permeability was observed to occur somewhere in the range of

$$2 \le \rho'_{3D} \le 4. \tag{8.16}$$

Once percolation occurs, permeabilities increase by around two orders of magnitude, for the case in which the average fracture to matrix permeability was about $k_f/k_m \approx 10^4$. This "bi-modal" distribution of permeabilities is a function of the permeability ratio, and it rapidly collapses into a single monotonically increasing permeability distribution as k_f/k_m decreases to below about 100.

The permeability of a fractured rock mass can be studied both in two and three dimensions, as a function of fracture density. In a study of the relationship between the permeabilities computed for two-dimensional and three-dimensional models, Lang *et al.* (2014) showed that stochastically generated two-dimensional systems will systematically produce lower predictions of permeability, as compared to the three-dimensional systems

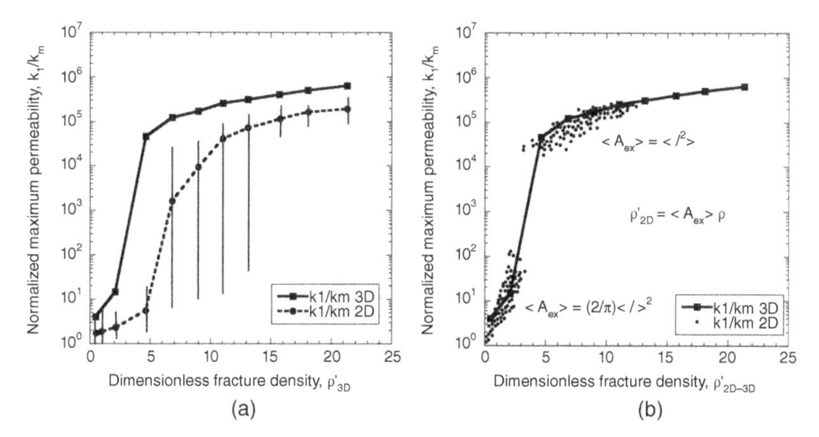

Figure 8.10 (a) Maximum principal permeability of a three-dimensional fractured rock mass (solid line) compared to the permeabilities of extracted cut-plane two-dimensional models of the same networks (dashed line). Matrix permeability is 1×10^{-15} m^2, and all fractures have apertures of 0.001 m. The vertical bars show the variation in the 2D permeabilities, and the dashed line indicates the average values at each fracture density. (b) The same results, plotted against a re-defined two-dimensional fracture density, using a modified excluded volume definition for high-density networks; see text for details.

from which these 2D planes were extracted. Figure 8.10a shows the predicted maximum principal permeability for the 2D and 3D models. These authors proposed a redefinition of the excluded volume for high-density datasets, so as to better compare datasets of the same density across dimensions. Figure 8.10b displays the permeabilities plotted as a function of fracture density when considering the density-dependent excluded area. It is clear that there is great variation between the computed permeabilities for different two-dimensional networks having the same fracture density. This variance is much more prominent for networks with dimensionless fracture densities $\rho'_{3D} < 15$ and less pronounced for networks with higher fracture densities. Two-dimensional planes extracted from three-dimensional fracture networks fail to capture the intricate fracture intersection topologies that form in three-dimensional systems, which serve to significantly enhance the flow of fluids (Lang *et al.*, 2014; Wang *et al.*, 2022).

Figure 8.11 compares the computed permeabilities of three hundred thirty-six stochastically generated networks and one hundred ninety three geomechanically generated networks, in each case assuming the same aperture in each fracture. The geomechanical networks correspond to six growing networks, and each dataset records a different stage of growth (Thomas *et al.*, 2020b); these datasets can be downloaded from Thomas and Paluszny (2020). As compared to the stochastically generated networks, the geomechanical networks are more anisotropic, are more permeable at higher densities, and have a tendency to percolate at higher densities. It can be noted that both k_{xx} and k_{yy} are much larger than k_{zz}, and the match between the "stochastic" and "geomechanical" permeabilities is better in the x and y directions, although there is more variation within the stochastic set. This is due to variations in the orientation of the fractures, which in about a third of the cases are organized into (often three) distinct sets. In the z-direction, the permeability is lowest; this is captured by a number of the stochastic models, but not all.

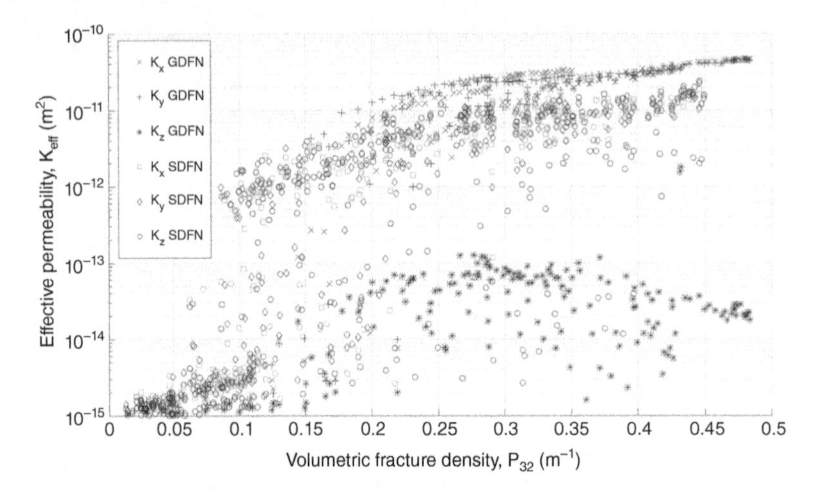

Figure 8.11 Effective permeabilities of stochastically generated Discrete Fracture Networks (SDFNs) and geomechanically generated Discrete Fracture Networks (GDFNs), as computed by Thomas *et al.* (2020b). In both cases, the effective permeabilities are anisotropic, with k_{xx}, $k_{yy} \gg k_{zz}$. Source: Adapted from Thomas *et al.* (2020b).

8.8 Effect of *In Situ* Stresses on Equivalent Permeability

In some fractured rock masses, the orientation of the global permeability tensor may be controlled more by dissolution and precipitation processes than by the remote stress field. However, in many cases, the permeability of a rock mass is strongly controlled by the current stress state because of the alteration of the local apertures due to the far-field stresses. Field observations of fractured rock permeability often reveal pronounced anisotropy, as a result of both the network geometry and the state of stress. Field studies have also shown that fracture-fault systems may develop higher permeabilities due to structural changes of the network in the direction of fracture-fault intersections orthogonal to fault slip vectors (Lang *et al.*, 2018). Such observations have led to empirical relationships that relate fracture hydraulic apertures to the physical process of shear displacement and the state of compression, *i.e.*, the acting normal stress.

The coupling between spatial variation of the geometry of fractures and their mechanical behavior has been shown to influence stress-dependent flow patterns and the resulting permeability of the rock mass (Baghbanan and Jing, 2008). Specifically, for stochastic models having length-dependent fracture apertures and permeabilities, large fractures with higher permeabilities tend to localize flow significantly, as compared to cases in which apertures are assumed to be the same among all fractures. When considering fractures at depth, both measurements and numerical simulations show that the fracture network permeability decreases with depth, as a function of confining stresses acting to close the fracture apertures. Permeability reduces down to a minimum threshold value, after which the permeability does not further decrease (Zhang and Sanderson, 1996). The rate of decrease depends both on the network geometry and the applied stress directions. The permeability of individual fractures continues to decrease with depth, down to a residual value, after which the mechanisms that control permeability at the fracture scale are much more dependent on roughness (*cf*., Sisavath *et al.*, 2003). It is worth noting that a change in the stress state can cause reactivation of preexisting fractures and channeling of flow in critically stressed fractures, as observed in previous two-dimensional studies that do not take into account small-scale effects (Latham *et al.*, 2013).

The ratio of the maximum to minimum far-field principal stresses has been shown to affect the magnitude and direction of the macroscopic permeability. If σ_{max} is the maximum remote principal stress, and σ_{min} is the minimum remote principal stress, the stress ratio can be defined as

$$m \equiv \sigma_{max} / \sigma_{min} . \tag{8.17}$$

It has been observed in two-dimensional numerical simulations of the permeability of stochastically generated networks (Baghbanan and Jing, 2008) that for low stress ratios, flow is dominated by a smaller number of large fractures having higher permeability. For higher stress ratios, $m > 3$, dilation caused by shear of predominantly smaller fractures becomes an important contribution to flow channeling and permeability. In fact, in models where the relationship between fracture length and aperture is not considered, this effect causes permeability to decrease more significantly than in the more realistic geomechanical case.

As an alternative to using the full permeability tensor to investigate the relationship between the directionality of the permeability tensor and the far-field stresses, Zhang and Sanderson (1996) proposed the following definition of the average deviation of the angle between the maximum principal permeability and the maximum far-field stress:

$$A(m) \equiv \frac{1}{N} \sum_{i=1}^{N} |\alpha_i|, \tag{8.18}$$

where α_i is the angle between the far-field major compressive remote stress and the direction of the maximum principal permeability, the sum is taken over all numerical experiments, and N is total number of numerical experiments. This average deviation angle was computed for various values of the stress ratio m. Using this parameter, they found that the maximum principal permeability direction often deviates from the direction of the applied major stress, both numerically and in the field. In particular, when considering fractures that have fixed apertures, networks that are composed of sets of fractures that can be regarded as being more "organized" or "systematic," will have values of $A(m)$ that are significantly greater than zero, whereas networks that are not organized and have macroscopically isotropic properties tend to yield values of $A(m) \approx 0$.

Remote stresses preferentially activate specific fractures that have a favorable orientation to the far-field stress. These fractures can form high-conductivity pathways that span the entire domain, and thereby dominate the macroscopic permeability. Previous two-dimensional studies (Baghbanan and Jing, 2008; Jing *et al.*, 2013) have found that the direction of maximum permeability of a fractured rock mass is aligned with the orientation of the maximum principal stress. However, these models assume *isotropic* transmissivity of individual fractures and, being two-dimensional models, cannot account for the effects of the intermediate principal stress. Most numerical studies of the mechanical deformation of fractured rocks have traditionally been two-dimensional, while most three-dimensional studies of permeability have been restricted to fluid flow and have not included mechanical effects. A few later models were extended to three dimensions, but ignored shear-induced dilation, or relied upon constitutive models that account for shear-induced dilation in an isotropic manner.

Lang *et al.* (2018) investigated the permeability of three-dimensional fractured rocks as a function of network geometry and remote compressive stresses. Fractures were assumed to be stochastically generated disks, and focus was placed on investigating the effect of the assumptions regarding individual fracture permeability on the overall properties of the fracture network. This three-dimensional analysis included the evolution of the individual fracture transmissivities, as a result of shear deformation. The simulations captured processes both at the mesoscale (meters) and at the small-scale (millimeters). Fractures were represented discretely at the mesoscale using a finite element-based continuum mechanics approach, which resolves the compressive stresses acting along the fracture plane, while taking into account friction along the fracture surfaces (Nejati *et al.*, 2016), using an Augmented-Lagrangian gap-based approach that assumes linear elastic constitutive behavior of the rock. This approach does not rely on arbitrarily imposed penalties for "interpenetration" of the opposing fracture walls, but instead relies on only three material properties: Young's modulus, Poisson's ratio, and a macroscopic friction coefficient for the fracture surfaces.

The volume of the rock was represented by isoparametric tetrahedral elements, and the fracture surfaces were discretized with isoparametric triangular elements (Paluszny and Zimmerman, 2013). The numerical model simulated the compression of the fractured rock mass, by first distributing the stresses along the triangular elements of the fracture surfaces. Subsequently, a small-scale model computed the resulting local fracture stiffness and local anisotropic fracture transmissivities, for each individual fracture, as a function of the contact stresses. The fracture was assumed to be a self-affine composite surface that acts as an elastic body under confining stress. Therefore, the method explicitly modeled changes to fracture roughness, using small-scale methods originally developed for the study of compaction, dissolution, and precipitation effects on fracture permeability (Lang *et al.*, 2015). Stiffness and transmissivity were computed by numerically modeling local shear-induced dilation and elastic compression. These mechanical and flow fracture properties were then used to populate the mechanical and flow properties of the fractures at the mesoscale, in order to evaluate the mesoscale permeability tensor of the fractured rock mass (Lang *et al.*, 2014).

As part of a network-scale study, 2120 fractures having different (initial) random rough surfaces were deformed. Fracture aperture distribution and permeability were found to be anisotropic, as a result of progressive deformation of the initially random small-scale fracture roughness. Simulations consistently yield anisotropic permeability of individual fractures, which influence the resulting macroscopic maximum permeability direction, which were found to align with the remote intermediate principal compressive stress.

When assuming isotropic fracture transmissivities, this model can reproduce commonly reported values of maximum permeability parallel to the maximum remote compressive stresses (Min *et al.*, 2004; Baghbanan and Jing, 2008; Jing *et al.*, 2013), because there is a preferential mechanical activation of fractures aligned with the maximum compressive stress. In contrast, when fracture permeability is modeled mechanically at a small scale,

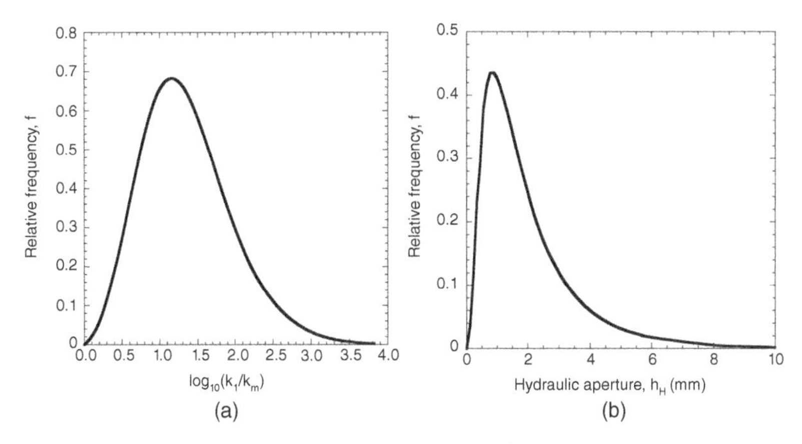

Figure 8.12 (a) Distribution of maximum principal permeability of the fractured rock mass, normalized with respect to the matrix permeability, over two hundred sixty-five DFM realizations. (b) Distribution of the hydraulic aperture of individual fractures, for the same realizations. The frequency functions on the vertical axes are to be interpreted such that, for example, the fraction of hydraulic apertures lying between h_H and $h_H + dh_H$ is given by $f(h_H)dh_H$. Curves represent distribution functions fitted to the simulated data. Source: Adapted from Lang *et al.* (2018).

as in the work of Lang *et al.* (2018), the anisotropic transmissivities induced by shear displacement will modify the macroscopic permeability tensor. Figure 8.12a illustrates how this anisotropy affects the macroscopic permeability of the entire system, for 265 different stochastically generated fracture networks. Figure 8.12b shows the corresponding average aperture distributions for the 2120 fractures simulated in the study. This distribution of anisotropic apertures in response to the applied stresses causes the fracture network's maximum permeability to be consistently aligned in the direction of the *intermediate* remote compressive stress.

The somewhat oversimplified, but essentially correct, explanation for this result is as follows. The greatest amount of shear displacement will be expected to occur on fractures subjected to a low normal stress (*i.e.*, lying in a plane normal to the direction of the minimum principal stress, σ_3), and this displacement will be aligned with the direction of the maximum principal stress, σ_1. As explained in Section 3.6, fractures subjected to significant shear displacement will develop higher permeability in the direction *perpendicular* to the shear direction than in the direction *parallel* to it (Auradou *et al.*, 2005). Hence, the permeability will be mainly enhanced in the direction within this fracture plane, perpendicular to the direction of maximum principal stress. Since this direction is therefore normal to both σ_1 and σ_3, it will necessarily lie in the direction of the intermediate principal stress, σ_2.

These findings are supported by field observations. In a study by Mattila and Follin (2019) that analyzed 193,698 fractures mapped from deep surface-based holes drilled in the proposed locations of nuclear waste repositories in Finland and Sweden, approximately six thousand fractures were detected to be flowing at steady state, and in general, more critically stressed fractures had a tendency to be more conductive. These researchers observed that in shallow rocks, at depths less than 1 km, permeability was controlled by dilation due to low normal stresses.

8.9 Channels and Preferential Flow Pathways

Fluid flow channels are observed in the field and can localize flow within a specific subset of fractures in a network (Tsang and Neretnieks, 1998). Geological observations show that flow intensely localizes in specific areas of fracture networks, leading to the formation of preferential pathways that strongly affect the permeability of the system and the manner in which this permeability is distributed among the fractures within the rock mass. Due to the lack of spatial correlation between more conductive and less conductive fractures, stochastic models do not usually reproduce the channeling that is characteristic of fractured aquifers (Maillot *et al.*, 2016). Geomechanical models are capable of doing so, in particular if they consider mechanically computed aperture distributions. Flow localization is systematically observed in geomechanically generated fracture networks, forming as a function of the geometry and topology of the connected network, as well as being enhanced by changes in applied stress (Thomas *et al.*, 2020b). In the simulations shown in Fig. 8.13, flow distributes itself evenly in well-connected geomechanically grown fracture networks (right) but tends to form strong channels when subjected to stresses (left). In fact, the more geomechanical realism that is included in the simulations, the more it is seen that flow patterns self-organize into flow channels.

A detailed analysis of channeling in fracture networks, within otherwise impermeable rock masses, *i.e.*, $k_m = 0$, was carried out by Maillot *et al.* (2016). They investigated both stochastically generated networks and "kinematic" networks that were generated according to heuristic kinematic rules, such as those briefly discussed in Section 8.2. They quantified the distribution of flow among the various fractures by a "ratio of participation" parameter. This concept was originally introduced by Bell and Dean (1970) to quantify the number of atoms that participate in a specific normal mode of vibration in a solid material. Following this idea, Maillot *et al.* (2016) defined a flow channeling density indicator, d_Q [1/m], as follows:

$$d_Q = \frac{1}{V} \frac{\sum (S_i Q_i)^2}{\sum S_i Q_i^2}, \tag{8.19}$$

where S_i [m^2] is the surface area of fracture i, Q_i [m^3/s] is volumetric flow through that fracture, and the summations are taken over all fractures. This parameter is a measure of the portion of the total fracture surface area for which the flow is significant. According to Maillot *et al.* (2016), $1/d_Q$ is representative of the average distance between main flow paths. Their analysis was carried out in terms of dimensionless variables, with the characteristic length taken to essentially be the minimum fracture radius, as defined in eq. (8.1), and the characteristic permeability is taken to be the log-mean transmissivity of the individual fractures divided by the characteristic length. Hence, in dimensionless form, small values of d_Q are associated with networks exhibiting highly channelized flow, since the distance between major flow paths is much larger than the typical distance between nearby fractures.

This definition can be extended to incorporate variations within the fracture when represented by surface meshes in three dimensions. The channeling indicator d_Q captures the relative flow distribution and is strongly correlated to the permeability of the system (Maillot *et al.*, 2016). For most of the high-density networks, a linear relationship was observed

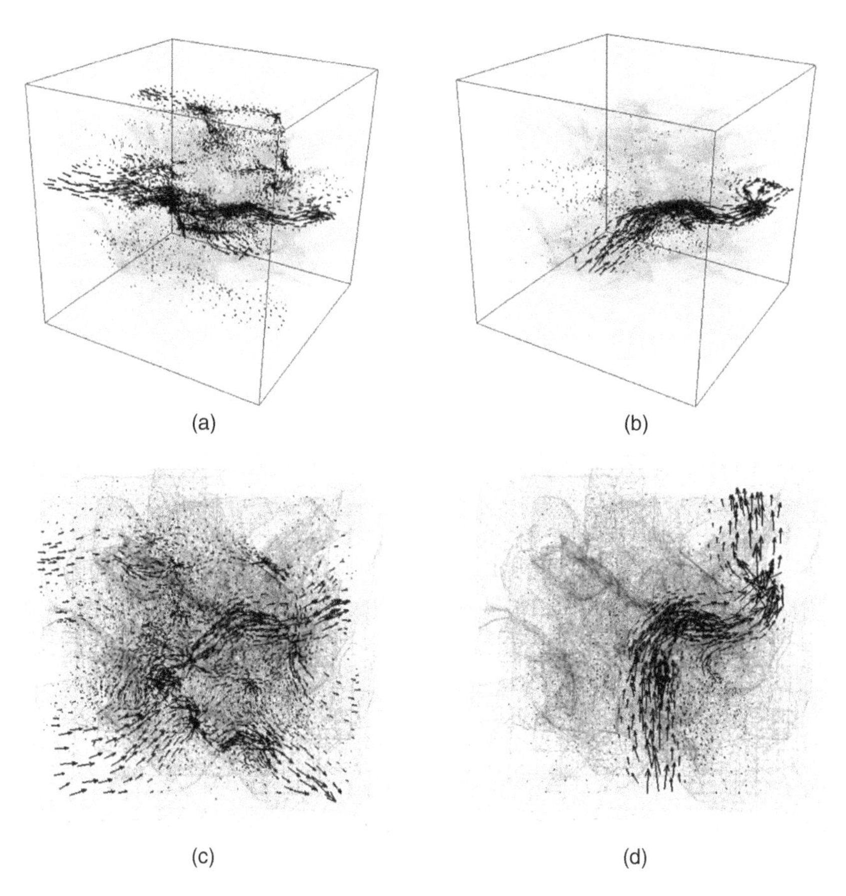

Figure 8.13 Flow velocity vectors in geomechanically grown fracture networks, with geomechanically generated apertures. The effect of fracture interaction on flow velocity vectors manifests as an emerging channelization of the flow, a direct result of fracture interaction during growth and its effect on stress-dependent aperture distributions. Images (a) and (b) are three-dimensional "side views," whereas (c) and (d) are top views of (a) and (b), respectively.

between the permeability and the flow channeling density indicator, which was given in dimensionless form as

$$k_{eq} = 1.4(d_Q - 0.25). \tag{8.20}$$

At lower fracture densities, the following quadratic relationship between the permeability and the normalized flow channeling density indicator was found to hold:

$$k_{eq} = 1.5d_Q^2. \tag{8.21}$$

After considering four sets of kinematic models of varying densities, the following relationship was found between the permeability and the fracture density parameter P_{32}:

$$k_{eq} = C_k(P_{32} - P_{Qc}), \tag{8.22}$$

where C_k is a proportionality constant, and P_{Qc} is the density of the "non-active" fractures that do not contribute to the macroscopic permeability of the system. Eliminating the permeability between eqs. (8.22) and (8.21), which is more accurate than eq. (8.20) at low densities, the following relationship can be found between the flow channeling density indicator d_Q and P_{32}:

$$d_Q = \sqrt{\frac{C_k}{15}(P_{32} - P_{Qc})}. \tag{8.23}$$

This conceptual model therefore predicts that lower fracture densities will lead to smaller values of d_Q, which implies more pronounced channeling. In stochastic models having a Poisson distribution of fractures, flow tends to essentially occupy less than half of the total fracture surface area, whereas for kinematically generated models, the proportion is even smaller, indicating more channeling in the latter as compared to the stochastic models. This is related to the spatial organization and intersections in the models, which for the kinematic models are a function of the mechanically informed nucleation rules that lead to a smaller number of intersections, by as much as a factor two, compared to Poisson models. This yields, in turn, lower permeabilities for the kinematic models. When comparing stochastic and geomechanically generated fracture networks having the same density, the number of intersections of stochastically generated datasets tends to be much larger than their geomechanical counterparts.

Problems

8.1 Consider a 3D cubic domain of size $1\,\mathrm{m}^3$, containing a stochastically generated network of disk-like fractures having a power-law distribution of radii. What is the dimensionless density P_{32} for eight sets that contain either 100, 500, 1000, or 10,000 fractures, and have a power-law exponent of either $n = -0.5$ or $n = -2.0$; in all eight cases, $\{r_{\min} = 0.1, \ r_{\max} = 0.2\}$. What are the maximum values of $r_{\max}$ that these datasets can take?

8.2 Consider the permeability of three mutually orthogonal sets of fractures of increasing density, assuming a matrix permeability of $10^{-15}\,\mathrm{m}^2$, with all fractures having uniform apertures of 0.001 m. Using the methods described in Chapter 7, compute the macroscopic permeabilities $\{k_{xx}, k_{yy}, k_{zz}\}$, and compare them against the SDFN and GDFN data shown in Fig. 8.11. Which networks are the most permeable?

8.3 Consider the fracture network datasets in Thomas and Paluszny (2020). Write a script to load the datasets and compute their surface areas. Solve the flow equations using the finite element method, finite difference method, or a suitable alternative, and thereby compute the flow field that results from applying fixed boundary conditions of 100 and 10 Pa fluid pressure on opposing faces. Assume that the intrinsic matrix permeability is $1 \times 10^{-15}\,\mathrm{m}^2$, and that each fracture has an aperture of 0.001 m. Compute the channeling indicator d_Q of the different datasets. How does the channeling indicator of the SDFNs compare to that of the GDFNs?

8.4 (Follows from Problem 8.3) Compute the directional permeabilities $\{k_{xx}, k_{yy}, k_{zz}\}$. Repeat the calculations for uniform apertures of 10^{-3} m, 10^{-4} m, and 10^{-5} m, and also consider a case in which the apertures take one of these three values in a random manner. How does the channeling indicator d_Q vary? How does the directional permeability compare to the results presented in Fig. 8.11?

8.5 (Follows from Problem 8.4) Implement the computation of the full permeability tensor of the DFN presented in Problem 8.3. After finding the full permeability tensor, compute the three principal permeabilities, by finding the three eigenvalues of the tensor. How do the principal permeabilities, $\{k_1, k_2, k_3\}$, compare to the diagonal components of the permeability tensor, $\{k_{xx}, k_{yy}, k_{zz}\}$?

References

Adler, P. M., Thovert, J.-F., and Mourzenko V. V. 2013. *Fractured Porous Media*. Oxford University Press, Oxford.

Auradou, H., Drazer, G., Hulin, J. P., and Koplik, J. 2005. Permeability anisotropy induced by the shear displacement of rough fracture walls. *Water Resources Research*, 41(9), W09423.

Azizmohammadi, S. and Matthäi, S. K. 2017. Is the permeability of naturally fractured rocks scale dependent? *Water Resources Research*, 53(9), 8041–63.

Baghbanan, A. and Jing, L. 2008. Stress effects on permeability in a fractured rock mass with correlated fracture length and aperture. *International Journal of Rock Mechanics and Mining Sciences*, 45(8), 1320–34.

Bear, J. 1988. *Dynamics of Fluids in Permeable Media*, Dover Publications, Mineola, N.Y.

Bell, R. J. and Dean, P. 1970. Atomic vibrations in vitreous silica. *Discussions of the Faraday Society*, 50, 55–61.

Bisdom, K., Bertotti, G., and Nick, H. M. 2016. The impact of different aperture distribution models and critical stress criteria on equivalent permeability in fractured rocks. *Journal of Geophysical Research: Solid Earth*, 121(5), 4045–63.

Bogdanov, I. I., Mourzenko, V. V., Thovert, J.-F., and Adler, P. M. 2003. Effective permeability of fractured porous media in steady state flow. *Water Resources Research*, 39(1), 1023.

Bogdanov, I., Mourzenko, V. V., Thovert, J.-F., and Adler P. M. 2007. Effective permeability of fractured porous media with power-law distribution of fracture sizes. *Physical Review E*, 76(3), 036309.

Bonneau, F., Caumon, G., and Renard, P. 2016. Impact of a stochastic sequential initiation of fractures on the spatial correlations and connectivity of discrete fracture networks. *Journal of Geophysical Research: Solid Earth*, 121(8), 5641–58.

Davy, P., Le Goc, R., Darcel, C., Bour, O., *et al.* 2010. A likely universal model of fracture scaling and its consequence for crustal hydromechanics. *Journal of Geophysical Research: Solid Earth*, 115(B10), B10411.

Durlofsky, L. J. 1991. Numerical calculation of equivalent grid block permeability tensors for heterogeneous porous media. *Water Resources Research*, 27(5), 699–708.

Goehring, L. 2013. Evolving fracture patterns: columnar joints, mud cracks and polygonal terrain. *Philosophical Transactions of the Royal Society A*, 371(2004), 20120353.

Holden, L. and Lia, O. 1992. A tensor estimator for the homogenization of absolute permeability. *Transport in Porous Media*, 8(1), 37–46.

Huseby, O., Thovert, J.-F., and Adler, P. M. 1997. Geometry and topology of fracture systems. *Journal of Physics A: Mathematical and General*, 30(5), 1415–44.

Jaeger, J. C., Cook, N. G. W., and Zimmerman, R. W. 2007. *Fundamentals of Rock Mechanics*, 4th ed., Wiley-Blackwell, Oxford.

Jing, L., Min, K.-B., Baghbanan, A., and Zhao, Z. 2013. Understanding coupled stress, flow and transport processes in fractured rocks. *Geosystem Engineering*, 16(1), 2–25.

Juanes, R., Samper, J., and Molinero, J. 2002. A general and efficient formulation of fractures and boundary conditions in the finite element method. *International Journal for Numerical Methods in Engineering*, 54(12), 1751–74.

Kim, J.-G. and Deo, M. D. 2000. Finite element, discrete-fracture model for multiphase flow in porous media. *AIChE Journal*, 46(6), 1120–30.

Lang, P. S., Paluszny, A., and Zimmerman, R. W. 2014. Permeability tensor of three-dimensional fractured porous rock and a comparison to trace map predictions. *Journal of Geophysical Research: Solid Earth*, 119(8), 6288–307.

Lang, P. S., Paluszny, A., and Zimmerman, R. W. 2015. Hydraulic sealing due to pressure solution contact zone growth in siliciclastic rock fractures. *Journal of Geophysical Research: Solid Earth*, 120(6), 4080–101.

Lang, P. S., Paluszny, A., Nejati, M., and Zimmerman, R. W. 2018. Relationship between the orientation of maximum permeability and intermediate principal stress in fractured rocks. *Water Resources Research*, 54, 8734–55.

Latham, J. P., Xiang, J., Belayneh, M., Nick, H. M., *et al.* 2013. Modelling stress-dependent permeability in fractured rock including effects of propagating and bending fractures. *International Journal of Rock Mechanics and Mining Sciences*, 57, 100–12.

Lavoine, E., Davy, P., Darcel, C., and Munier, R. 2020. A discrete fracture network model with stress-driven nucleation: impact on clustering, connectivity, and topology *Frontiers in Physics*, 8, article 9.

Lesueur, M., Guével, A., and Poulet, T. 2022. Reconciling asymmetry observations in the permeability tensor of digital rocks with symmetry expectations. *Advances in Water Resources*, 170, 104334.

Leung, C. T. O. and Zimmerman, R. W. 2012. Estimating the hydraulic conductivity of two-dimensional fracture networks using network geometric properties. *Transport in Porous Media*, 97, 777–97.

Long, J. C. S., Remer, J. S., Wilson, C. R., and Witherspoon, P. A. 1982. Porous media equivalents for networks of discontinuous fractures. *Water Resources Research*, 18(3), 645–58.

Maillot, J., Davy, P., Le Goc, R., Darcel, C., and de Dreuzy, J. R. 2016. Connectivity, permeability, and channeling in randomly distributed and kinematically defined discrete fracture network models. *Water Resources Research*, 52(11), 8526–45.

Matsumoto, M. and Nishimura, T. 1998. Mersenne twister: A 623-dimensionally equidistributed uniform pseudo-random number generator. *ACM Transactions on Modeling and Computer Simulation*, 8(1), 3–30.

Matthäi, S. K. and Belayneh, M. 2004. Fluid flow partitioning between fractures and a permeable rock matrix. *Geophysical Research Letters*, 31(7), L07602.

Matthäi, S. K., Mezentsev, A. A., Pain, C. C., and Eaton, M. D. 2005. A high-order TVD transport method for hybrid meshes on complex geological geometry. *International Journal for Numerical Methods in Fluids*, 47(10–11), 1181–87.

Mattila, J. and Follin, S. 2019. Does *in situ* state of stress affect fracture flow in crystalline settings? *Journal of Geophysical Research: Solid Earth*, 124(5), 5241–53.

Min, K.-B., Rutqvist, J., Tsang, C.-F., and Jing, L. 2004. Stress-dependent permeability of fractured rock masses: a numerical study. *International Journal of Rock Mechanics and Mining Sciences*, 41(7), 1191–1210.

Monteagudo, J. E. P. and Firoozabadi, A. 2004. Control-volume method for numerical simulation of two-phase immiscible flow in two- and three-dimensional discrete-fractured media. *Water Resources Research*, 40(7), 07405.

Mourzenko, V., Thovert, J.-F., and Adler, P. M. 2005. Percolation of three-dimensional fracture networks with power-law size distributions. *Physical Review E*, 72(3), 036103.

Nejati, M., Paluszny, A., and Zimmerman, R. W. 2016. A finite element framework for modeling internal frictional contact in three-dimensional fractured media using unstructured tetrahedral meshes. *Computer Methods in Applied Mechanics and Engineering*, 306, 123–50.

Odling, N. E. 1997. Scaling and connectivity of joint systems in sandstones from western Norway. *Journal of Structural Geology*, 19(10), 1257–71.

Olson, J. and Pollard, D. D. 1989. Inferring paleostresses from natural fracture patterns: a new method. *Geology*, 17(4), 345–48.

Paluszny, A. and Matthäi S. K. 2010. Impact of fracture development on the effective permeability of porous rocks as determined by 2-D discrete fracture growth modeling. *Journal of Geophysical Research: Solid Earth*, 115(B2), B02203.

Paluszny, A. and R. W. Zimmerman 2011. Numerical simulation of multiple 3D fracture propagation using arbitrary meshes. *Computer Methods in Applied Mechanics and Engineering*, 200(9–12), 953–66.

Paluszny, A. and Zimmerman, R. W. 2013. Numerical fracture growth modeling using smooth surface geometric deformation. *Engineering Fracture Mechanics*, 108, 19–36.

Reichenberger, V., Jakobs, H., Bastian, P., and Helmig, R. 2006. A mixed-dimensional finite volume method for two-phase flow in fractured porous media. *Advances in Water Resources*, 29(7), 1020–36.

Renshaw, C. E. and Pollard, D. D. 1994. Numerical simulation of fracture set formation: a fracture mechanics model consistent with experimental observations. *Journal of Geophysical Research: Solid Earth*, 99(B5), 9359–72.

Sisavath, S., Al-Yaarubi, A., Pain, C. C., and Zimmerman, R. W. 2003. A simple model for deviations from the cubic law for a fracture undergoing dilation or closure. *Pure and Applied Geophysics*, 160, 1009–22.

Thomas, R. and Paluszny, A. 2020. Geomechanical and discrete fracture networks in Thomas et al. 2020, *Journal of Geophysical Research: Solid Earth. British Geological Survey*. (Dataset). 10.5285/47593fbe-95da-48ac-a03f-773b7abe1d97.

Thomas, R. N., Paluszny, A., and Zimmerman, R. W. 2020a. Growth of three-dimensional fractures, arrays, and networks in brittle rocks under tension and compression. *Computers and Geotechnics*, 121, 103447.

Thomas, R. N., Paluszny, A., and Zimmerman, R. W. 2020b. Permeability of three-dimensional numerically grown geomechanical discrete fracture networks with evolving geometry and mechanical apertures. *Journal of Geophysical Research: Solid Earth*, 125(4), e2019JB018899.

Trefethen, L. N. and Bau, D. 1997. *Numerical Linear Algebra*. Society of Industrial and Applied Mathematics, Philadelphia.

Tsang, C.-F. and Neretnieks, I. 1998. Flow channeling in heterogeneous fractured rocks. *Reviews of Geophysics*, 36(2), 275–98.

Wen, X. H., Durlofsky, L. J., and Edwards, M. G. 2003. Use of border regions for improved permeability upscaling. *Mathematical Geology*, 35(5), 521–47.

Wang, X. H., Zheng, J., and Sun, H. Y. 2022. A method to identify the connecting status of three-dimensional fractured rock masses based on two-dimensional geometric information. *Journal of Hydrology*, 614, 128640.

Wu, H. and Pollard, D. D. 1992. Propagation of a set of opening-mode fractures in layered brittle materials under uniaxial strain cycling. *Journal of Geophysical Research: Solid Earth*, 97(B3), 3381–96.

Wu, X. H., Efendiev, Y., and Hou, T. Y. 2003. Analysis of upscaling absolute permeability. *Discrete and Continuous Dynamical Systems Series B*, 2(2), 185–204.

Zhang, X. and Sanderson, D. J. 1996. Effects of stress on the two-dimensional permeability tensor of natural fracture networks. *Geophysical Journal International*, 125, 912–24.

Zoback, M. L., Zoback, M. D., Adams, J., Assumpção, M., *et al*. 1989. Global patterns of tectonic stress. *Nature*, 341(6240), 291–98

9

Dual-Porosity Models for Fractured-Porous Rocks

9.1 Introduction

Chapter 7 discussed analytical models for estimating the macroscopic-scale permeability of a rock mass that contains a large number of interconnected, hydraulically transmissive fractures situated in an otherwise permeable rock mass. As a special case, the host rock might be considered, for all practical purposes, impermeable. In any case, the macroscopic permeability of such a fractured rock mass is mainly due to the interconnected fracture network. In these models, the fractures are typically idealized as being planar features, with planforms that are usually of simple shapes such as circular, elliptical, and square. On the other hand, Chapter 8 presented numerical simulations of fluid flow through fractured-porous rock masses, in which the fractures are explicitly represented as "thin," highly permeable but not necessarily regularly shaped fluid conduits, emplaced in a porous and permeable host rock.

Another commonly used approach to modeling fluid flow through fractured-porous rock masses, which is particularly suitable for transient processes that occur during production of fluids from subsurface water, geothermal, or oil and gas reservoirs, is the "dual-porosity" model, originally proposed by Barenblatt *et al.* (1960). In the dual-porosity model, the "macroscopic-scale" permeability of the rock mass is provided solely by the fracture network. Fluid is assumed to be able to flow on a macroscopic scale, for example, to a well or into an underground tunnel, only through this fracture network. However, the regions of rock between the fractures, which in the context of dual-porosity models are called "matrix blocks," contain most of the fluid. These porous and permeable matrix blocks feed fluid into, or out of, the fracture network but do not directly transmit fluid over macroscopic distances.

The effective-medium permeability models described in Chapter 7 take the porous matrix rock as their starting point and then compute the permeability perturbation caused by the introduction of fractures into this porous medium. In contrast, dual-porosity models take as their starting point an idealized medium that contains *only* fractures, situated in an *impermeable* rock mass. It is assumed that this fractured rock mass can be modeled as a porous and permeable continuum with an appropriate macro-scale permeability and porosity that account for the flow and transport properties of the fracture network. Transient fluid flow through this fractured rock mass is modeled by a traditional pressure diffusion equation, which will be derived in Section 9.2, which is similar to the equation that governs

Fluid Flow in Fractured Rocks, First Edition. Robert W. Zimmerman and Adriana Paluszny.

Companion website: www.wiley.com/go/zimmerman/fluidflowinfracturedrocks

flow in a non-fractured porous rock mass. The matrix blocks are then assumed to provide a "source–sink" term for fluid that may enter or leave the fractured continuum. In a dual-porosity model, the fractures are never modeled explicitly, and their effect is felt only through their contribution to the permeability and porosity of the fractured continuum.

The fracture-matrix flow interaction term that governs the flow of fluid between the fracture network and the matrix blocks, in the context of Barenblatt-type dual-porosity models, is discussed in some detail in Section 9.3, where it is shown to be a quasi-steady-state model that is accurate only at long times. In Section 9.4, the fracture-matrix interaction equation derived in Section 9.3 is combined with the pressure diffusion equation for the fracture continuum from Section 9.2, yielding the two coupled equations of the standard dual-porosity model. The Warren–Root solution for the important problem of flow to a vertical well in a laterally unbounded reservoir is presented and analyzed in Section 9.5. A more accurate fully transient model for the fracture-matrix interaction term is presented in Section 9.6, and a simple nonlinear fracture-matrix flow model is derived and discussed in Section 9.7. Finally, extensions of the dual-porosity model to account for two-phase flow, gravity, and other effects are briefly discussed in Section 9.8.

9.2 Pressure Diffusion Equation for the Fractured Continuum

The development of the classical dual-porosity model for fractured rock masses begins with the conceptual model of a network of fractures emplaced in an *impermeable* host rock. On a macroscopic scale, larger than the typical spacing between fractures, this system can be thought of as a porous medium, with permeability and porosity that are due solely to the fracture network. This permeability could be estimated by one of the analytical or semi-analytical methods discussed in Chapter 7 or could be treated as a fitting parameter whose value is obtained by analyzing well-test data. In any case, the next step is to model transient fluid flow in this "fractured continuum" by a standard pressure diffusion equation, such as that used for modeling non-fractured porous media. The derivation of this pressure diffusion equation, which can be found in many textbooks and monographs (de Marsily, 1986; Bear, 1988), can be summarized as follows, based on the derivation given by Zimmerman (2018).

Consider a fluid having density ρ [kg/m^3], flowing at a volumetric flowrate per unit area q [m/s], through a one-dimensional slab of porous rock having cross-sectional area A [m], as in Fig. 9.1. Although the following derivation does not require any specification of the pore geometry, the pore space in this rock is assumed to consist of an *interconnected fracture network*. Applying the law of conservation of mass to the portion of rock lying between x and $x+\Delta x$, over the time interval between t and $t+\Delta t$, leads to the following mass balance equation:

$$[A(x)\rho(x)q(x) - A(x+\Delta x)\rho(x+\Delta x)q(x+\Delta x)]\Delta t + S(t)\Delta t = m(t+\Delta t) - m(t), \quad (9.1)$$

in which $A(x)\rho(x)q(x)\Delta t$ is the mass entering the "control volume" from the left, $A(x+\Delta x)\rho(x+\Delta x)q(x+\Delta x)\Delta t$ is the mass exiting from the right, and $m(t)$ [kg] is the amount of fluid mass stored in the control volume at time t. The term $S(t)$ [kg/s] is a generic "source/sink" term that quantifies the rate at which fluid mass enters (or leaves, if $S(t)$ is negative) the control volume through mechanisms other than Darcy-like flow,

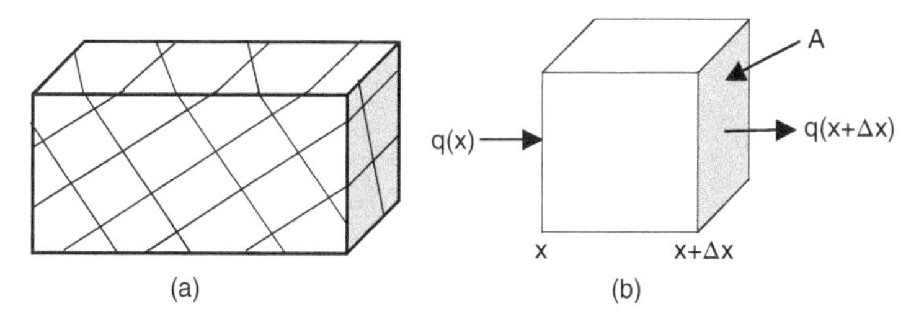

Figure 9.1 (a) Fractured-porous rock mass with an interconnected network of fractures that break up the matrix rock into discrete porous and permeable matrix blocks. (b) One-dimensional porous slab used in the derivation of an equation that represents conservation of mass. At this stage of the derivation, the pore space in this slab of rock is assumed to consist solely of the *interconnected fracture network*.

such as through a well that penetrates the slab in this region, for example. Factoring out A, dividing both sides by Δt, and letting $\Delta t \to 0$, yields

$$-A[\rho q(x+\Delta x) - \rho q(x)] + S(t) = \lim_{\Delta t \to 0} \frac{m(t+\Delta t) - m(t)}{\Delta t} = \frac{\partial m}{\partial t}, \tag{9.2}$$

where the product ρq is temporarily treated as a single entity. The mass stored in the pore space of this region of the rock is given by $m = \rho V_p$, where V_p is the pore volume of this region, which is, by definition, equal to the macroscopic volume, $A\Delta x$, multiplied by the porosity, ϕ. Hence, $m = \rho V_p = \rho\phi V = \rho\phi A\Delta x$, in which case eq. (9.2) leads to

$$-A[\rho q(x+\Delta x) - \rho q(x)] + S(t) = \frac{\partial(\rho\phi)}{\partial t} A\Delta x. \tag{9.3}$$

Dividing through by $A\Delta x$, and letting $\Delta x \to 0$, yields

$$-\frac{\partial(\rho q)}{\partial x} + s = \frac{\partial(\rho\phi)}{\partial t}, \tag{9.4}$$

where $s(t) = S(t)/A\Delta x$, with units of [kg/m^3 s], is the source/sink term of added mass, per unit volume of the porous medium.

Applying the product rule and chain rule on the right-hand side of eq. (9.4) gives

$$\begin{aligned} \frac{\partial(\rho\phi)}{\partial t} &= \rho\frac{\partial\phi}{\partial t} + \phi\frac{\partial\rho}{\partial t} = \rho\frac{d\phi}{dp}\frac{\partial p}{\partial t} + \phi\frac{d\rho}{dp}\frac{\partial p}{\partial t} \\ &= \rho\phi\left[\left(\frac{1}{\phi}\frac{d\phi}{dp}\right) + \left(\frac{1}{\rho}\frac{d\rho}{dp}\right)\right]\frac{\partial p}{\partial t} = \rho\phi(c_\phi + c_F)\frac{\partial p}{\partial t}, \end{aligned} \tag{9.5}$$

where c_F [1/Pa] is the compressibility of the pore fluid, and c_ϕ [1/Pa] is the "pore compressibility" of the rock mass. If the volumetric flow per unit area, q, is assumed to obey Darcy's law, eq. (2.13), the derivative term on the left-hand side of eq. (9.5) takes the form

$$\begin{aligned} -\frac{\partial(\rho q)}{\partial x} &= -\frac{\partial}{\partial x}\left[\frac{-\rho k}{\mu}\frac{\partial p}{\partial x}\right] = \frac{k}{\mu}\left[\rho\frac{\partial^2 p}{\partial x^2} + \frac{\partial\rho}{\partial x}\frac{\partial p}{\partial x}\right] \\ &= \frac{k}{\mu}\left[\rho\frac{\partial^2 p}{\partial x^2} + \frac{d\rho}{dP}\frac{\partial p}{\partial x}\frac{\partial p}{\partial x}\right] = \frac{\rho k}{\mu}\left[\frac{\partial^2 p}{\partial x^2} + \left(\frac{1}{\rho}\frac{d\rho}{dP}\right)\left(\frac{\partial p}{\partial x}\right)^2\right] \\ &= \frac{\rho k}{\mu}\left[\frac{\partial^2 p}{\partial x^2} + c_F\left(\frac{\partial p}{\partial x}\right)^2\right]. \end{aligned} \tag{9.6}$$

Inserting eqs. (9.5) and (9.6) into eq. (9.4) yields

$$\frac{k}{\mu}\left[\frac{\partial^2 p}{\partial x^2} + c_F\left(\frac{\partial p}{\partial x}\right)^2\right] + \frac{s}{\rho} = \phi(c_\phi + c_F)\frac{\partial p}{\partial t}. \tag{9.7}$$

For liquids, the second term inside the brackets on the left will be negligible compared to the first term (Matthews and Russell, 1967; Zimmerman, 2018), thereby yielding the following one-dimensional, linearized form of the pressure diffusion equation:

$$\frac{k}{\mu}\frac{\partial^2 p}{\partial x^2} + \frac{s}{\rho} = \phi c\frac{\partial p}{\partial t}, \tag{9.8}$$

in which $c = c_\phi + c_F$ is the *total compressibility* of the rock-fluid system. In the fully three-dimensional case, the pressure diffusion equation takes the form

$$\frac{\partial p}{\partial t} = \frac{k}{\phi\mu c}\left(\frac{\partial^2 p}{\partial x^2} + \frac{\partial^2 p}{\partial y^2} + \frac{\partial^2 p}{\partial z^2}\right) + \frac{s}{\rho\phi c} = \frac{k}{\phi\mu c}\nabla^2 p + \frac{s}{\rho\phi c}. \tag{9.9}$$

This equation has the form of a classical diffusion equation, with a source/sink term, and is mathematically analogous to the equation that governs heat flow through a heat-conducting solid (Carslaw and Jaeger, 1959). In the present case of fluid flow through a porous medium, the pressure plays a role analogous to the temperature, and the *hydraulic diffusivity*, $k/\phi\mu c$, plays the role of the thermal diffusivity.

The pressure that appears in eq. (9.9) is the pressure of the fluid *in the fracture system* at a given location (x, y, z) in the rock mass. In general, the fluid pressure, as well as parameters such as the porosity and permeability, will be different in the rock matrix than in the fracture network, and so the subscript f will be used to denote the variables that pertain to the fractures (not to be confused with the subscript F that is used in this chapter to denote the fluid), in which case eq. (9.9) is written as

$$\frac{\partial p_f}{\partial t} = \frac{k_f}{\mu\phi_f c_f}\nabla^2 p_f + \frac{s}{\rho\phi_f c_f}. \tag{9.10}$$

Although this equation is written here in the traditional mathematical form in which the material parameters multiply the spatial derivative terms, a form that is physically more revealing can be obtained by multiplying through by the *storativity*, $\phi_f c_f$, *i.e.*, reverting to a three-dimensional version of eq. (9.8):

$$\phi_f c_f\frac{\partial p_f}{\partial t} = \frac{k_f}{\mu}\nabla^2 p_f + \frac{s}{\rho}. \tag{9.11}$$

In this form, the term on the left represents the local rate of increase in storage of fluid on a volumetric basis, with units of [m^3/m^3 s], which should be interpreted as "cubic meters of fluid, per cubic meter of porous medium, per second." The first term on the right is the negative of the volumetric divergence of fluid out of the local region, and the second term on the right is the volumetric source/sink term, both of which also have units of [m^3/m^3 s].

9.3 Fracture/Matrix Fluid Interaction Term

The pressure diffusion equation (9.11) is quite general, in the sense that the mass source/sink term s, and its corresponding volumetric source/sink term s/ρ, can be due to any number of processes, such as injection or extraction of fluid through well, for

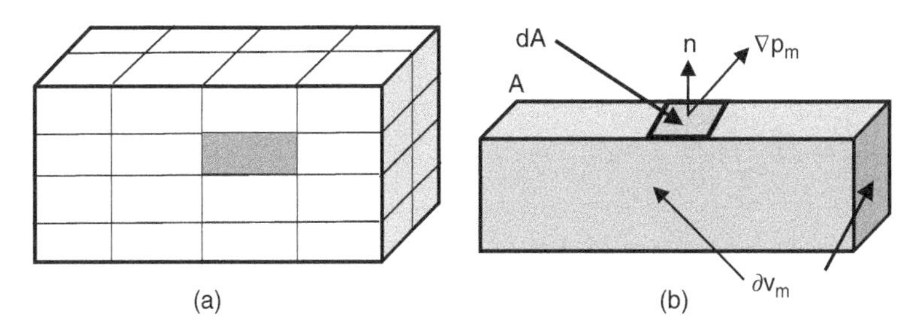

Figure 9.2 (a) Fractured-porous rock mass, showing a "representative volume element" that can be thought of as a computational gridblock in a numerical simulator. This computational gridblock contains numerous (in this case, 32) physical matrix blocks. (b) One of the matrix blocks that contributes to the source/sink term for fluid flowing into the fracture network. The matrix-fracture flow is computed by integrating Darcy's law over the outer surfaces of all the physical matrix blocks within the computational gridblock of the dual-porosity medium.

example. In the context of a dual-porosity model, this term represents fluid that is fed into the fracture network from the porous and permeable matrix blocks, in which case it is conventional to denote s/ρ by q_{mf} (Fig. 9.2). In order to derive an explicit expression for q_{mf}, it is convenient to consider a "gridblock" that occupies a region of space that has macroscopic volume V_f.

It is important to bear in mind that such a computational gridblock is *not* coincident with a matrix block (see Fig. 9.2) and actually must contain a large number of matrix blocks in order to justify the homogenization process that allows the fractured rock mass to be represented by a porous and permeable continuum. Since the fractures occupy a fraction ϕ_f of the total volume, the matrix blocks occupy a fraction $1-\phi_f$ of the total volume, and so the volume occupied by matrix blocks is $(1-\phi_f)V_f$. The number of matrix blocks contained in the gridblock is therefore equal to $(1-\phi_f)V_f/v_m$, where v_m is the volume of an individual matrix block.

According to the sign convention used in deriving eq. (9.11), the source term q_{mf} is *positive* if fluid flows *into* the fracture network. The flowrate of fluid from an individual matrix block into the fracture network can therefore be found by integrating the outward flux over the entire outer boundary of that matrix block (Zimmerman *et al.*, 1993):

$$Q(\text{single matrix block}) = -\int_{\partial v_m} \frac{k_m}{\mu}\frac{\partial p_m}{\partial n}dA, \tag{9.12}$$

where k_m is the permeability of the matrix block, p_m is the fluid pressure within the matrix block, ∂v_m is the outer boundary of the matrix block, n is the coordinate in the direction of the outward unit normal vector of this boundary, and dA is the differential of area on the boundary (Fig. 9.2b). The derivative $\partial p_m/\partial n$ can be computed as $\nabla p_m \bullet \mathbf{n}$, where ∇p_m is the pressure gradient in the matrix block and $\mathbf{n}$ is the outward unit normal vector to the surface ∂v_m.

The total inflow term for the entire fractured-medium gridblock is then found by multiplying the flow from each matrix block by the number of matrix blocks within the fracture gridblock, $(1-\phi_f)V_f/v_m$, yielding

$$Q_{mf} = \frac{-(1-\phi_f)V_f}{v_m}\int_{\partial v_m} \frac{k_m}{\mu}\frac{\partial p_m}{\partial n}dA. \tag{9.13}$$

The volumetric influx to the fracture gridblock, per unit macroscopic volume, is therefore given by

$$q_{mf} = \frac{Q_{mf}}{V_f} = \frac{-(1-\phi_f)}{v_m} \int_{\partial v_m} \frac{k_m}{\mu} \frac{\partial p_m}{\partial n} dA. \tag{9.14}$$

If the matrix block is assumed to be homogeneous, then the permeability and viscosity can be taken outside of the integral, to yield

$$q_{mf} = \frac{-(1-\phi_f)k_m}{\mu v_m} \int_{\partial v_m} \frac{\partial p_m}{\partial n} dA. \tag{9.15}$$

The integral in the above equation could be evaluated by solving the pressure diffusion equation inside the matrix block and then computing the integral exactly, as has been done by de Swaan (1976), Kazemi *et al.* (1976), and others; this approach is discussed in more detail in Section 9.6. A simpler model, proposed by Barenblatt *et al.* (1960), can be derived from the following order-of-magnitude considerations and approximations. First, assume that the integrand has the same value over the entire boundary of the matrix block; this would be true for a spherical matrix block and is an approximation for matrix blocks of other shapes. The pressure gradient term, taken in the direction pointing out of the matrix block, will be roughly given by $-(\overline{p}_m - p_f)/l$, where $\overline{p}_m$ is the mean pressure in the matrix block, and l is some length scale that will be on the order of, but necessarily somewhat less than, the "radius" of the matrix block. The matrix-fracture flux term can then be approximated as

$$q_{mf} = \frac{(1-\phi_f)k_m}{\mu v_m} \frac{(\overline{p}_m - p_f)A_m}{l}, \tag{9.16}$$

where A_m is the outer surface area of an individual matrix block. A check of the units of the terms in eq. (9.16) will verify that q_{mf} has the correct units of [m^3/m^3 s], or [1/s]. The combination A_m/lv_m, with dimensions of length^{-2}, is referred to as the *shape factor* and is usually denoted by α, although sometimes by σ, leading to the following form of the matrix-fracture transfer term:

$$q_{mf} = \frac{(1-\phi_f)\alpha k_m}{\mu}(\overline{p}_m - p_f). \tag{9.17}$$

The relationship between the numerical value of the shape factor and the geometry of the matrix block is discussed in detail in Chapter 10.

9.4 Equation for the Evolution of the Mean Pressure in the Matrix Blocks

Equation (9.17), or more generally, eq. (9.15), provides the source term that is needed in the pressure diffusion equation for the fractures, eq. (9.11). These two equations contain three time-varying unknowns: $\{p_f, \overline{p}_m, q_{mf}\}$. A third equation, which is needed in order to yield a properly closed mathematical problem, can be found by considering conservation of mass *within the matrix blocks*. Consider again an individual matrix block, having volume v_m. Flow within this matrix block is governed by an equation of the same form as eq. (9.11), except that there is no *local* source/sink term within the matrix block, and subscripts m are now used to denote properties of the matrix block:

$$\phi_m c_m \frac{\partial p_m}{\partial t} = \frac{k_m}{\mu} \nabla^2 p_m. \tag{9.18}$$

This equation can be integrated over the entire matrix block:

$$\iiint_{v_m} \phi_m c_m \frac{\partial p_m}{\partial t} dV = \iiint_{v_m} \frac{k_m}{\mu} \nabla^2 p_m dV. \tag{9.19}$$

Applying the divergence theorem to the integral on the right-hand side yields

$$\iiint_{v_m} \phi_m c_m \frac{\partial p_m}{\partial t} dV = \iint_{\partial v_m} \frac{k_m}{\mu} \frac{\partial p_m}{\partial n} dA, \tag{9.20}$$

where ∂v_m represents the outer boundary of the matrix block, and $\partial p_m / \partial n$ is the derivative of the pressure in the direction of the outward unit normal vector at the outer surface of the matrix block.

The standard assumption that the porosity and compressibility are each uniform throughout the matrix block, and constant in time, allows these two terms to be taken outside of the integral on the left side of eq. (9.20). Since mechanical deformation of the rock mass is *not* accounted for in this model, the region occupied by the matrix block will not vary with time, and so the time derivative can also be taken outside of the integral on the left-hand side. After multiplying and dividing the left-hand side by v_m, the result is

$$\phi_m c_m v_m \frac{\partial}{\partial t} \left[\frac{1}{v_m} \iiint_{v_m} p_m dV \right] = \iint_{\partial v_m} \frac{k_m}{\mu} \frac{\partial p_m}{\partial n} dA. \tag{9.21}$$

The bracketed term is, by definition, equal to the mean pore fluid pressure within the matrix block, $\overline{p}_m$. Hence,

$$\phi_m c_m v_m \frac{\partial \overline{p}_m}{\partial t} = \iint_{\partial v_m} \frac{k_m}{\mu} \frac{\partial p_m}{\partial n} dA. \tag{9.22}$$

Comparison of eqs. (9.22) and (9.14) immediately shows that (Zimmerman *et al.*, 1993)

$$q_{mf} = -\phi_m c_m (1 - \phi_f) \frac{\partial \overline{p}_m}{\partial t}. \tag{9.23}$$

The standard dual-porosity model, therefore, consists of three coupled equations. Equation (9.23) embodies the mass balance of the pore fluid in the matrix blocks. Equation (9.17) represents the matrix-fracture flow interaction term, according to the approximate matrix-fracture interaction model of Barenblatt *et al.* (1960). Equation (9.11) embodies the mass balance of the pore fluid in the fracture network. For convenience and future reference, this latter equation is re-written as follows, using the explicit identification of the generic source/sink term s/ρ with the matrix-fracture interaction term, q_{mf}:

$$\phi_f c_f \frac{\partial p_f}{\partial t} = \frac{k_f}{\mu} \nabla^2 p_f + q_{mf}. \tag{9.24}$$

The matrix-fracture flow interaction term q_{mf} can be eliminated from explicitly appearing in the governing equations by inserting eq. (9.17) into eqs. (9.24) and (9.23), leading to the following pair of coupled differential equations that govern the evolution of the pore pressure in both the matrix blocks and the fractures (Warren and Root, 1963):

$$\phi_f c_f \frac{\partial p_f}{\partial t} + \phi_m c_m \frac{\partial \overline{p}_m}{\partial t} = \frac{k_f}{\mu} \nabla^2 p_f, \tag{9.25}$$

$$\phi_m c_m \frac{\partial \overline{p}_m}{\partial t} = \frac{\alpha k_m}{\mu}(p_f - \overline{p}_m). \tag{9.26}$$

9.5 Warren–Root Solution for Flow to a Well in a Dual-Porosity Medium

One of the earliest problems to be solved within the context of the dual-porosity model was that of the flow of fluid to a vertical well in an infinite dual-porosity reservoir at a constant flowrate (Warren and Root, 1963; see Fig. 9.3). This is the most important flow problem in dual-porosity reservoirs, both historically, and in terms of practical importance. It also provides a clear and concrete example of the different "flow regimes" that occur during flow processes in fractured reservoirs, thereby highlighting the differences between porous-medium and dual-porosity-medium behavior.

The first step in presenting the Warren–Root solution is to rewrite the governing eqs. (9.25) and (9.26) in terms of radial coordinates, which is convenient for modeling flow to a vertical well. When using radial coordinates for flow on the scale of the reservoir, eq. (9.26) remains unchanged in form, but eq. (9.25) takes the form (Zimmerman, 2018)

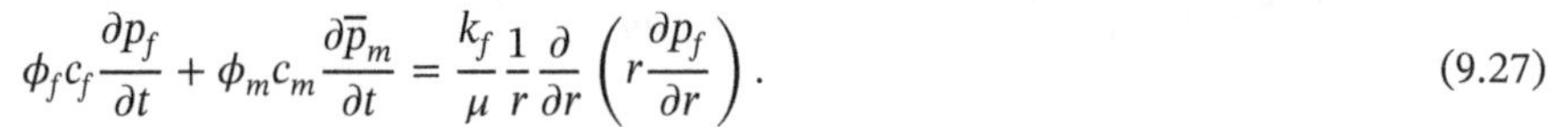

$$\phi_f c_f \frac{\partial p_f}{\partial t} + \phi_m c_m \frac{\partial \overline{p}_m}{\partial t} = \frac{k_f}{\mu} \frac{1}{r} \frac{\partial}{\partial r}\left(r \frac{\partial p_f}{\partial r} \right). \tag{9.27}$$

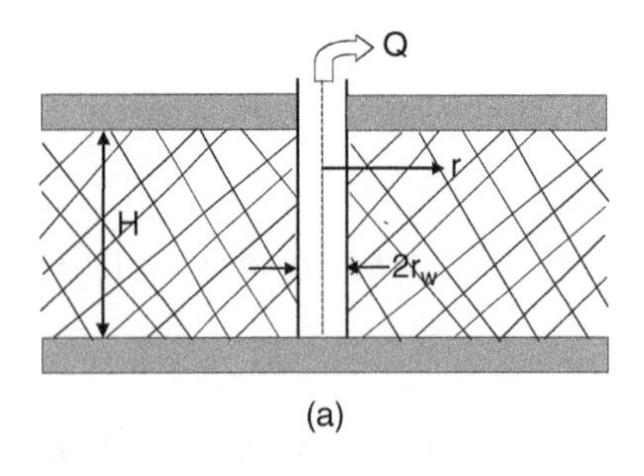

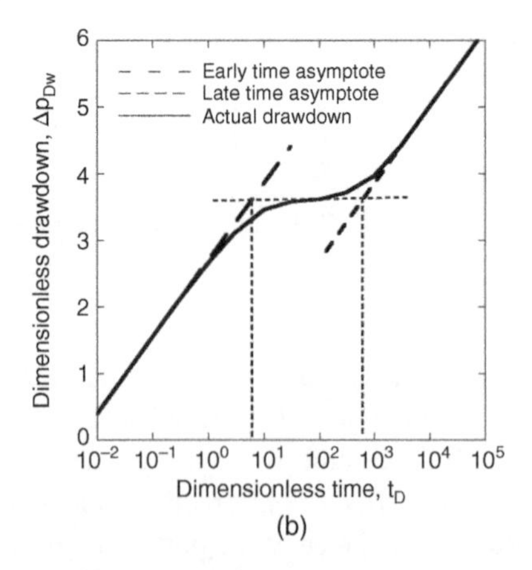

Figure 9.3 (a) Vertical well of radius r_w in a laterally unbounded dual-porosity reservoir of thickness H, extracting fluid from the reservoir at a volumetric flowrate Q. Shaded regions represent the impermeable overburden and underburden rock layers. (b) Dimensionless pressure drawdown at the well, as a function of dimensionless time, according to the analytical solution, eq. (9.38), derived by Warren and Root (1963), for the case $\omega = 0.01$ and $\lambda = 0.001$. The transition times t_{D1} and t_{D2} are indicated by the vertical dotted lines; see text for further details.

The two governing differential equations for this problem can be put into *dimensionless* form, by defining the following dimensionless variables. The dimensionless time is defined as

$$t_D = \frac{k_f t}{(\phi_f c_f + \phi_m c_m)\mu\, r_w^2}, \tag{9.28}$$

in which r_w is the wellbore radius. The dimensionless radius for points in the reservoir that are located at a distance r from the centerline of the wellbore is defined as

$$r_D = \frac{r}{r_w}. \tag{9.29}$$

The dimensionless pressure in the fractures is defined as

$$p_{Df} = \frac{2\pi\, k_f H(p_i - p_f)}{\mu Q}, \tag{9.30}$$

in which Q is the volumetric flowrate into the well, H is the thickness of the reservoir, and p_i is the initial pressure in the reservoir, before any fluid is produced. The dimensionless pressure in the matrix blocks is defined as

$$p_{Dm} = \frac{2\pi\, k_f H(p_i - \overline{p}_m)}{\mu Q}. \tag{9.31}$$

The two governing differential equations for dual-porosity flow can now be written in terms of these dimensionless variables (see Problem 9.1), as follows:

$$\frac{\phi_f c_f}{\phi_f c_f + \phi_m c_m}\frac{\partial p_{Df}}{\partial t_D} + \frac{\phi_m c_m}{\phi_f c_f + \phi_m c_m}\frac{\partial p_{Dm}}{\partial t_D} = \frac{1}{r_D}\frac{\partial}{\partial r_D}\left(r_D\frac{\partial p_{Df}}{\partial r_D}\right), \tag{9.32}$$

$$\frac{\phi_m c_m}{\phi_f c_f + \phi_m c_m}\frac{\partial p_{Dm}}{\partial t_D} = \frac{\alpha k_m r_w^2}{k_f}(p_{Df} - p_{Dm}). \tag{9.33}$$

The equations can be further simplified in appearance by defining two additional dimensionless parameters, which characterize the reservoir, and are independent of the particular production conditions (*i.e.*, independent of Q or p_i). The first parameter, the ratio of the storativity of the fracture network to the total storativity of the fracture-plus-matrix system, is the "storativity ratio," ω:

$$\omega = \frac{\phi_f c_f}{\phi_f c_f + \phi_m c_m}. \tag{9.34}$$

The second dimensionless parameter, the transmissivity ratio, λ, is the ratio of matrix permeability to fracture permeability, multiplied by a dimensionless geometrical factor, αr_w^2, which is essentially a measure of the ratio of the areal size of the wellbore to the areal size of a typical matrix block:

$$\lambda = \frac{\alpha k_m r_w^2}{k_f}. \tag{9.35}$$

Since the fracture porosity is typically much less than the matrix porosity, and the fracture permeability is much greater than the matrix permeability, in practice, both of these dimensionless parameters are very small. In most fractured reservoirs, it is the case that $\omega < 0.1$, and $\lambda < 0.001$.

In terms of the dimensionless parameters ω and λ, the two governing equations for radial flow in a dual-porosity reservoir, eqs. (9.32) and (9.33), take the form (da Prat, 1990, p. 64):

$$\omega\frac{\partial p_{Df}}{\partial t_D} + (1-\omega)\frac{\partial p_{Dm}}{\partial t_D} = \frac{1}{r_D}\frac{\partial}{\partial r_D}\left(r_D\frac{\partial p_{Df}}{\partial r_D}\right), \tag{9.36}$$

$$(1-\omega)\frac{\partial p_{Dm}}{\partial t_D} = \lambda(p_{Df} - p_{Dm}). \tag{9.37}$$

If the reservoir were actually a "single-porosity" reservoir, then the storativity ratio ω would equal 1, the transmissivity ratio λ would equal 0, eq. (9.36) would then reduce to the standard pressure diffusion equation, and eq. (9.37) would reduce to the identity $0 = 0$.

Warren and Root (1963) solved the problem of a vertical well producing at a constant flowrate Q in a laterally unbounded dual-porosity reservoir (Fig. 9.3a), using Laplace transforms. Noting that, according to the dual-porosity concept, fluid can only flow to the well through the fracture network, and so the pressure of the fluid in the well will essentially be equal to the pressure of the fluid in the fracture network at the well, where $r_D = r/r_w = 1$. Hence, the dimensionless "pressure drawdown" in the well can be found by evaluating $p_{Df}(r_D, t_D)$ at the value $r_D = 1$.

For dimensionless times that satisfy the condition $t_D > 100\omega$, which usually covers all times of practical interest, the dimensionless pressure drawdown at the well, often denoted by Δp_{Dw}, was found to be given by

$$\Delta p_{Dw} \equiv p_{Df}(r_D = 1, t_D) = \frac{1}{2}\left\{\ln t_D + 0.8091 + Ei\left[\frac{-\lambda t_D}{\omega(1-\omega)}\right] - Ei\left[\frac{-\lambda t_D}{(1-\omega)}\right]\right\}, \tag{9.38}$$

where Ei is the exponential integral function, defined by Matthews and Russell (1967)

$$-Ei(-x) = \int_x^\infty \frac{e^{-u}}{u}du. \tag{9.39}$$

The awkward minus signs are artifacts of the original mathematical definition of this function. Although some hydrologists (*cf*., de Marsily, 1986) remove both minus signs on the left side of eq. (9.39) and redefine the resulting function as the "Theis function" or the "well function," $W(x)$, the notation of eq. (9.39) is standard in the petroleum engineering and mathematics literature. Numerical values of the Ei function can be found in de Marsily (1986) or Zimmerman (2018). For $x < 0.01$, the following highly accurate logarithmic approximation, $-Ei(-x) \approx -0.5772 - \ln x$, can be used. For large values of x, it can be shown (see Problem 9.2) that $-Ei(-x)$ goes to zero exponentially fast as $x \to \infty$.

In the early-time regime, when $t_D > 100\omega$, but is not yet "too large," the variables inside the Ei functions in eq. (9.38) will be small enough to allow the use of the aforementioned logarithmic approximation, and eq. (9.38) reduces to

$$\Delta p_{Dw} = \frac{1}{2}[\ln(t_D/\omega) + 0.8091]. \tag{9.40}$$

Reverting to *dimensional* form, the pressure drawdown at the well, defined as $\Delta p_w = p_i - p_w$, takes the following form during this early-time regime:

$$\Delta p_w = \frac{\mu Q}{4\pi k_f H}\left\{\ln\left[\frac{k_f t}{\phi_f \mu c_f r_w^2}\right] + 0.8091\right\}, \tag{9.41}$$

which is precisely the drawdown that would occur in a single-porosity system consisting only of fractures with no matrix blocks (Matthews and Russell, 1967). The explanation of this result is that at early times, fluid has not yet had sufficient time to flow out of the matrix blocks due to their relatively low permeability, and so the matrix storativity is irrelevant.

At "late" times, both of the *Ei* terms in eq. (9.38) will effectively be zero, and so the drawdown is given by

$$\Delta p_{Dw} = \frac{1}{2}(\ln t_D + 0.8091), \tag{9.42}$$

which in dimensional form is

$$\Delta p_w = \frac{\mu Q}{4\pi k_f H}\left\{\ln\left[\frac{k_f t}{(\phi_f c_f + \phi_m c_m)\mu r_w^2}\right] + 0.8091\right\}. \tag{9.43}$$

This is precisely the drawdown that would occur in a single-porosity reservoir whose permeability is that of the fracture system but whose storativity is that of the fractures *plus* the matrix blocks. This reflects the fact that, at late times, fluid has had sufficient time to travel from the matrix blocks into the fracture network.

When plotted on a semi-log plot, both the early-time solution given by eq. (9.41) and the late-time solution given by eq. (9.43), yield straight lines, with exactly the same slope:

$$\frac{d\Delta p_w}{d\ln t} = \frac{\mu Q}{4\pi k_f H}. \tag{9.44}$$

The vertical offset of these two lines, denoted by $\delta\Delta p_{Dw}$, which can be found by comparing eqs. (9.40) and (9.42), is given by

$$\delta\Delta p_{Dw} = -\frac{1}{2}\ln\omega = \frac{1}{2}\ln(1/\omega). \tag{9.45}$$

In dimensional form, the vertical offset is (Streltsova, 1988):

$$\delta\Delta p_w = \frac{\mu Q}{4\pi k_f H}\ln(1/\omega). \tag{9.46}$$

The full curve for the drawdown in a well in an infinite dual-porosity reservoir is shown in Fig. 9.3b. According to the Warren–Root solution, which is based on the Barenblatt model for matrix-fracture flow, there is a transition regime between the two semi-log straight lines, during which the pressure drawdown changes very gradually. Bourdet and Gringarten (1980) showed that if a horizontal line is drawn through the drawdown curve at the midpoint (horizontally) between the two semi-log straight lines, this line will intersect the early-time and late-time semi-log straight lines at times given by

$$t_{D1} = \frac{\omega}{\gamma\lambda} \text{ and } t_{D2} = \frac{1}{\gamma\lambda}, \tag{9.47}$$

where $\gamma = 1.781$ is Euler's number. The horizontal offset of the two sloping straight lines is therefore given by $\ln t_{D2} - \ln t_{D1} = \ln(1/\omega)$. Recalling the definition of λ given in eq. (9.35), the expression given above for t_{D2} shows that the time required to reach full "double-porosity" behavior, wherein fluid is produced from both the fractures and the matrix blocks, is inversely proportional to the matrix permeability, as would be expected.

9.6 Fully Transient model for Matrix-to-Fracture Flow

Several extensions have been made to the Barenblatt model, to remedy the deficiency that the matrix-fracture flow interaction equation, eq. (9.17), is only accurate at "late" times during the depletion of fluid from a matrix block. Some of these fully transient models will be derived and discussed in this section and Section 9.7. In passing, these derivations will shed light on why the Barenblatt fracture/matrix flow model is often referred to as a "quasi-steady-state" model and is asymptotically accurate only at sufficiently large times. Rasoulzadeh and Kuchuk (2019) have recently presented a detailed and rigorous derivation of dual-porosity models, using homogenization theory (Allaire, 1992). Their discussion sheds further light on the range of validity of the Warren–Root model, including the influence of the fracture/matrix permeability and porosity ratios.

As mentioned in Section 9.4, fluid flow *within* a given matrix block is governed by the following partial differential equation:

$$\frac{\partial p_m(\mathbf{x}, t)}{\partial t} = \frac{k_m}{\phi_m \mu c_m} \nabla^2 p_m(\mathbf{x}, t), \tag{9.48}$$

where $\mathbf{x}$ is the position vector of a point within the block. Consider now the problem in which the pressure within the matrix block is initially at some uniform value, p_i,

$$p_m(\mathbf{x}, t = 0) = p_i, \tag{9.49}$$

after which the pressure at the entire outer boundary is instantaneously changed to some constant value, p_f:

$$p_m(\mathbf{x} \in \partial v_m, t > 0) = p_f, \tag{9.50}$$

where ∂v_m represents the outer boundary of the matrix block, and $\mathbf{x} \in \partial v_m$ denotes points on the boundary of the block.

For any specific block shape, the solution of this problem will always be of the following form (Courant and Hilbert, 1953, p. 312; Zimmerman and Bodvarsson, 1995):

$$p_m(\mathbf{x}, t) = p_f + (p_i - p_f) \sum_{n=1}^{\infty} C_n F_n(\mathbf{x}) \exp(-\lambda_n k_m t / \phi_m \mu c_m), \tag{9.51}$$

where the eigenfunctions, $F_n(\mathbf{x})$, and eigenvalues, λ_n, numbered such that $0 < \lambda_1 < \lambda_2 \ldots$, satisfy the equation $\nabla^2 F_n(\mathbf{x}) = -\lambda_n F_n(\mathbf{x})$ and the boundary condition $F_n(\mathbf{x} \in \partial v_m) = 0$. In mathematical terms, λ_n are the "eigenvalues of the Laplacian operator for the region occupied by the matrix block, with Dirichlet-type boundary conditions." These eigenvalues are unrelated to the transmissivity ratio λ; unfortunately, the notation λ is standard usage in both cases. Comparison of eqs. (9.49) and (9.51) shows that the Fourier coefficients C_n must be chosen such that

$$\sum_{n=1}^{\infty} C_n F_n(\mathbf{x}) = 1. \tag{9.52}$$

The eigenfunctions, eigenvalues, and Fourier coefficients can be found explicitly for simple shapes such as spheres, cylinders, and rectangular parallelepipeds (Carslaw and Jaeger, 1959; Crank, 1975). However, it is easier and more instructive to carry out the discussion and analysis in the following fully general context, which does not require explicit knowledge of these functions or coefficients.

First, note that since $0 < \lambda_1 < \lambda_2 \ldots$, the series in eq. (9.51) will eventually be dominated by the first term in the series, and so, at sufficiently late times,

$$p_m(\mathbf{x}, t) \approx p_f + (p_i - p_f)C_1F_1(\mathbf{x})\exp(-\lambda_1 k_m t/\phi_m \mu c_m). \tag{9.53}$$

This equation can be integrated over the matrix block, as in eq. (9.21), to find the average pressure in the matrix block, $\overline{p}_m(t)$:

$$\overline{p}_m(t) \approx p_f + (p_i - p_f)C_1\langle F_1(\mathbf{x})\rangle \exp(-\lambda_1 k_m t/\phi_m \mu c_m), \tag{9.54}$$

where $\langle F_1(\mathbf{x})\rangle$ denotes the spatial average, taken over the entire matrix block. Differentiating this expression with respect to time yields

$$\frac{d\overline{p}_m(t)}{dt} \approx \frac{\lambda_1 k_m}{\phi_m \mu c_m}(p_f - p_i)C_1\langle F_1(\mathbf{x})\rangle \exp(-\lambda_1 k_m t/\phi_m \mu c_m). \tag{9.55}$$

Elimination of the time-dependent exponential terms between eqs. (9.54) and (9.55) shows that the mean pressure in the matrix block satisfies the following first-order ordinary differential equation (Zimmerman *et al.*, 1993):

$$\frac{d\overline{p}_m(t)}{dt} \approx \frac{\lambda_1 k_m}{\phi_m \mu c_m}(p_f - \overline{p}_m), \tag{9.56}$$

where the $\approx$ symbol should be interpreted as meaning "with increasing accuracy as time increases." This equation has precisely the same form as eq. (9.26), and these two equations will become identical if the shape factor α is identified with the first (*i.e.*, the smallest) eigenvalue, λ_1. Finally, comparison with eq. (9.23), which was exact, shows that the matrix-fracture flow term is given by

$$q_{mf} \approx \frac{(1 - \phi_f)\alpha k_m}{\mu}(\overline{p}_m - p_f), \tag{9.57}$$

which is precisely the matrix-fracture flow term, eq. (9.17), which was suggested by Barenblatt *et al.* (1960). The preceding arguments provide a rational means of defining the shape factor and also show that the Barenblatt model for matrix-fracture flow will be accurate *only* at sufficiently large times.

If the pressure of the fluid in the surrounding fractures changes in a step-function manner, eq. (9.51) would provide the exact pressure distribution within the matrix block, and the matrix-fracture flow term could be computed exactly by integrating the fluid flux over the outer surface. In realistic subsurface flow situations, the pressure in the fractures surrounding the matrix block will change gradually, rather than instantaneously. This situation can be handled by performing a convolution integral (Carslaw and Jaeger, 1959; Zimmerman, 2018), which contains in its integrand the flowrate that occurs for the case of a unit step-function change in the fracture pressure, from $p_i = 0$ to $p_f = 1$. In the case of this step-function change in the pressure at the outer boundary of the matrix block, the pressure distribution (9.51) takes the form

$$p_m(\mathbf{x}, t) = 1 - \sum_{n=1}^{\infty} C_n F_n(\mathbf{x})\exp(-\lambda_n k_m t/\phi_m \mu c_m). \tag{9.58}$$

Substituting this expression into eq. (9.48), and integrating over the entire matrix block, yields

$$\frac{d\overline{p}_m}{dt} = \frac{1}{v_m}\iiint_{v_m} \frac{k_m}{\phi_m \mu c_m} \nabla^2 p_m(\mathbf{x}, t) dV$$

$$= \frac{k_m}{\phi_m \mu c_m v_m}\iiint_{v_m} \nabla^2 \sum_{n=1}^{\infty} C_n F_n(\mathbf{x}) \exp(-\lambda_n k_m t/\phi_m \mu c_m) dV. \tag{9.59}$$

But $\nabla^2 F_n(\mathbf{x}) = -\lambda_n F_n(\mathbf{x})$, by definition, and so

$$\frac{d\overline{p}_m}{dt} = \frac{-k_m}{\phi_m \mu c_m}\sum_{n=1}^{\infty} \lambda_n C_n \langle F_n(\mathbf{x})\rangle \exp(-\lambda_n k_m t/\phi_m \mu c_m). \tag{9.60}$$

Combining this result with eq. (9.23) yields

$$q_{umf}(t) = \frac{k_m(1-\phi_f)}{\mu}\sum_{n=1}^{\infty} \lambda_n C_n \langle F_n(\mathbf{x})\rangle \exp(-\lambda_n k_m t/\phi_m \mu c_m), \tag{9.61}$$

where, following Najurieta (1980), the subscript u is used to indicate that this equation pertains *only* to the case of a unit step-function increase of the fracture pressure, from 0 to 1 (in the appropriate units).

In the general case in which the pressure of the fluid in the surrounding fractures, $p_f(t)$, changes in an arbitrary manner, the volumetric flux from the matrix block into the fracture network, at time t, can be found by evaluating the following *convolution integral*:

$$q_{mf}(t) = \int_0^t q_{umf}(t-\tau)\frac{dp_f(\tau)}{d\tau} d\tau, \tag{9.62}$$

in which $q_{umf}(t)$ is given by eq. (9.61), and τ is a dummy variable of integration. The domain of integration ranges from $\tau = 0$ at the start of the process, up to the current time, when $\tau = t$. Dual-porosity models in which the matrix-fracture flux is calculated using convolution have been described by, among others, de Swaan (1976), Najurieta (1980), Lu and Connell (2011), and Rasoulzadeh and Kuchuk (2019). When implemented into a numerical simulator, this approach is more accurate than a Warren–Root type approach, but also more computationally demanding, due to the need to store all past fracture pressures, and to evaluate the convolution integral in each computational fracture gridblock, at each time step.

9.7 Nonlinear Matrix-Fracture Transfer Model

Zimmerman *et al.* (1993) developed a dual-porosity model that is more accurate than the Warren–Root approach at early times of the matrix-fracture flow period, but which retains the simplicity of using a single ordinary differential equation, rather than a convolution integral, to model the evolution of the mean matrix pressure. To be concrete, consider a dual-porosity medium containing spherical matrix blocks of radius a. Under the same step-function boundary conditions as used in Section 9.6, the mean pressure in the matrix block is given by (see Problem 9.5)

$$\overline{p}_m(t) = p_f - (p_f - p_i)\frac{6}{\pi^2}\sum_{n=1}^{\infty} \frac{\exp(-n^2\pi^2 k_m t/\phi_m \mu c_m a^2)}{n^2}. \tag{9.63}$$

At late times, this expression reduces to

$$\overline{p}_m(t) = p_f - (p_f - p_i)\frac{6}{\pi^2}\exp(-\pi^2 k_m t/\phi_m \mu c_m a^2). \tag{9.64}$$

As explained in Section 9.6, differentiation of this expression, and elimination of the time variable between eq. (9.64) and its derivative, leads to a Warren–Root-type matrix-fracture interaction equation,

$$\frac{d\overline{p}_m(t)}{dt} = \frac{\pi^2 k_m}{\phi_m \mu c_m a^2}(p_f - \overline{p}_m). \tag{9.65}$$

Since the full series given by eq. (9.63) is more accurate than eq. (9.64), over all times, it might be thought that a more accurate transfer function could be derived by differentiating eq. (9.63). Unfortunately, it is not possible to eliminate t between eq. (9.63) and its derivative, to arrive at an ordinary differential equation that depends solely on the matrix and fracture pressures, and does not explicitly depend on time. However, the following simple equation (Vermeulen, 1953) provides a good approximation to eq. (9.63), in all time regimes:

$$\overline{p}_m(t) = p_i + (p_f - p_i)[1 - \exp(-\pi^2 k_m t/\phi_m \mu c_m a^2)]^{1/2}. \tag{9.66}$$

Elimination of t between this expression and its time derivative leads to the following nonlinear ordinary differential equation for the matrix pressure:

$$\frac{d\overline{p}_m(t)}{dt} = \frac{\pi^2 k_m}{\phi_m \mu c_m a^2}\frac{(p_f - p_i)^2 - (\overline{p}_m - p_i)^2}{2(\overline{p}_m - p_i)}. \tag{9.67}$$

If this differential equation is taken to govern the variation of matrix pressure with time, for the case of a step-change in fracture pressure, it integrates (by construction) to eq. (9.66). Integration of the Warren–Root equation, (9.65), for the case of a step-change in fracture pressure, and imposing the initial condition $\overline{p}_m(t = 0) = p_i$, leads to

$$\overline{p}_m(t) = p_i + (p_f - p_i)[1 - \exp(-\pi^2 k_m t/\phi_m \mu c_m a^2)]. \tag{9.68}$$

This expression is not quite the same as eq. (9.64), because the late-time approximation to the exact pressure function does not satisfy the initial conditions. Comparison of the exact expression for the matrix pressure, eq. (9.63), the expression given by the Vermeulen model, eq. (9.66), and the expression given by the Warren–Root model, eq. (9.68), clearly shows that the Warren–Root model yields inaccurate matrix pressures, and consequently inaccurate flowrates, until times at which the depletion process is nearly complete (Fig. 9.4).

As with the Warren–Root model, although the differential equation (9.67) for the matrix pressure was derived for the case of a step-function change in fracture pressure, the assumption is made that it can be used for arbitrary changes in the fracture pressure with time. Zimmerman *et al.* (1993) verified that eq. (9.67) is much more accurate than the Warren–Root equation, for linear, power-law, and exponential variations of fracture pressure with time. It was also found to have reasonable accuracy for the case when the fracture pressure oscillates sinusoidally in time, provided that absolute value signs are inserted into eq. (9.67) in an appropriate manner.

The accuracy of the nonlinear dual-porosity model was demonstrated by Zimmerman *et al.* (1993), for the problem of fluid infiltration into a semi-infinite dual-porosity half-space, containing spherical matrix blocks of radius a (Fig. 9.5a). They considered a

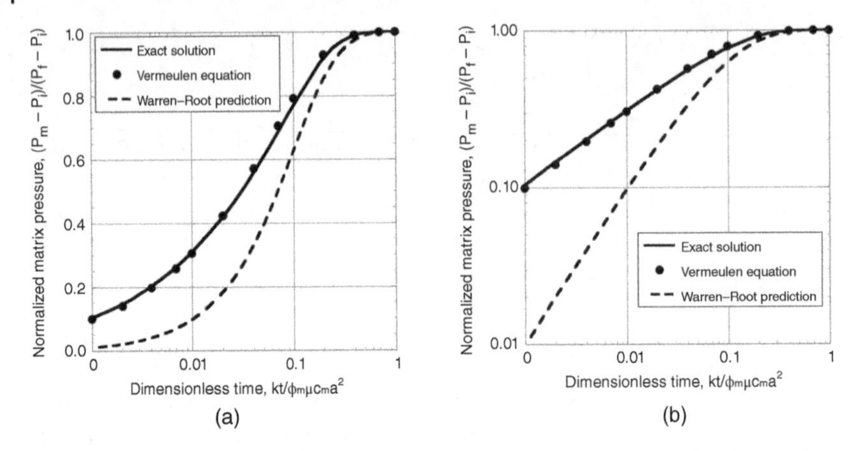

Figure 9.4 Normalized average pressure in a spherical matrix block of radius a, whose outer boundary is subjected to a step-function increase in pressure from p_i to p_f, according to the exact solution, the Vermeulen approximation, and the Warren–Root model, (a) in semi-log format, and (b) in log–log format.

rock formation that was initially at pressure p_i, after which the pressure at the boundary was abruptly changed to some value p_o. The parameters of the fractured-porous rock mass were chosen as $k_m = 10^{-15}$ m^2, $k_f = 10^{-18}$ m^2, $\phi_m = 0.1$, $\phi_f = 0.001$, and $a = 10$ m. The pore fluid was assumed to be water at 20 °C, which has a viscosity of $\mu = 0.001$ Pa s, a density of $\rho = 1000$ kg/m^3, and a compressibility of $c = 4.5 \times 10^{-10}$/Pa. The pore compressibilities of the fractures and matrix were set to zero, so that the total compressibilities of the fracture and matrix domains reflected only that of the fluid, *i.e.*, $c_f = c_m = 4.5 \times 10^{-10}$/Pa. The initial and boundary pressures were taken to be $p_i = 10$ MPa and $p_o = 11$ MPa. This problem was solved by implementing the nonlinear fracture-matrix transfer model, as given by eqs. (9.67) and (9.26), into an early version of the TOUGH simulator (Transport of Unsaturated Groundwater and Heat; Pruess, 1987).

The predicted flux into the formation from the pressurized boundary is plotted in Fig. 9.5b, using both the nonlinear fracture-matrix transfer function and the Warren–Root transfer function. Also shown in this figure are the early-time, intermediate-time, and late-time asymptotes that can be derived from the analytical solution to this problem that was obtained using Laplace transform methods by Nitao and Buscheck (1991). These three regimes are analogous to the three regimes that occur during the radial flow problem discussed in Section 9.5, although in that case, the external boundary condition was one of fixed flowrate, rather than fixed pressure. The early-time flux is

$$q(t) = (p_o - p_i)\left(\frac{\phi_f c_f k_f}{\pi \mu t}\right)^{1/2} = 3.78 \times 10^{-7}\, t^{-1/2}\ \text{m/s} \tag{9.69}$$

where, in this and subsequent numerical expressions on the right side, t must be in seconds. This flux is equal to the flux that would occur in a porous medium whose permeability and storativity are due only to the fracture network. The late-time flux is

$$q(t) = (p_o - p_i)\left[\frac{(\phi_f c_f + \phi_m c_m)k_f}{\pi \mu t}\right]^{1/2} = 3.80 \times 10^{-6}\, t^{-1/2}\ \text{m/s}, \tag{9.70}$$

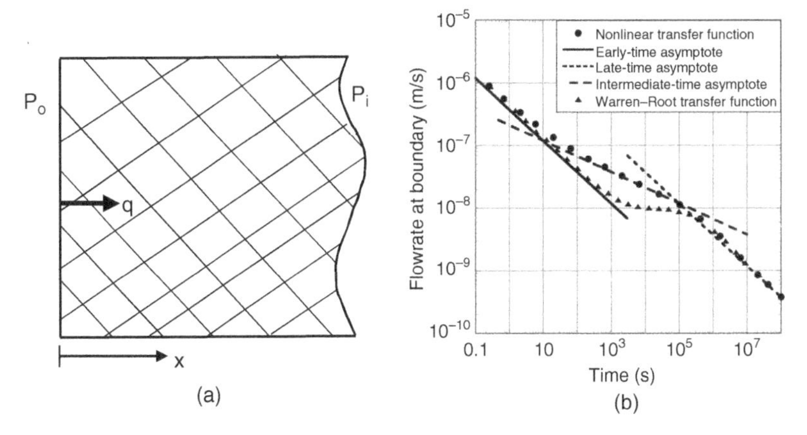

Figure 9.5 (a) Problem of a semi-infinite dual-porosity rock mass, with a step-function increase in the pressure at the boundary, without gravity. (b) Instantaneous flux at the boundary, as computed using the nonlinear transfer function and the Warren–Root transfer function, along with the asymptotes derived from the analytical solution of Nitao and Buscheck (1991). See text, and Zimmerman *et al.* (1993), for further details. Source: Adapted from Zimmerman *et al.* (1993).

which is equal to the flux that would occur in a porous medium whose permeability is due only to the fracture network, but whose storativity is due to both the fractures and the matrix blocks. In the intermediate-time regime, which is dominated by highly transient flow interactions between the fractures and matrix blocks, the boundary influx is given by

$$q(t) = (p_o - p_i)\left(\frac{144\phi_m k_m k_f^2 c_f^2}{\pi^3 \mu^3 c_m a^2 t}\right)^{1/4} = 2.14\times 10^{-7}\, t^{-1/4}\ \text{m/s}. \tag{9.71}$$

As can be seen in Fig. 9.5b, the nonlinear dual-porosity model provides accurate results over all three-time regimes. The flux obtained using the Warren–Root model is inaccurate during the intermediate regime when imbibition from the fractures into the matrix starts to become important. Since the Warren–Root model underestimates the flow from the fractures into the matrix blocks during the initial stage of fracture-to-matrix flow, as illustrated in Fig. 9.4, it consequently underestimates the total flux into the formation and predicts a boundary flux that tracks the early-time Nitao–Buscheck asymptote for too long a time. Eventually, during the intermediate regime, the Warren–Root model predicts a roughly constant flux into the formation, which can be observed in the graph between about 10^4 and 10^5 seconds. In this regime, the analytical solution, and the nonlinear transfer function model, both predict a boundary flux that decreases according to $t^{-1/4}$. As expected based on the discussion given in Section 9.6, the Warren–Root model becomes accurate again in the late-time regime.

9.8 Multi-Phase Flow, Gravity Effects, and Other Extensions

Although all preceding sections of this chapter have focused on fractured-porous media that contain only a single fluid phase, the dual-porosity concept has also been widely used to model multi-phase flow processes. The simplest example is the case of “unsaturated flow” in air–water systems. In subsurface rocks or soils located above the water table, the pore space is typically filled with a mixture of water and air. The fraction of the pore space that is filled with water is denoted by the (water) saturation, s. The difference in pressure between the water phase and the air phase is described by a *capillary pressure function*, which is a material property that depends on s.

In general, both the air and water phases will be able to flow in response to pressure gradients, with Darcy's law for each phase modified by the inclusion of a *relative permeability function*, $0 \leq k_r(s) \leq 1$ (see Chapter 12). In the commonly used Richards' model (Hillel, 1980), the air phase is assumed to be "infinitely" mobile, due to its very low viscosity. Consequently, the pressure in the air phase is assumed to be uniform (or determined by hydrostatic equilibrium), and a flow equation for the air phase is not needed. This assumption leads to flow equations for the water phase that are similar to the equations for single-phase water flow, except that they are intrinsically nonlinear, due to the nonlinear dependence of the relative permeability and capillary pressure functions on the saturation.

Although such a highly nonlinear model does not lend itself to analytical solutions of boundary value problems, such as flow to a well, flow into a half-space, *etc.*, dual-porosity models based on the Richards' equation can be implemented numerically. Gerke and van Genuchten (1993a,b) presented a dual-porosity model for unsaturated flow, and examined different methods of computing the effective permeability to be used in the fracture-matrix transfer function. Zimmerman *et al.* (1996) extended the single-phase nonlinear dual-porosity model of Zimmerman *et al.* (1993) to unsaturated flow and validated the predictions of their new model against fine-grid simulations of vertical and horizontal infiltration, in which the matrix blocks were explicitly discretized. Lewandowska *et al.* (2005) presented a computational dual-porosity model for unsaturated flow and validated it against laboratory experiments on vertical infiltration under gravity.

Dual-porosity models are frequently used to model the two-phase flow of oil and water in petroleum reservoirs. Kazemi *et al.* (1976) generalized the Warren–Root transfer function in an obvious way, by writing separate flow terms for oil and water phases, and incorporating the appropriate relative permeability factor for each phase. Gilman and Kazemi (1983) incorporated gravity effects by adding a gravity-related component of the driving force to the fracture/matrix pressure difference. Quandalle and Sabathier (1989) improved this model by treating vertical flow and horizontal flow differently. Lu *et al.* (2008) developed a "general transfer function" that accounts for fluid expansion, capillary imbibition, gravity drainage, and diffusion, using functional forms that reduce to the correct behavior in the early-time and late-time regimes. Abushaikha and Gosselin (2008) provided a detailed review of these various models, along with numerous validation simulations. van Heel *et al.* (2008) presented a fully transient dual-porosity model that accounted for gravity and thermal effects and used it to study gas/oil gravity drainage during steam injection. Lu and Connell (2007) developed a dual-porosity model for gas production from coal seams that accounted for the adsorption/desorption of gas on the outer surface of the matrix blocks. Huyakorn *et al.* (1994) developed a numerical dual-porosity model for three-phase flow problems involving water, air, and a nonaqueous-phase liquid for application to contaminant remediation.

The dual-porosity models discussed above each assume that all of the matrix blocks in the rock mass, or at least in a given computational gridblock, have the same shape factor. Particularly in light of the fact that the "shape factor" actually reflects the *size* of the matrix block more than its geometric "shape" (see Chapter 10), it is probably more realistic to assume the existence of a distribution of matrix block sizes and therefore a distribution of shape factors. Although it would be convenient if a distribution of block sizes and shapes could be rigorously represented by an "equivalent" shape factor, this is unfortunately not the case (Zimmerman and Bodvarsson, 1995; Lu and Connell, 2011). The simplest way to understand the difficulty that arises is to consider a fractured/porous medium containing spherical matrix blocks of two different sizes, each having a mean pressure described by eq. (9.64). The average matrix pressure, computed as a volume-weighted average of these two exponential functions, cannot be represented by a single exponential function, regardless of the choice of the equivalent radius.

Lu and Connell (2011) developed a computational dual-porosity model in which the matrix blocks were essentially assumed to have the same shape, but a distribution of sizes, and considered in detail the specific cases in which the block sizes were distributed according to a uniform distribution and a gamma distribution. Ding (2019) extended the MINC method (Pruess and Narashimhan, 1985) so that it can be used for reservoirs having a distribution of block sizes, and studied different methods for creating the nested shells that are needed in the discretized matrix blocks. Cinco-Ley *et al.* (1985) investigated the problem of flow to a vertical well in a dual-porosity reservoir containing a distribution of matrix block sizes, using both the Warren–Root model and the fully transient model for matrix-fracture flow. They concluded that the matrix block size that would be inferred from well-test analysis corresponds to a volume-weighted harmonic mean of the individual block sizes, whereas the inferred value of λ (see eq. (9.35)) would correspond to a volume-weighted arithmetic mean of the individual block sizes.

In some fractured media, mineralization along the faces of the fracture may lead to a mechanical "skin" that impedes the flow of fluid into or out of the matrix block. Moench (1984) developed an analytical dual-porosity model that accounted for a thin, low-permeability skin region surrounding each matrix block. If the fracture skin permeability is very low, most of the pressure drop during matrix-fracture flow occurs across the fracture skin, and the matrix block itself will be nearly in a state of uniform pressure – thereby essentially satisfying an implicit assumption of the Warren–Root approach. Furthermore, by replacing the matrix permeability with the skin zone permeability in the definition of λ in eq. (9.35), Moench found that the new model coincides with the original Warren–Root model. This analysis provides a rational justification for the use of the Warren–Root model for fractured-porous reservoirs containing fracture skin.

Problems

9.1 Supply all of the steps required to transform the governing equations for flow to a vertical well in a dual-porosity reservoir, eqs. (9.25) and (9.26), into dimensionless form, eqs. (9.36) and (9.37).

9.2 Starting from the definition of the Ei function given by eq. (9.39), prove that, for any positive value of x, $-Ei(-x) < e^{-x}/x$. Then, show that for $x > 1$, it follows that $-Ei(-x) < e^{-x}$, and so $-Ei(-x)$ goes to zero exponentially fast as $x \to \infty$.

9.3 Describe and sketch the way that the drawdown curve in Fig. 9.3b would change if the storativity ratio ω increased (or decreased) by a factor of 10, and if the transmissivity ratio λ increased (or decreased) by a factor of 10.

9.4 Imagine a vertical well of radius 0.1, producing water at a rate of 0.001 m^3/s, from an unbounded, 15 m thick, naturally fractured aquifer. The water has a viscosity of 1×10^{-3} Pa s. The fracture permeability is 100 mD, the matrix permeability is 0.1 mD, the fracture porosity is 0.002, the matrix porosity is 0.2, and total compressibility of the fracture system and matrix system is, in each case, 10×10^{-6}/psi. The matrix blocks can be treated as cubes of length 1 m. Plot the drawdown at the well, as a function of time, starting at one hour after the start of production, and continuing until 100 days after the start of production.

9.5 For the problem of a unit-step increase in the pressure at the outer boundary of a *spherical* matrix block of radius a, the eigenvalues are $\lambda_n = n^2\pi^2/a^2$, the eigenfunctions are $F_n(r) = (a/r)\sin(n\pi r/a)$, and the Fourier coefficients are $C_n = -2(-1)^n/n\pi$ (Najurieta, 1980). Using these results, along with eq. (9.60), find an explicit expression for $d\overline{p}_m/dt$. The resulting expression should be consistent with eq. (9.38) of Carslaw and Jaeger (1959), which states that

$$\overline{p}_m(t) = 1 - \frac{6}{\pi^2}\sum_{n=1}^{\infty}\frac{\exp(-n^2\pi^2 k_m t/\phi_m \mu c_m a^2)}{n^2}$$

References

Abushaikha, A. S. A. and Gosselin, O. R. 2008. Matrix-fracture transfer function in dual-porosity medium flow simulation: review, comparison, and validation, in *SPE Europec Annual Conference and Exhibition*, Rome, 9–12 June 2008, Paper SPE 113890.

Allaire, G. 1992. Homogenization and two-scale convergence. *SIAM Journal of Mathematical Analysis*, 23(6), 1482–1518.

Barenblatt, G. I., Zheltov, Iu. P., and Kochina, I. N. 1960. Basic concepts in the theory of seepage of homogeneous liquids in fissured rocks (strata). *Journal of Applied Mathematics and Mechanics*, 24(5), 852–64.

Bear, J. 1988. *Dynamics of Fluids in Porous Media*, Dover, Mineola, N.Y.

Bourdet, D. and Gringarten, A. C. 1980. Determination of fissure volume and block size in fractured reservoirs by type-curve analysis, in *SPE Annual Fall Technical Conference and Exhibition*, Dallas, 21–24 September 1980, Paper SPE 9293.

Carslaw, H. S., and Jaeger, J. C. 1959. *Conduction of Heat in Solids*, 2nd ed., Clarendon Press, Oxford.

Cinco-Ley, H., Samaniego, F., and Kucuk, F. 1985. The pressure transient behavior for naturally fractured reservoirs with multiple block sizes, in *SPE Fall Annual Technical Conference and Exhibition*, Las Vegas, 22–25 September 1985, Paper SPE 14168.

Courant, R., and Hilbert, D. 1953. *Methods of Mathematical Physics*, vol. 1, Interscience Publishers, New York.

Crank, J. 1975. *The Mathematics of Diffusion*, Clarendon Press, Oxford.

da Prat, G. 1990. *Well Test Analysis for Fractured Reservoir Evaluation*, Elsevier, Amsterdam.

de Marsily, G. 1986. *Quantitative Hydrogeology*, Academic Press, San Diego.

de Swaan, A. 1976. Analytic solution for determining naturally fractured reservoir properties by well testing. *Society of Petroleum Engineers Journal*, 16(2), 117–22.

Ding, D. Y. 2019. Modeling of matrix/fracture transfer with nonuniform-block distributions in low-permeability fractured reservoirs, *Society of Petroleum Engineers Journal*, 24(6), 2653–70.

Gerke, H. H., and van Genuchten, M. T. 1993a. A dual-porosity model for simulating the preferential movement of water and solutes in structured porous media. *Water Resources Research*, 29(2), 305–19.

Gerke, H. H., and van Genuchten, M. T. 1993b. Evaluation of a first-order water transfer term for variably saturated dual-porosity flow models. *Water Resources Research*, 29(4), 1225–38.

Gilman J. R., and Kazemi, H. 1983. Improvement in simulation of naturally fractured reservoirs. *Society of Petroleum Engineers Journal*, 23(4), 695–707.

Hillel, D. 1980. *Fundamentals of Soil Physics*, Academic Press, New York.

Huyakorn, P. S., Panday, S., and Wu, Y. S. 1994. A three-dimensional multiphase flow model for assessing NAPL contamination in porous and fractured media, 1. Formulation. *Journal of Contaminant Hydrology*, 16, 109–30.

Kazemi, H., Merrill, L. S., Porterfield, K. L., and Zeman, P. R. 1976. Numerical simulation of water-oil flow in naturally fractured reservoirs. *Society of Petroleum Engineers Journal*, 16(4), 317–26.

Lewandowska, J., Szymkiewicz, A., Gorczewska, W., and Vauclin, M. 2005. Infiltration in a double-porosity medium: Experiments and comparison with a theoretical model, *Water Resources Research*, 41, W02022.

Lu, H. Y., Di Donato, G., and Blunt, M. J. 2008. General transfer functions for multiphase flow in fractured reservoirs. *Society of Petroleum Engineers Journal*, 13(3), 289–97.

Lu, M., and Connell, L. D. 2007. A dual-porosity model for gas reservoir flow incorporating adsorption behaviour—part I. Theoretical development and asymptotic analyses. *Transport in Porous Media*, 68, 153–73.

Lu, M., and Connell, L. D. 2011. A statistical representation of the matrix–fracture transfer function for porous media. *Transport in Porous Media*, 86, 777–803.

Matthews, C. S., and Russell, D. G. 1967. *Pressure Buildup and Flow Tests in Wells*, Society of Petroleum Engineers, Dallas.

Moench, A. F. 1984. Double-porosity models for a fissured groundwater reservoir with fracture skin. *Water Resources Research*, 20(7), 831–46.

Najurieta, H. L. 1980. A theory for pressure transient analysis in naturally fractured reservoirs. *Journal of Petroleum Technology*, 269, 1241–50.

Nitao, J. J., and Buscheck, T. A. 1991. Infiltration of a liquid front in an unsaturated, fractured porous medium. *Water Resources Research*, 27(8), 2099–2112.

Pruess, K. 1987. *TOUGH User's Guide. Report LBL-20700*, Lawrence Berkeley National Laboratory, Berkeley, CA.

Pruess, K., and Narashimhan, T. N. 1985. A practical method for modelling fluid and heat flow in fractured porous media. *Society of Petroleum Engineers Journal*, 25(1), 14–26.

Quandalle, P., and Sabathier, J. C. 1989. Typical features of a multipurpose reservoir simulator. *SPE Reservoir Engineering*, 4(4), 75–80.

Rasoulzadeh, M., and Kuchuk, F. J. 2019. Pressure transient behavior of high-fracture-density reservoirs (dual-porosity models). *Transport in Porous Media*, 121(3), 901–40.

Streltsova, T. D. 1988. *Well Testing in Heterogeneous Formations*, Wiley, New York.

van Heel, A. P. G., Boerrigter, P. M., and van Dorp, J. J. 2008. Thermal and hydraulic matrix-fracture interaction in dual-permeability simulation. *SPE Reservoir Engineering and Evaluation*, 11(4), 735–49.

Vermeulen, T. 1953. Theory of irreversible and constant-pattern solid diffusion. *Industrial and Engineering Chemistry*, 45(8), 1644–70.

Warren J. E., and Root, P. J. 1963. The behaviour of naturally fractured reservoirs. *Society of Petroleum Engineers Journal*, 3(3), 245–55.

Zimmerman, R. W. 2018. *Fluid Flow in Porous Media*, World Scientific, Singapore and London.

Zimmerman, R. W., and Bodvarsson, G. S. 1995. Effective block size for imbibition and absorption in dual-porosity media. *Geophysical Research Letters*, 22(11), 1461–64.

Zimmerman, R. W., Hadgu, T., and Bodvarsson, G. S. 1996. A new lumped-parameter model for flow in unsaturated dual-porosity media. *Advances in Water Resources*, 19(5), 317–27.

Zimmerman, R. W., Chen, G., Hadgu, T., and Bodvarsson, G. S. 1993. A numerical dual-porosity model with semi-analytical treatment of fracture/matrix flow. *Water Resources Research*, 29(7), 2127–37.

10

Matrix Block Shape Factors

10.1 Introduction

The traditional dual-porosity model introduced by Barenblatt *et al.* (1960) and Warren and Root (1963) assumes that the volumetric fluid flow from the matrix blocks to the fracture network, per unit volume of matrix block, can be described by the following linear expression:

$$q = \frac{k_m \alpha}{\mu}(\overline{p}_m - p_f). \tag{10.1}$$

This expression is equivalent to eq. (9.17), in which the flux, denoted there by q_{mf}, was normalized with respect to the bulk (matrix plus fracture) volume. For the purposes of this chapter, it is more convenient to normalize the flux against the matrix block volume, although in practice these two volumes are always very nearly equal to each other, differing only by a multiplicative factor of $(1 - \phi_f)$. The traditional dual-porosity model therefore requires an estimate of the matrix block shape factor, α. Although in well-test analysis the shape factor is often subsumed into the transmissivity ratio parameter λ, as defined in eq. (9.35), which is then treated as a fitting parameter, a full physical interpretation of the results requires at least an approximate knowledge of the relationship between α and the geometry of the matrix blocks. The nonlinear model for matrix-fracture flow introduced by Zimmerman *et al.* (1993), in which eq. (10.1) is replaced by a more accurate interaction equation, which also involves the same shape factor parameter. Hence, implementation of either of these dual-porosity models, and the proper interpretation of their predictions, require an understanding of the relationship between the geometry of the matrix blocks and the numerical value of the shape factor.

Various approaches that have been proposed to estimate the shape factor of a given matrix block are discussed in Section 10.2. For the commonly applied approach in which the shape factor is chosen so that the mean pressure in the matrix block as predicted by the Warren–Root model agrees with the exact pressure distribution at late times, for a block with a constant-pressure boundary condition, some exact results, general theorems, and upper/lower bounds are discussed in Sections 10.3 and 10.4. For block shapes for which analytical results are not available, a method to compute the shape factor using a numerical simulator is presented in Section 10.5. Approximate scaling laws to estimate

Fluid Flow in Fractured Rocks, First Edition. Robert W. Zimmerman and Adriana Paluszny.

Companion website: www.wiley.com/go/zimmerman/fluidflowinfracturedrocks

the shape factor based on simple geometrical attributes such as surface area and volume are presented in Section 10.6. Some researchers have suggested that the shape factor should be computed based on the problem of constant flux out of the matrix block rather than constant pressure at the outer boundary. This approach is discussed in detail in Section 10.7, along with a few specific results. Finally, an analytical expression for the constant-flux shape factor of a general brick-like matrix block, having sides of length $\{L_x, L_y, L_z\}$, is presented in Section 10.8.

10.2 Approaches to Choosing the Shape Factor

In the paper that introduced the dual-porosity model, Barenblatt *et al.* (1960) did not discuss the numerical value of the shape factor, aside from pointing out that it was inversely proportional to the square of some "characteristic length" of the matrix block. Since then, numerous equations for the shape factor have been proposed for blocks of various shapes. The most commonly assumed shape has been a cube of length L. For cubical blocks, Warren and Root (1963) suggested $\alpha = 60/L^2$, without offering any derivation. (Unfortunately, Fig. 1b of Warren and Root (1963) seems to have led to a common misunderstanding that the dual-porosity model requires or assumes a "sugar-cube" matrix block geometry, which is not true). Kazemi *et al.* (1976) suggested $\alpha = 12/L^2$, based on a finite-difference approximation to the flow equations, with the entire matrix block represented by a single finite-difference cell. Coats (1989) derived the value $\alpha = 24/L^2$, starting from the claim that, at sufficiently long times, "$\partial P/\partial t$ is independent of position within the matrix." This assumption has the effect of converting the governing equation for pressure within the matrix block into a steady-state problem with a constant "source term." However, this starting assumption may be correct for some reservoir-scale processes, but not for all.

de Swaan (1990) suggested that α should be chosen so that the Warren–Root model will correctly predict the time at which the mean pressure change in the matrix block attains 50% of its eventual value in the case of a step-change in the fracture pressure. He thereby found, for cubical blocks, that this criterion leads to the same shape factor as found by Warren and Root, $\alpha = 60/L^2$, but leads to a different shape factor than that proposed by Warren and Root for a slab-like block: $15/L^2$, instead of $12/L^2$.

Gerke and van Genuchten (1993) and Landereau *et al.* (2001) analyzed the problem of flow into a cubical matrix block using Laplace transforms and concluded that $\alpha = 49.62/L^2$. This result was obtained by starting with the basic problem of a step-function change in the fracture pressure, and then assuming that the late-time behavior of the pressure in the matrix block corresponds to its behavior in the Laplace domain as the Laplace variable s goes to zero. Although this assumption has been made frequently in the hydrology literature, in general, it is incorrect (see Carslaw and Jaeger, 1949, p. 280; Doetsch, 1961, p. 213; Chen and Stone, 1993). The late-time behavior of a function is actually governed by the behavior of its Laplace transform in the neighborhood of the pole in the complex plane that has the largest real part, which in the problem of a step-function change in the pressure at the outer boundary of the matrix block does *not* occur at $s = 0$ (see Mathias and Zimmerman, 2003).

A few researchers have studied the shape factor for the case of constant flux at the outer boundary of the matrix block rather than constant pressure at the outer boundary (Coats, 1989; Hassanzadeh and Pooladi-Darvish, 2006; Mora and Wattenbarger, 2009). Taking this approach, Coats (1989) found a value of $\alpha = 49.58/L^2$ for a cubical block, and Mora and Wattenbarger (2009) computed $\alpha = 49.48/L^2$. These values differ slightly due to numerical roundoff and essentially agree with the values obtained by Gerke and van Genuchten (1993) and Landereau *et al.* (2001). But although a "constant flux" boundary condition can be applied at the wellbore in an infinite reservoir, it cannot pertain indefinitely to the *depletion* of fluid from a finite-sized matrix block. Consequently, most researchers have, either implicitly or explicitly, assumed that the constant pressure boundary condition *for individual matrix blocks* is the most appropriate one to use when constructing dual-porosity models.

The wide range of the values that have been proposed for the shape factor of a cubical block can in part be attributed to the fact that eq. (10.1) for the flow between the fractures and matrix blocks is highly inaccurate at early times (de Swaan, 1976). Hence, since eq. (10.1) is not valid over all times, there is no obvious objective criterion for defining the most appropriate value of α. Zimmerman *et al.* (1993) suggested that α should be chosen so as to render eq. (10.1) asymptotically accurate at long times for the case in which the fracture pressure undergoes an instantaneous "step-change" by some finite amount. This criterion for choosing α has subsequently been used by Lim and Aziz (1995), Mora and Wattenbarger (2009), and others. As shown by Zimmerman and Bodvarsson (1995), this definition is equivalent to choosing α to be the smallest eigenvalue of the Laplacian operator in the interior of the matrix block under constant pressure boundary conditions (known as "Dirichlet" boundary conditions in the mathematics literature).

The identification of the shape factor with the smallest eigenvalue of the Laplace operator under constant-pressure boundary conditions provides a consistent framework in which to study this problem and allows various known exact results, general theorems, and upper/lower bounds to be taken from the mathematics and physics literature. However, two pertinent facts must be borne in mind. Firstly, no choice of the shape factor will eliminate the defect that the Warren–Root model is grossly inaccurate at early times. Secondly, in the context of the Warren–Root model, no single choice of shape factor for a matrix block of a given size and shape is optimal for all possible boundary conditions (see Hassanzadeh and Pooladi-Darvish, 2006).

As an example, consider a spherical matrix block of radius a that is initially in equilibrium with the surrounding fractures, at some pressure p_i, after which the fracture pressure at the outer boundary of the matrix block decreases linearly with time, according to $p_f = p_i - Bt$, where the constant B has dimensions of pressure/time. Although spherical matrix blocks do not provide a very physically realistic model for fractured rocks, they are highly relevant to soils, for which the "matrix block" is a soil aggregate surrounded by macropores (van Genuchten and Dalton, 1986). Additionally, the large number of existing mathematical solutions to the diffusion equation in spherical coordinates makes this geometry very mathematically convenient. The boundary condition of a linearly increasing pressure can be viewed as the opposite "extreme case" of the step-function boundary condition, in which the pressure first changes abruptly and then remains constant. The exact solution for the mean matrix block pressure in the case of a linearly increasing boundary pressure is (Crank, 1975, p. 93)

$$\overline{p}_m(t) = p_i - Bt + \frac{\phi_m \mu c_m a^2 B}{15k_m} - \frac{6}{\pi^4}\sum_{n=1}^{\infty}\frac{1}{n^4}\exp(-n^2\pi^2 k_m t/\phi_m \mu c_m a^2). \tag{10.2}$$

At sufficiently late times, all of the exponential terms will have died out, and the mean matrix block pressure will be given by

$$\overline{p}_m(t) = p_i - Bt + \frac{\phi_m \mu c_m a^2 B}{15k_m}. \tag{10.3}$$

The first two terms on the right side are equal to the fracture pressure, and so the pressure difference is $\overline{p}_m - p_f = \phi_m \mu c_m a^2 B/15k_m$. The time derivative of the matrix pressure is $d\overline{p}_m/dt = -B$, and so in this problem, the late-time behavior corresponds to constant flux out of the matrix block rather than constant pressure at the outer boundary. Combining the two previous equations yields $d\overline{p}_m/dt = 15k_m(p_f - \overline{p}_m)/\phi_m \mu c_m a^2$. Imposing the condition that this equation should agree with eq. (9.26), which governs the evolution of the matrix pressure according to the Warren–Root model, leads to the conclusion that the shape factor should be chosen to be $\alpha = 15/a^2$. This same result was found by Hassanzadeh and Pooladi-Darvish (2006), by numerically inverting the solution for pressure evolution in the matrix block that was obtained using Laplace transforms and examining the late-time behavior of the fracture-matrix flux. These authors examined numerous different boundary conditions and found that if the fracture pressure changes abruptly and/or rapidly, the appropriate shape factor is π^2/a^2, whereas if the fracture pressure changes slowly and/or continually, the appropriate shape factor is $15/a^2$.

10.3 Some Specific Results and General Theorems

By Identifying the shape factor with the smallest eigenvalue of the Laplacian operator, Zimmerman *et al.* (1993) derived the result $\alpha = 3\pi^2/L^2$ for a cubical block. This value was confirmed by Lim and Aziz (1995) and was subsequently re-derived by Mathias and Zimmerman (2003) in the Laplace domain.

Using this definition of the matrix block shape factor, values can be found for a few other simple but important shapes (Zimmerman *et al.*, 1993; Lim and Aziz, 1995; Zimmerman and Bodvarsson, 1995; Mora and Wattenbarger, 2009). These values include $3\pi^2/L^2$ for a cube of side L, $2\pi^2/L^2$ for an infinitely long "matchstick-shaped" prism having two sides of length L, and π^2/L^2 for a sheet-like matrix block having thickness L. These values are all special cases of the general expression for a brick-like block having sides L_x, L_y, and L_z (Lim and Aziz, 1995):

$$\alpha = \pi^2\left(\frac{1}{L_x^2} + \frac{1}{L_y^2} + \frac{1}{L_z^2}\right). \tag{10.4}$$

The shape factor of a sphere of radius a is π^2/a^2, and the shape factor of an infinitely long prismatic block having a circular cross-section of radius a is $(2.405)^2/a^2 = 5.784/a^2$, where 2.405 is the first zero of J_0, the Bessel function of the first kind, of order 0 (Zimmerman *et al.*, 1993).

Based on the identification of the shape factor of a block of a given geometry as being the smallest eigenvalue of the Laplacian operator in that region, under conditions of constant pressure at the outer boundary (which can mathematically be taken to be zero), some general results can be obtained by appealing to theorems available in the mathematics literature. For example, if the shape factor for an infinitely long "matchstick-shaped" block having a given cross-sectional shape is denoted by α_{2D}, the shape factor of a block having this same cross-sectional shape, but a *finite* length L, will be given *exactly* by Gottlieb (1988)

$$\alpha_{3D} = \alpha_{2D} + \frac{\pi^2}{L^2}. \tag{10.5}$$

As an example of the use of this theorem, consider an infinitely long block having a square cross-section with sides l and l. This block has a shape factor of $2\pi^2/l^2$, and eq. (10.5) then correctly predicts that a cube of side l has a shape factor of $3\pi^2/l^2$. At the other extreme, by letting the block length L become very small, it follows from eq. (10.5) that the term π^2/L^2 will eventually dominate the term α_{2D}, thereby showing that a thin disk-like matrix block of thickness L has a shape factor of π^2/L^2, *regardless* of its cross-sectional shape in the plane normal to the thin direction.

Since the problem of calculating the eigenvalues of the Laplacian operator arises in numerous areas of applied mathematics, it might be thought that shape factors for blocks of many other shapes could easily be found in the mathematics literature. Unfortunately, although there is a vast literature on the asymptotic distribution of the *larger* eigenvalues (*i.e.*, Kac, 1966), there are surprisingly few exact results for the *smallest* eigenvalue for a given shape. In fact, the only other two-dimensional shapes for which exact results seem to be available are an equilateral triangle having three sides of length L, for which $\alpha = 16\pi^2/3L^2$ (Pinsky, 1980; Práger, 1998; Siudeja, 2007), and an isosceles right triangle having two sides of length L, for which $\alpha = 5\pi^2/L^2$ (Gottlieb, 1988).

10.4 Upper and Lower Bounds on the Shape Factor

Although exact results are not available for general polygonal-shaped matrix blocks in either two or three dimensions, useful upper and lower bounds are available. Specifically, the first eigenvalue of the Laplacian operator in a given region, with zero boundary conditions, must be smaller than the first eigenvalue of any sub-domain of that region (Courant and Hilbert, 1953, p. 409). Hence, the shape factor of a given matrix block is always less than the shape factor of any region that can be inscribed within the block and greater than the shape factor of any region that circumscribes the block. For a three-dimensional matrix block, a useful set of easily computable bounds can therefore be found by considering the *largest* sphere that can be *inscribed* within the block and the *smallest* sphere that *circumscribes* the block. For a two-dimensional block (*i.e.*, a long prismatic block with a uniform cross-section), the analogous bounds can be found by considering the largest circle that can be inscribed within the block and the smallest circle that circumscribes the block.

Another useful 2D theorem concerning the eigenvalues of the Laplacian operator under zero boundary conditions, first conjectured in 1877 by the physicist Rayleigh and proven in the 1920s by the mathematicians Faber and Krahn (see Szegö, 1954), is that, among all shapes having area A, the circle has the largest first (*i.e.*, smallest) eigenvalue. Since the area of a circle of radius a is $A = \pi a^2$, and the shape factor of that circle is $5.784/a^2$, this theorem implies that for any noncircular block of area A, a lower bound for the shape factor is $\alpha > 5.784\pi/A$. In most cases, this bound is more stringent, and therefore more useful, than the lower bound that can be obtained from the smallest circle that circumscribes the block.

As an example of the use of these bounds, consider a two-dimensional square matrix block of length L. The largest circle that can be inscribed within this square would have a radius $L/2$. Since the shape factor of a circle of radius a is $5.784/a^2$, the shape factor of this square must be less than $5.784/(L/2)^2 = 23.14/L^2$. On the other hand, the smallest circle that can be circumscribed around this square would have a radius $L/\sqrt{2}$, and so the shape factor of the square must be greater than $5.784/(L/\sqrt{2})^2 = 11.57/L^2$. Finally, since the area of the square is L^2, the Faber–Krahn bound shows that $\alpha > 5.784\pi/L^2 = 18.17/L^2$. This lower bound is much sharper than the "circumscribed circle" lower bound, $11.57/L^2$. Hence, it can be concluded that $18.17/L^2 < \alpha < 23.14/L^2$. The exact value for a square is $2\pi^2/L^2 = 19.74/L^2$, which indeed satisfies these bounds. The arithmetic mean of the inscribed-circle upper bound and the Faber–Krahn lower bound is $20.66/L^2$, which is only 5% larger than the exact value.

These bounds each have very simple physical explanations. In the case of a step-function change in the pressure in the fracture that surrounds the matrix block, the matrix block pressure eventually decays exponentially, as shown in eq. (9.51). The time constant for the pressure decay, which is to say the time constant for fluid depletion, is therefore inversely proportional to α, and so *smaller* values of α correspond to *slower* depletion. If a block can be inscribed within another block, this inscribed block will obviously drain faster and therefore have a larger shape factor. Conversely, if a block can be circumscribed around another block, this circumscribed block will obviously drain more slowly and therefore have a smaller shape factor.

The Rayleigh bound, on the other hand, follows heuristically from the observation that the "last drop" of fluid that will drain from a circular block of radius a will drain from the center, since this point is the farthest from any point on the boundary of the block. If this circular shape is distorted in any way, while maintaining the same total area, every point within the distorted block will be no farther than a from some point on the boundary, and so this distorted block will presumably drain faster than the original circular block and therefore will have a larger shape factor.

10.5 Methodology for Numerical Calculation of the Shape Factor

Equation (10.1) gives the volumetric flowrate of fluid out of a matrix block according to the standard Warren–Root model. By performing a simple mass balance on the matrix block, this flowrate can be equated to the change in the mean pressure inside the block, yielding the following differential equation for the mean pressure, eq. (9.26), repeated here for convenience:

$$\phi_m c_m \frac{d\overline{p}_m}{dt} = \frac{\alpha k_m}{\mu}(p_f - \overline{p}_m). \tag{10.6}$$

If the matrix block is initially at some uniform pressure p_i, the change in the (mean) matrix pressure caused by instantaneously changing the fracture pressure to some constant value $p_f \neq p_i$ can be found by solving eq. (10.6) subject to the initial condition

$$\overline{p}_m(t = 0) = p_i, \tag{10.7}$$

which yields (Streltsova, 1988)

$$\overline{p}_m(t) = p_f + (p_i - p_f)\exp\left(\frac{-\alpha k_m t}{\phi_m \mu c_m}\right). \tag{10.8}$$

Hence, according to the Warren–Root model, the mean matrix block pressure will decay exponentially, from p_i to p_f.

The mean pressure in the matrix block, with initial uniform pressure p_i and a constant pressure p_f maintained at its outer boundary, can also be computed *exactly* by solving the full pressure diffusion equation inside the matrix block (eq. 9.48) to find the local pressure distribution $p_m(\mathbf{x}, t)$ and then averaging the pressure over the entire block. The governing equation and boundary/initial conditions for this exact model are

$$\frac{\partial p_m(\mathbf{x}, t)}{\partial t} = \frac{k_m}{\phi_m \mu c_m}\nabla^2 p_m(\mathbf{x}, t), \tag{10.9}$$

$$p_m(\mathbf{x} \in \partial v_m, t > 0) = p_f, \tag{10.10}$$

$$p(\mathbf{x}, t = 0) = p_i, \tag{10.11}$$

where ∂v_m denotes the outer boundary of the matrix block, and $\mathbf{x} \in \partial v_m$ denotes points on the boundary of the block.

For matrix block shapes for which an analytical solution is *not* available, this problem can be solved using a numerical simulator with a sufficiently fine grid (see Rostami *et al.*, 2020). The mean pressure $\overline{p}_m$ in the matrix block can be computed at each time step by averaging the pressures over all the numerical cells within the matrix block. By requiring the Warren–Root model and the "exact" solution to asymptotically agree at large times, the numerically computed normalized mean pressure $[\overline{p}_m(t) - p_f]/[p_i - p_f]$, at *late times*, can be fitted to an exponential of the form $A\exp(-\beta t)$, where A and β are fitting constants. Comparison with eq. (10.8) shows that $\beta = \alpha k_m/\phi_m \mu c_m$, from which it follows that the shape factor is given by

$$\alpha = \frac{(\phi_m \mu c_m)\beta}{k_m}. \tag{10.12}$$

Wuthicharn and Zimmerman (2011) used this approach to compute shape factors for several two- and three-dimensional block shapes, using the black-oil simulator Eclipse-100 to conduct the required fine-grid simulations. As an example, consider a two-dimensional square matrix block of length $L = 1000$ ft (=304.8 m), initially saturated with water at 900 psi, surrounded by a fracture saturated with water at a pressure of 1000 psi.

The pore fluid was taken to have a viscosity of 1 cP, and a compressibility of 3.40×10^{-6}/psi. The matrix block was discretized into a 10×10 square grid. In order to simulate the constant-pressure boundary condition at the outer boundary of the matrix block, the matrix cells were surrounded by a ring of "fracture" cells that have a very high permeability (1000 mD), so as to maintain the "fracture" at a uniform pressure; see Mora and Wattenbarger (2009). The porosity of the fracture cells was chosen to be very high (0.9), and the porosity of the matrix block was taken to be much lower (0.001), so that the total pore volume of the "fracture" was much larger than that of the matrix block. This helped to ensure that the fracture would remain (nearly) at constant pressure as fluid flowed from the fracture into the matrix block. The matrix permeability was taken to be 1 mD, and the formation pore compressibility was taken to be 3.40×10^{-6}/psi. The grid had a thickness of one cell in the third direction, so as to maintain two-dimensional flow.

For this scenario, the pore fluid flows from the ring of fracture cells into the matrix block until the two regions reach pressure equilibration at a final equilibrium pressure of 999.75 psi. This represents a change in the fracture pressure of less than 0.03%, confirming that the fracture cells did indeed essentially act as a constant-pressure boundary condition. The evolution of local cell pressures as a function of time is shown in Fig. 10.1, where it can be seen that the pressure diffuses very rapidly at early times, and the pressure profile is symmetric, as expected.

The mean matrix pressure is plotted as a function of time in Fig. 10.2a. In Fig. 10.2b, the mean pressure is plotted in normalized form, as in eq. (10.8), along with the exponential function that provides the best fit to the computed values at large times. From the best-fitting value of β = 16.886/day = 1.95×10^{-4}/sec, and using the values $\phi_m = 0.001$, μ = 1 cP = 0.001 Pa s, $c_m = c_p + c_F = 6.80\times10^{-6}$/psi = 9.86×10^{-10}/Pa, and k_m = 1 mD = 0.986×10^{-15} m^2, eq. (10.12) yields a shape factor of $\alpha = 1.95\times10^{-4}$ m^2, or $18.10/L^2$. This differs from the theoretical value of $2\pi^2/L^2 = 19.74/L^2$, due to the effect of grid coarseness. The theoretical value can be approached more closely by using a series of finer $N\times N$ meshes, calculating the shape factor for each value of N, plotting the shape factors against $1/N$, and extrapolating to the limit as $N\to\infty$, *i.e.*, $1/N\to0$, as explained by Yeo and Zimmerman (2000). The extrapolated value of $19.63/L^2$ obtained by Wuthicharn and Zimmerman (2011) agrees with the theoretical value of $19.74/L^2$ to within better than 1%. Similarly accurate results were obtained for the other two-dimensional shapes for which a theoretical result is available: a circle, an isosceles triangle, and a right

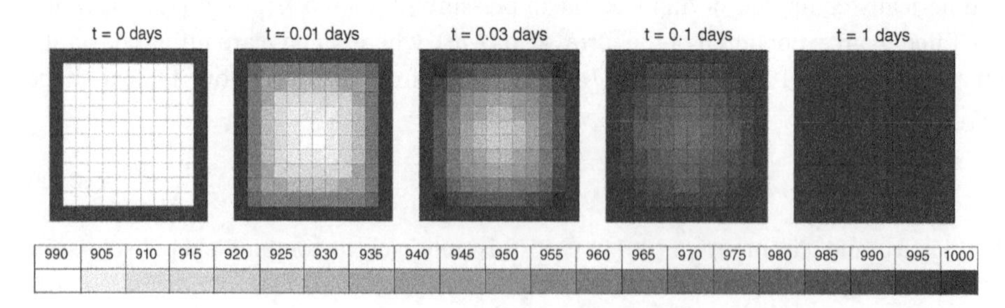

Figure 10.1 Evolution of local pressure for the problem of fluid flowing into a square two-dimensional matrix block discretized into a 10×10 grid, under constant-pressure boundary conditions. Gray scale shows the pressures in units of psi. See text for details.

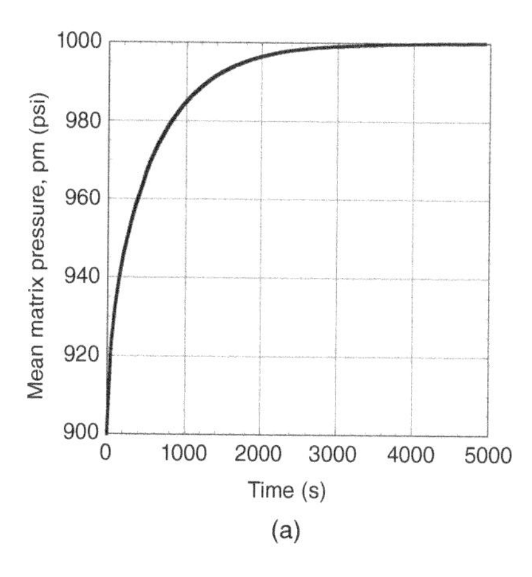

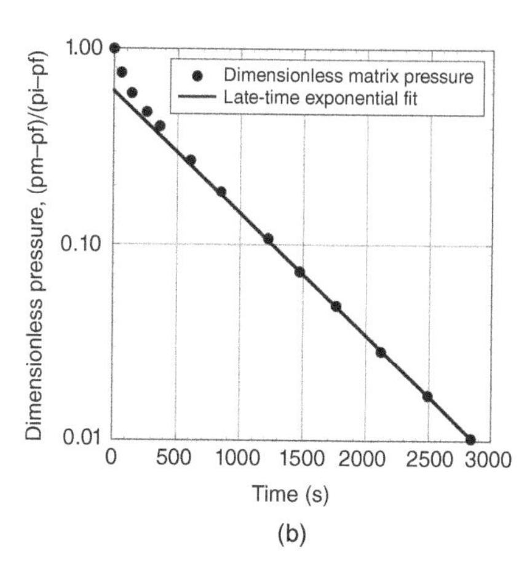

Figure 10.2 (a) Evolution of mean pressure in the matrix block for the problem of fluid flow into a square block, as shown in Fig. 10.1. (b) Semi-log plot of the normalized mean matrix block pressure as a function of time, along with the exponential function that gives the best fit at large times to the simulated pressures.

equilateral triangle. Their simulations also generally showed that using about 500 total cells is sufficient to obtain 2% accuracy for the shape factor of a two-dimensional matrix block.

10.6 Scaling Laws for Irregularly Shaped Matrix Blocks

Since it is not feasible to use the method described in Section 10.5 for every matrix block shape that might be encountered in a reservoir, it would be useful to develop an approximate scaling law that could be used to estimate the shape factor based on simple geometrical attributes. Consider first a prismatic matrix block that has a uniform cross-section, with area A and perimeter P, and is "infinitely long" in the third direction. As the shape factor must have dimensions of L^{-2}, it is necessarily inversely proportional to the square of some "characteristic length," which can be denoted by l. Assuming that the appropriate characteristic length depends only on the area, A, and the perimeter, P, possible candidates for l^2 are A, P^2, $(A/P)^2$, $A^{3/2}/P$, *etc.*, and the scaling law would imply that $\alpha = C/l^2$, where C is some dimensionless constant. The accuracy of any such scaling law can be tested by evaluating the product αl^2 for various shapes, and assessing whether or not it has nearly the same numerical value for most shapes.

Wuthicharn and Zimmerman (2011) tested the scaling laws based on the four abovementioned characteristic lengths against the known shape factors for a square, an isosceles right triangle, and a circle, and also for one irregularly shaped matrix block, shown in Table 10.1, that was chosen so as not to be too similar to a square or a triangle. The shape factor of this matrix block was computed numerically, following the procedure described in Section 10.5. For each of the four scaling laws and each of the four shapes, they computed the "constant" $C = \alpha l^2$, and then computed the standard deviation of C normalized against its mean value. The normalized standard deviation was 18% when using $l^2 = A$, 60% when using $l^2 = P^2$,

Table 10.1 Test of the shape factor scaling law $\alpha = CP/A^{3/2}$ for some prismatic blocks (*i.e.*, infinitely long in the third direction).

Matrix block shape		$\alpha A^{3/2}/P$	Deviation from mean value
Square		4.91	0%
Isosceles right triangle		5.10	4%
Circle		5.07	3%
Irregular shape		4.55	−7%
Mean	4.91		
Standard deviation (SD)	0.26		
SD/Mean	5%		

24% when using $l^2 = (A/P)^2$, and 5% when using $l^2 = A^{3/2}/P$. The scaling law $\alpha = CP/A^{3/2}$ was therefore the most accurate by far of the four tested laws, and the mean value of C for this scaling law for the four tested shapes was 4.91. They suggested that, for simplicity, the scaling law for prismatic matrix blocks be taken as $\alpha = 5P/A^{3/2}$.

Based on additional scaling arguments and approximations to the integral that appears in eq. (9.15), Wuthicharn and Zimmerman (2011) suggested that the appropriate extension of this scaling law to *three-dimensional* matrix blocks would be $\alpha = 5S/V^{4/3}$, where V is the volume of the block and S is its surface area. When applied to a cube of length L, this scaling law predicts a shape factor of $\alpha = 30/L^2$, which differs by only 1.3% from the exact result $\alpha = 3\pi^2/L^2$. When applied to a sphere of radius a, this scaling law gives $\alpha = 9.31/a^2$, which differs by only 6% from the exact result $\alpha = \pi^2/a^2$. They also tested the three-dimensional scaling law $\alpha = 5S/V^{4/3}$ against two other shapes for which analytical expressions for the shape factor are not available. The first (Fig. 10.3a) is a pyramid with a square base of side L, and height L. The volume of this block is $V = L^3/3$, and its surface area is $S = (1 + \sqrt{5})L^2$. The scaling law predicts $\alpha = 70.0/L^2$, whereas their fine-grid simulation yielded a value of $\alpha = 78.0/L^2$, showing that the scaling law underpredicts this shape factor by 10%. The second shape (Fig. 10.3b) was an irregular block shaped like a slightly deformed cube, bounded by six nonparallel faces. In this case, the fine-grid simulation yielded a shape factor of $\alpha = 5.02S/V^{4/3}$, which is only 0.4% greater than the value predicted by the scaling law.

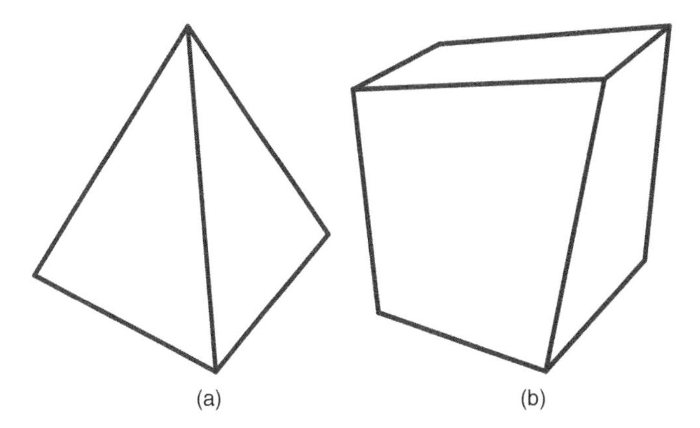

Figure 10.3 (a) Pyramidal-shaped matrix block, with a square base of side L, and height L. (b) Slightly irregular three-dimensional matrix block.

10.7 Shape Factor Under Constant-Flux Boundary Conditions

As mentioned in Section 10.2, a few researchers have studied the shape factor for the case of constant flux at the outer boundary of the matrix block, rather than constant pressure (Coats, 1989; Hassanzadeh and Pooladi-Darvish, 2006; Mora and Wattenbarger, 2009). An argument for using this definition of the shape factor was given by Hassanzadeh *et al.* (2009), who used Laplace transforms to solve the dual-porosity equations for flow to a vertical well in an unbounded reservoir, using the fully transient pressure diffusion equation within the matrix blocks, eq. (9.48). They imposed a boundary condition of constant flux *at the wellbore*, in which case the "boundary condition" at the fracture-matrix interface emerges naturally from the solution. They then numerically fit the late-time matrix-fracture flux to the Warren–Root equation, and found that, for the cases of slab-like, cylindrical, and spherical matrix blocks, the best-fitting shape factor at late times was in fact the one that corresponds to *constant flux* at the outer boundary of the matrix block. They concluded that the "constant-flux" shape factor should be used in this particular reservoir problem.

For slab-like, cylindrical, and spherical matrix blocks, the constant-flux shape factor can be computed *exactly* based on analytical solutions to the diffusion equation that are available in the literature (Carslaw and Jaeger, 1959; Crank, 1975). For example, for a spherical matrix block of radius a, with a uniform and constant volumetric flux Q [$\mathrm{m}^3/\mathrm{m}^2\,\mathrm{s}$] over the outer surface, the pressure distribution is given by (Crank, 1975, p. 91)

$$p_m(r,t) = p_i - \frac{Q\mu a}{k_m}\left[\frac{3k_m t}{\phi_m \mu c_m a^2} + \frac{r^2}{2a^2} - \frac{3}{10} - \frac{2a}{r}\sum_{n=1}^{\infty}\frac{\sin(\beta_n r/a)}{\beta_n^2 \sin\beta_n}\exp\left(\frac{-\beta_n^2 k_m t}{\phi_m \mu c_m a^2}\right)\right], \tag{10.13}$$

in which the normalized eigenvalues β_n are the positive roots of the equation $\beta = \tan\beta$. At sufficiently large times, all of the exponential terms in the series will have died off, and the pressure distribution will be given by

$$p_m(r,t) = p_i - \frac{Q\mu a}{k_m}\left(\frac{3k_m t}{\phi_m \mu c_m a^2} + \frac{r^2}{2a^2} - \frac{3}{10}\right). \tag{10.14}$$

The average value of $r^2/2a^2$ over the entire sphere is 3/10, and so the mean pressure in the matrix block is $\overline{p}_m(t) = p_i - (3Qt/\phi_m c_m a)$. By setting $r = a$ in eq. (10.14), the pressure at the outer surface of the block, which must equal the fracture pressure, is found to be given by $p_f(t) = p_i - (3Qt/\phi_m c_m a) - (Q\mu a/5k_m)$. Hence, $\overline{p}_m - p_f = Q\mu a/5k_m$. Since the outer surface area of the sphere is $4\pi a^2$, and the volume of the sphere is $4\pi a^3/3$, the volumetric flux per unit volume of matrix block, which is the flux that appears in the Warren–Root matrix-fracture transfer function, eq. (10.1), is $q = 3Q/a$. Hence, $\overline{p}_m - p_f = q\mu a^2/15k$. Insertion of this expression into eq. (10.1) yields $\alpha = 15/a^2$ (Mora and Wattenbarger, 2009), which is higher than the constant-pressure shape factor, $\alpha = \pi^2/a^2$. The value $\alpha = 15/a^2$ was also shown in Section 10.2 to be the shape factor that pertains to the case in which the fracture pressure decreases (or increases) linearly with time.

Analyses similar to that given above, based on solutions available in Crank (1975), reveal that the constant-flux shape factor is $\alpha = 12/L^2$ for a thin slab of thickness L, and $\alpha = 8/a^2$ for a long cylinder of radius a. These values agree very closely with the values found numerically by Hassanzadeh *et al.* (2009). Both of these constant-flux shape factors are greater than the constant-pressure shape factors, which are $\alpha = \pi^2/L^2$ for a thin slab, and $\alpha = 5.784/a^2$ for a long cylinder.

For matrix blocks of other shapes, a subtle problem arises when attempting to derive the shape factor by applying a constant and spatially uniform flux over the outer surface of the matrix block, as was done in the aforementioned analytical solutions. For any block that has "less symmetry" than a thin slab, long cylinder, or sphere, imposing a spatially uniform constant flux over the outer boundary does *not* lead to a spatially uniform boundary pressure. But the dual-porosity model assumes that the fracture pressure is *uniform* over the entire outer boundary. Therefore, for matrix blocks of other shapes, the "constant-flux" shape factor cannot be computed by imposing a spatially uniform constant flux at the outer boundary. However, the "constant-flux" shape factor can be computed by applying a boundary condition in which the *pressure* is spatially uniform over the entire outer boundary, and decreases linearly with time, as in the problem discussed at the end of Section 10.2, as this problem leads to a late-time *total* flux, integrated over the entire outer boundary, that is constant in time.

A general methodology for computing the constant-flux shape factor can therefore be formulated as follows. The governing equation and boundary/initial conditions are

$$\frac{dp_m(\mathbf{x},t)}{dt} = \frac{k_m}{\phi_m \mu c_m}\nabla^2 p_m(\mathbf{x},t), \tag{10.15}$$

$$p_m(\mathbf{x} \in \partial v_m, t) = p_i - Bt, \tag{10.16}$$

$$p_m(\mathbf{x}, t = 0) = p_i, \tag{10.17}$$

where ∂v_m denotes the outer boundary of the block, and $\mathbf{x} \in \partial v_m$ denotes all points on the outer boundary. Guided by the form of the solution for a sphere (Crank, 1975, p. 93), the solution for a block of *any* shape can be constructed as the sum of four terms:

$$p_m(\mathbf{x},t) = p_i - Bt + F(\mathbf{x}) + G(\mathbf{x},t). \tag{10.18}$$

The first two terms on the right account for the boundary condition (10.16) and the initial condition (10.17), but do not satisfy the pressure diffusion equation (10.15). The third term,

$F(\mathbf{x})$, must be added so that the first three terms, taken together, satisfy eq. (10.15). Since the first two terms, taken together, satisfy the initial condition (10.17), the fourth term, $G(\mathbf{x}, t)$, must then be added to cancel out the *nonzero* initial condition due to $F(\mathbf{x})$.

Specifically, insertion of the first three terms of eq. (10.18) into the pressure diffusion equation (10.15), yields the following differential equation for $F(\mathbf{x})$:

$$\nabla^2 F(\mathbf{x}) = \frac{-B\phi_m \mu c_m}{k_m}, \tag{10.19}$$

If the first three terms, taken together, are required to satisfy the boundary condition (10.16), then $F(\mathbf{x})$ must satisfy zero-pressure boundary conditions, *i.e.*,

$$F(\mathbf{x} \in \partial v_m) = 0. \tag{10.20}$$

Inserting the full pressure distribution as given by eq. (10.18) into eqs. (10.15)–(10.17) shows that $G(\mathbf{x}, t)$ must satisfy the diffusion equation (10.15), rewritten here for convenience,

$$\frac{dG(\mathbf{x}, t)}{dt} = \frac{k_m}{\phi_m \mu c_m} \nabla^2 G(\mathbf{x}, t), \tag{10.21}$$

along with the following boundary/initial conditions:

$$G(\mathbf{x} \in \partial v_m, t) = 0, \tag{10.22}$$

$$G(\mathbf{x}, t = 0) = -F(\mathbf{x}). \tag{10.23}$$

Many general properties and bounds for the constant-flux shape factor can be derived *without* fully solving the above equations. The term $G(\mathbf{x}, t)$ will always take the form of an infinite series of eigenfunctions, each multiplied by an appropriate Fourier coefficient, and by a term that decays exponentially with time. At late times, these terms will have died off, the pressure in the matrix block will be described by $p_m(\mathbf{x}, t) = p_i - Bt + F(\mathbf{x})$, and the average matrix block pressure will be $\overline{p}_m(t) = p_i - Bt + \langle F(\mathbf{x})\rangle$, where the angle brackets denote the average taken over the entire block. The pressure in the surrounding fractures is $p_f(t) = p_i - Bt$, and so the pressure difference between the fractures and matrix block is given by $\overline{p}_m - p_f = \langle F(\mathbf{x})\rangle$. Comparison of this expression with the Warren–Root transfer function, eq. (10.1), shows that $\alpha = (q\mu/k_m)/\langle F(\mathbf{x})\rangle$. In other words, knowledge of the function $F(\mathbf{x})$ is sufficient to determine the shape factor; knowledge of $G(\mathbf{x}, t)$ is not needed.

The volumetric flux per unit volume out of the matrix block is related to the mean matrix pressure by $q = -\phi_m c_m (d\overline{p}_m/dt)$. The expression for $\overline{p}_m(t)$ given in the previous paragraph implies that $d\overline{p}_m/dt = -Bt$, and so it follows that $B = q/\phi_m c_m$, where q is the volumetric flux *at late times*. The governing equation for $F(\mathbf{x})$, eq. (10.19), then takes the form $\nabla^2 F(\mathbf{x}) = -q\mu/k_m$. Comparison with the result $\alpha = (q\mu/k_m)/\langle F(\mathbf{x})\rangle$ suggests that it is convenient to define a new normalized function, $\widehat{F}(\mathbf{x}) = (k_m/q\mu)F(\mathbf{x})$, which will satisfy the equation

$$\nabla^2 \widehat{F}(\mathbf{x}) = -1, \tag{10.24}$$

along with zero boundary conditions, as in eq. (10.20). Since $\alpha = (q\mu/k_m)/\langle F(\mathbf{x})\rangle$ and $F(\mathbf{x}) = (q\mu/k_m)\widehat{F}(\mathbf{x})$, it follows that $\alpha = 1/\langle \widehat{F}(\mathbf{x})\rangle$. The formalism outlined above leads to the same result, eq. (10.24), as does the "closure problem" suggested by Quintard and Whitaker (1996) in the context of their "volume averaging" approach to modeling dual-porosity media.

The governing equation and boundary conditions for $\widehat{F}(\mathbf{x})$, namely eqs. (10.24) and (10.20), comprise the "Poisson equation with Dirichlet boundary conditions," which occurs in many areas of mathematical physics. In two dimensions, this problem governs the torsion of prismatic beams (Sokolnikoff, 1956, pp. 114–34; Landau and Lifshitz, 1986, pp. 59–64), as well as the laminar flow of a viscous fluid along a pipe of uniform cross-section (Berker, 1963). The analogy between these three problems extends further, in that the "figure of merit" in each case is essentially the mean value of the function $\widehat{F}(\mathbf{x})$. Term-by-term comparison of the equations given above for the shape factor problem, and the equations that govern pipe flow, eqs. (2.8) and (2.13), show that the permeability of the pipe is given by $k = \langle\widehat{F}(\mathbf{x})\rangle$, where the physical interpretation of $\widehat{F}(\mathbf{x})$ is that of a normalized local fluid velocity. Hence, since $\alpha = 1/\langle\widehat{F}(\mathbf{x})\rangle$, the constant-flux shape factor for a prismatic matrix block of a given cross-sectional shape is *exactly* given by $\alpha = 1/k$, where k is the permeability of a pipe having that same cross-section.

As an illustration of this conclusion, note that the permeability of a circular pipe of radius a is $a^2/8$ (White, 2006), whereas the constant-flux shape factor of a cylindrical matrix block is $8/a^2$ (Hassanzadeh and Pooladi-Darvish, 2006). Constant-flux shape factors for long prismatic blocks of several simple shapes, such as circles, squares, and equilateral triangles, can therefore be found directly from the literature on laminar pipe flow (Berker, 1963; Sisavath *et al.*, 2001; White, 2006), and are listed in Table 10.2. The constant-flux shape factor for a given geometry is typically about 20–60% larger than the corresponding constant-pressure shape factor.

Bounds on the constant-flux shape factor immediately follow from known mathematical results concerning the Poisson equation (10.24). These bounds are analogous to the bounds that hold for the constant-pressure shape factor. For example, the constant-flux shape factor of a block of a given shape is larger than the shape factor of any shape that circumscribes the block and is smaller than the shape factor of any shape that can be

Table 10.2 Shape factors for some geometrically simple block shapes.

Matrix block shape		"Constant-pressure" shape factor	"Constant-flux" shape factor
Thin slab of thickness L	L	π^2/L^2	$12/L^2$
Long square prism of $L \times L$ cross-section	L, L	$2\pi^2/L^2$	$28.45/L^2$
Long triangular prism of sides L	L	$16\pi^2/3L^2$	$80/L^2$
Long circular prism of radius a	2a	$5.784/a^2$	$8/a^2$
Sphere of radius a	2a	π^2/a^2	$15/a^2$
Cube of side L	L	$3\pi^2/L^2$	$49.62/L^2$

inscribed within that block. Furthermore, the constant-flux shape factor of a 3D nonspherical block of volume V is necessarily greater than the constant-flux shape factor of a sphere having that same volume. Since the constant-flux shape factor of a sphere is $15/a^2$, and its volume is $V = 4\pi a^3/3$, the shape factor of any nonspherical block of volume V must be greater than $15(4\pi/3V)^{2/3} = 38.98V^{-2/3}$.

As an example of the use of these bounds, consider a cubical block of side L. The largest inscribed sphere would have radius $L/2$, and the smallest circumscribed sphere would have radius $L\sqrt{3}/2$. The inscribed/circumscribed sphere bounds imply that the shape factor of this cube must satisfy $20/L^2 < \alpha < 60/L^2$. The "isoperimetric bound" based on the sphere whose volume is equal to L^3 implies that $\alpha > 38.98/L^2$. The most stringent set of bounds are therefore $38.98/L^2 < \alpha < 60/L^2$. The mean value of these two upper and lower bounds, $49.49/L^2$, is very close to the exact value, $49.62/L^2$ (Quintard and Whitaker, 1996).

10.8 Constant-Flux Shape Factor for a Brick-like Matrix Block

The most important and flexible matrix block shape for practical applications is the brick-like "rectangular parallelepiped," such as would be formed by three orthogonal sets of fractures, each set having its own spacing, $\{L_x, L_y, L_z\}$. Following the formalism outlined in Section 10.7, and making use of standard textbook results for solving the Poisson and diffusion equations using the eigenfunction method, the pressure distribution in a three-dimensional brick-like matrix block, subject to boundary and initial conditions (10.16) and (10.17), can easily be found to be given by

$$p_m(x,y,z,t) = p_i - Bt + \frac{64\phi_m \mu c_m B}{k_m \pi^3} \sum_{l=1,3,}^{\infty} \sum_{m=1,3,}^{\infty} \sum_{n=1,3,}^{\infty} \frac{\sin\left(\frac{l\pi x}{L_x}\right)\sin\left(\frac{m\pi y}{L_y}\right)\sin\left(\frac{n\pi z}{L_z}\right)\left[1-\exp\left(\frac{-\lambda_{lmn}^2 k_m t}{\phi_m \mu c_m}\right)\right]}{lmn\,\lambda_{lmn}^2}, \tag{10.25}$$

where $$\lambda_{lmn}^2 = \frac{l^2\pi^2}{L_x^2} + \frac{m^2\pi^2}{L_y^2} + \frac{n^2\pi^2}{L_z^2}, \tag{10.26}$$

and the three sums are taken over all odd positive integers.

The mean value of a term such as $\sin(n\pi z/L_z)$ over the interval from 0 to L_z is $2/n\pi$, and so the mean pressure in the matrix block is given by

$$\overline{p}_m(t) = p_i - Bt + \frac{512\phi_m \mu c_m B}{\pi^8 k_m} \sum_{l=1,3,}^{\infty} \sum_{m=1,3,}^{\infty} \sum_{n=1,3,}^{\infty} \frac{1-\exp^{-\lambda_{lmn}^2 k_m t/\phi_m \mu c_m}}{l^2 m^2 n^2 \left[(l^2/L_x^2) + (m^2/L_y^2) + (n^2/L_z^2)\right]}. \tag{10.27}$$

At late times, all of the exponential terms will have died out. Recalling from Section 10.7 that the volumetric flux *at late times* is given by $q = \phi_m c_m B$, in this time regime the mean matrix pressure can be written as

$$\overline{p}_m(t) = p_i - Bt + \frac{512\mu q}{\pi^8 k_m} \sum_{l=1,3,}^{\infty} \sum_{m=1,3,}^{\infty} \sum_{n=1,3,}^{\infty} \frac{1}{l^2 m^2 n^2 \left[(l^2/L_x^2) + (m^2/L_y^2) + (n^2/L_z^2)\right]}. \tag{10.28}$$

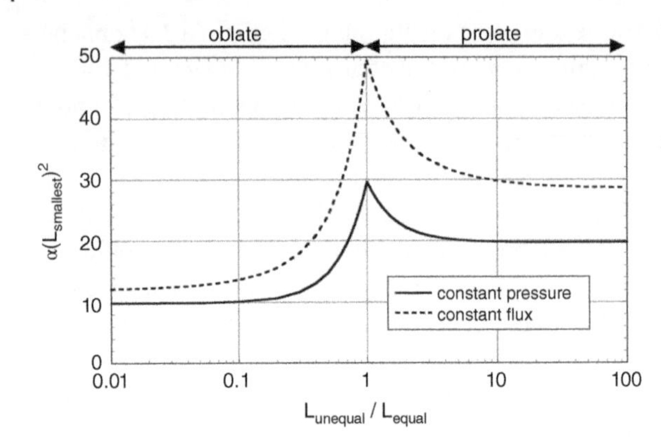

Figure 10.4 Constant-pressure and constant-flux shape factors of "oblate" and "prolate" brick-like matrix blocks, normalized with respect to the square of the smallest dimension. Oblate blocks have $L_x \leq L_y = L_z$, and prolate blocks have $L_x \geq L_y = L_z$; the horizontal axis denotes L_x/L_y in both cases.

Aside from the factor $\mu q/k_m$, the third term on the right side of eq. (10.28) is precisely the term identified as $\langle \widehat{F}(\mathbf{x}) \rangle$ in Section 10.7, and so the constant-flux shape factor of a general brick-like matrix block is given by Coats (1989)

$$\alpha = \frac{\pi^8}{512}\left\{\sum_{l=1,3,}^{\infty}\sum_{m=1,3,}^{\infty}\sum_{n=1,3,}^{\infty}\frac{1}{l^2m^2n^2\left[\left(l^2/L_x^2\right)+\left(m^2/L_y^2\right)+\left(n^2/L_z^2\right)\right]}\right\}^{-1}. \tag{10.29}$$

The triple summation in eq. (10.29) must be evaluated numerically. For a thin slab-like block having thickness L, which corresponds to the case $L_x = L \ll L_y = L_z$, the numerical values converge to the result $12/L^2$ that can also be obtained by starting with a one-dimensional flow problem. A long prismatic block having a square cross-section of side L, which corresponds to $L_x \gg L_y = L_z = L$, has a shape factor of $28.45/L^2$. A cube of side L, which corresponds to $L_x = L_y = L_z = L$, has a shape factor of $49.62/L^2$. In each of these three limiting cases, it can be noticed that the parameter L that appears in the expression for the shape factor refers to the *smallest* of the three sides, $\{L_x, L_y, L_z\}$. As pointed out by de Swaan (1990), the shape factor of a matrix block is primarily, although not solely, controlled by the smallest dimension of the block.

The shape factor of a brick-like matrix block that has two equal sides is plotted in Fig. 10.4, as a function of the aspect ratio of the block. These blocks are referred to here as "oblate," formed by compressing a cube along one axis, or "prolate," formed by elongating a cube along one axis. In each regime, the shape factor is normalized with respect to the *smallest* of the three lengths. The constant-pressure shape factors are given by eq. (10.4), and constant-flux shape factors are given by eq. (10.29). For all shapes, the constant-flux shape factor is larger than the constant-pressure shape factor. As might be expected, oblate matrix blocks having a thickness that is less than about one-tenth of the transverse dimensions essentially act like thin disks, and prolate blocks having a length that is greater than about ten times that of the smaller dimensions essentially act like infinitely long prisms. However, the constant-pressure shape factors approach their asymptotic limits much more rapidly than the constant-flux shape factors.

Problems

10.1 Using the fact that the constant-pressure shape factor of a sphere of radius a is π^2/a^2, along with the bounds discussed in Section 10.2, estimate the shape factor for a cube of side L. Compare this estimate with the known result, $3\pi^2/L^2$.

10.2 Consider a pyramidal block with a square base of side L, and height L, as in Fig. 10.3a. Use the various bounds discussed in Section 10.4, and the known results for the constant-pressure shape factors of cubes and spheres, to develop bounds on the constant-pressure shape factor of this pyramidal block. Compare these bounds with the fine-grid simulation of Wuthicharn and Zimmerman (2011), which yielded a value of $\alpha = 78.0/L^2$.

10.3 Assuming that the "matrix blocks" that can be identified in Fig. 1.9a each have long prismatic shapes extending into the third dimension, choose a few of these blocks, and estimate their constant-pressure shape factor using the methods discussed in this chapter.

10.4 Using eqs. (10.4) and (10.29), calculate the constant-pressure and constant-flux shape factors for a brick-like matrix block that has lengths L, $2L$, and $3L$.

References

Barenblatt, G. I., Zheltov, Iu. P., and Kochina, I. N. 1960. Basic concepts in the theory of seepage of homogeneous liquids in fissured rocks (strata). *Journal of Applied Mathematics and Mechanics*, 24(5), 852–64.

Berker, R. 1963. Intégration des équations du mouvement d'un fluide visqueux incompressible [Integration of the equations of motion of an incompressible viscous fluid]. *Handbuch der Physik*, VIII/2, Siegfried Flügge ed., Springer-Verlag, Berlin, pp. 1–384.

Carslaw, H. S., and Jaeger, J. C. 1949. *Operational Methods in Applied Mathematics*, 2nd ed. Oxford University Press, New York and Oxford.

Carslaw, H. S., and Jaeger, J. C. 1959. *Conduction of Heat in Solids*, 2nd ed. Oxford University Press, New York and Oxford.

Chen, C.-S., and Stone, W. D. 1993. Asymptotic calculation of Laplace inverse in analytical solutions of groundwater problems. *Water Resources Research*, 29(1), 207–09.

Coats, K. H. 1989. Implicit compositional simulation of single-porosity and dual-porosity reservoirs, in *SPE Symposium on Reservoir Simulation*, Houston, 6–8 February 1989, Paper SPE 18427.

Courant, R., and Hilbert, D. 1953. *Methods of Mathematical Physics*, vol. 1, Interscience Publishers, New York.

Crank, J. 1975. *The Mathematics of Diffusion*, 2nd ed., Clarendon Press, Oxford.

de Swaan, A. 1976. Analytic solution for determining naturally fractured reservoir properties by well testing. *Society of Petroleum Engineers Journal*, 16(2), 117–22.

de Swaan, A. 1990. Influence of shape and skin of matrix-rock blocks on pressure transients in fractured reservoirs. *SPE Reservoir Engineering and Formation Evaluation*, 5(4), 344–52.

Doetsch, G. 1961. *Guide to the Applications of Laplace Transforms*. Van Nostrand Reinhold, New York.

Gerke, H. H., and van Genuchten, M. T. 1993. Evaluation of a first-order water transfer term for variably saturated dual-porosity flow models. *Water Resources Research*, 29(4), 1225–38.

Gottlieb, H. P. W. 1988. Eigenvalues of the Laplacian for rectilinear regions. *Journal of the Australian Mathematical Society*, B29, 270–81.

Hassanzadeh, H., and Pooladi-Darvish, M. 2006. Effects of fracture boundary conditions on matrix-fracture transfer shape factor. *Transport in Porous Media*, 64, 51–71.

Hassanzadeh, H., Pooladi-Darvish, M., and Atabay, S. 2009. Shape factor in the drawdown solution for well testing of dual-porosity systems. *Advances in Water Resources*, 32, 1652–63.

Kac, M. 1966. Can one hear the shape of a drum? *American Mathematical Monthly*, 73, 1–23.

Kazemi, H., Merrill, L. S., Porterfield, K. L., and Zeman, P. R. 1976. Numerical simulation of water-oil flow in naturally fractured reservoirs. *Society of Petroleum Engineers Journal*, 16(4), 317–26.

Landau, L. D., and Lifshitz, E. M. 1986. *Theory of Elasticity*, 3rd ed., Elsevier, Amsterdam.

Landereau, P., Noetinger, B., and Quintard, M. 2001. Quasi-steady two-equation model for diffusive transport in fractured porous media: large-scale properties for densely fractured systems. *Advances in Water Resources*, 24, 863–76.

Lim, K. T., and Aziz, K. 1995. Matrix-fracture transfer shape factors for dual-porosity simulators. *Journal of Petroleum Science and Engineering*, 13(3, 4), 169–78.

Mathias, S. A., and Zimmerman, R.W. 2003. Laplace transform inversion for late-time behavior of groundwater flow problems. *Water Resources Research*, 39(10), paper 1283.

Mora, C. A., and Wattenbarger, R. A. 2009. Analysis and verification of dual porosity and CBM shape factors. *Journal of Canadian Petroleum Technology*, 48(2), 1–5.

Pinsky, M. A. 1980. The eigenvalues of an equilateral triangle. *SIAM Journal of Mathematical Analysis*, 11(5), 819–27.

Práger, M. 1998. Eigenvalues and eigenfunctions of the Laplace operator on an equilateral triangle. *Applications of Mathematics*, 43(4), 311–20.

Quintard, M., and Whitaker, S. 1996. Transport in chemically and mechanically heterogeneous porous media. II: comparison with numerical experiments for slightly compressible single-phase flow. *Advances in Water Resources*, 19(1), 49–60.

Rostami, P., Sharifi, M., and Dejam, M. 2020. Shape factor for regular and irregular matrix blocks in fractured porous media. *Petroleum Science*, 17, 136–52.

Sisavath, S., Jing, X. D., and Zimmerman, R. W. 2001. Laminar flow through irregularly-shaped pores in sedimentary rocks. *Transport in Porous Media*, 45(1), 41–62.

Siudeja, B. 2007. Sharp bounds for eigenvalues of triangles. *Michigan Journal of Mathematics*, 55, 243–54.

Sokolnikoff, I. S. 1956. *The Mathematical Theory of Elasticity*, 2nd ed., McGraw-Hill, New York.

Streltsova, T. D. 1988. *Well Testing in Heterogeneous Formations*, Wiley, New York.

Szegö, G. 1954. Inequalities for certain eigenvalues of a membrane of a given area. *Archive of Rational Mechanics and Analysis*, 3, 343–56.

van Genuchten, M. Th., and Dalton, F. N. 1986. Models for simulating salt movement in aggregated field soils. *Geoderma*, 38(1–4), 165–83.

Warren J. E., and Root, P. J. 1963. The behaviour of naturally fractured reservoirs. *Society of Petroleum Engineers Journal*, 3(3), 245–55.

White, F. M. 2006. *Viscous Fluid Flow*, 3rd ed., McGraw-Hill, New York.

Wuthicharn, K. and Zimmerman, R. W. 2011. Shape factors for irregularly shaped matrix blocks, in *SPE Reservoir Characterization and Simulation Conference*, Abu Dhabi, 9–11 October 2011, Paper SPE 148060.

Yeo, I. W., and Zimmerman, R. W. 2000. Accuracy of the renormalization method for computing effective conductivities of heterogeneous media. *Transport in Porous Media*, 45(1), 129–38.

Zimmerman, R. W., and Bodvarsson, G. S. 1995. Effective block size for imbibition or absorption in dual-porosity media. *Geophysical Research Letters*, 22(11), 1461–64.

Zimmerman, R. W., Chen, G., Hadgu, T., and Bodvarsson, G. S. 1993. A numerical dual-porosity model with semi-analytical treatment of fracture/matrix flow. *Water Resources Research*, 29(7), 2127–37.

11

Solute Transport in Fractured Rock Masses

11.1 Introduction

Subsurface fluids often contain dissolved chemical species, referred to as solutes. Often, these solutes are unwanted contaminants, whereas in some cases they are intentionally introduced into the water to serve as tracers. Modeling of solute transport through groundwater can be conducted to predict how these contaminants or tracers move in the subsurface due to natural or engineered processes. A fundamental understanding of fluid flow and solute transport in fractured rock masses is essential for analyzing mass and energy transport in fracture networks at the field scale. Applications of the analysis of the migration of solutes within the subsurface include contaminant transport, groundwater management, reservoir engineering, and exploration for deposits of minerals and metals.

The study of solute transport also includes the local or regional transport of chemical or nuclear materials relevant to disposal of waste in the subsurface. Transport in fractured media can also be used to reconstruct the history of the natural deposition of a mineral deposit and understand how porosity evolves in geological structures. During transport within a fractured network, a number of different physical processes take place, each of which influences the extent and the character of the migration of the solute mass. These include diffusion within the matrix, advection, dispersion, interface sorption, and chemical reactions (Berkowitz, 2002). Distributed fractures introduce heterogeneity into the rock, which affects the distribution of the flow velocities, which in turn affects how the material migrates and disperses. Fractured networks have a distribution of orientations, sizes, and apertures, which influence the permeability at multiple scales. This results in difficulties in identifying a suitable "representative elementary volume" that captures behavior at a larger scale. The assumption that the fractured rock mass is a porous continuum with permeability variations does not always serve to capture the influence of fractures on solute transport. This has led to the need to develop specific methods to model flow and transport in rock masses, taking specifically into account the effect of fractures on the transport process.

Field and laboratory experiments to study transport in fractured media have investigated the migration of solutes through fractured systems, showing that fractured systems behave differently than homogeneous porous media. As opposed to the classical Fickian transport, as discussed in single fractures in Chapter 6, fractured media often depart from this behavior on a larger scale. This behavior is referred to as "non-Fickian" or sometimes as "anomalous" transport.

Fluid Flow in Fractured Rocks, First Edition. Robert W. Zimmerman and Adriana Paluszny.

Companion website: www.wiley.com/go/zimmerman/fluidflowinfracturedrocks

In fact, depending on the properties of the fracture network and the current *in situ* conditions, fractures can considerably accelerate, channel, or impede the migration of solutes, depending on their aperture, their connectivity, their filling, and the stresses to which they are subjected. Although laboratory studies of transport within single fractures are feasible, upscaling the results of these studies to multiple fractures under *in situ* conditions is problematic, due to the difficulty of building and monitoring large and complex systems of fractures, natural or engineered, and monitoring how solutes advance within these systems. Therefore, a number of numerical techniques have emerged to model this behavior and develop an understanding of the interplay between the influence of *in situ* stresses, fracture network geometry, and fluid flow. The predictions of these numerical models have been compared to field observations and are used in the field to plan injection campaigns and predict migration paths.

Numerical and analytical models represent the fractured system with different degrees of realism. Some methods assume that fractures exist in two dimensions, or are organized into a perfectly regular lattice. Other methods assume that the movement of solutes through the system is random. In the real system, fracture apertures and permeability reflect a combined effect of several factors, such as the geometry of the fractures, chemical dissolution and precipitation processes, fluid pressure, and *in situ* stresses. The variation of the permeabilities of fractures within a network, their intersections, and the contrast between fracture and matrix permeability all play an important role in determining how solutes are transported through a fractured rock mass.

This chapter discusses the migration of a solute component dissolved in a single-phase fluid that is distributed within the entire pore space of a fractured rock. Section 11.2 presents the advection–dispersion equation (ADE) and defines the terms that appear in it. Methods for numerically solving this equation are reviewed in Section 11.3. Factors that cause solute transport to deviate from the ADE model, thereby leading to "anomalous transport," are briefly discussed in Section 11.4. Section 11.5 discusses the concept of channeling, which allows a powerful simplification of the modeling process. Sections 11.6 and 11.7 present additional transport modeling methodologies, such as the particle tracking method and the continuous time random walk (CTRW) approach. Finally, the effects of matrix permeability and *in situ* stresses on the observed solute transport properties of fractured media are discussed in Sections 11.8 and 11.9.

11.2 Advection–Dispersion and Solute Transport Equations

Solute transport is controlled by the variations in fluid flow velocity that occur in the medium. Velocities within a fracture network can vary by orders of magnitude when inspecting different scales (Matthäi, 2003). Modeling solute transport in a fracture network is generally subdivided into two steps: modeling the fluid flow problem, which depends on the geometry and permeability of the system, and solving the transport equations based on the fluid flow velocity distribution (Bear, 1988).

The transport of an inert solute can be modeled locally by the standard hyperbolic ADE, as derived in Chapter 6 (see also de Marsily, 1986):

$$\phi(\mathbf{x})\frac{\partial C}{\partial t} = \nabla \cdot [\mathbf{D}(\mathbf{x})\nabla C(\mathbf{x}, t) - \mathbf{u}(\mathbf{x})C(\mathbf{x}, t)], \tag{11.1}$$

where C [kg/m^3] is the concentration of the solute or tracer in the pore fluid, ϕ [–] is the local porosity, $\mathbf{u}$ [m/s] is the Darcy velocity, and $\mathbf{D}$ [m^2/s] is the dispersion tensor of the medium. The first term on the right is the dispersion term, which represents the spread of the solute due to molecular diffusion or dispersion due to local heterogeneities. The second term represents transport due to the solute advecting along with the flow.

If the x-axis of the coordinate system is aligned with the regional mean flow velocity U, which generally lies in a horizontal plane, the dispersion tensor will have a "longitudinal" component, $D_{xx} \equiv D_L$, and two "transverse" components in the horizontal ($D_{yy} \equiv D_{Th}$) and vertical ($D_{zz} \equiv D_{Tv}$) directions, respectively. In the case of a formation having an isotropic permeability field and assuming that the dispersion coefficients do not vary in space, eq. (11.1) takes the form

$$\phi \frac{\partial C}{\partial t} + U \frac{\partial C}{\partial x} = D_L \frac{\partial^2 C}{\partial x^2} + D_{Th} \frac{\partial^2 C}{\partial y^2} + D_{Tv} \frac{\partial^2 C}{\partial z^2}. \tag{11.2}$$

The dispersion coefficients will contain a component due to molecular diffusion and a component due to the mixing caused by local velocity field heterogeneities. For the velocities that are typically of practical interest, which is to say at large Peclet numbers (see Chapter 6), molecular diffusion is negligible, and it is found that the dispersion coefficients are proportional to the regional velocity. Consequently, they can be written as

$$D_L = \alpha_L U, \quad D_{Th} = \alpha_{Th} U, \quad D_{Tv} = \alpha_{Tv} U, \tag{11.3}$$

where the α terms are referred to as the *dispersivities*.

Transverse dispersivities are usually about 1–2 orders of magnitude smaller than the longitudinal values at a given site (Gelhar *et al.*, 1992), although there seem to be exceptional cases in which the transverse vertical dispersivity is larger than the horizontal value (Zech *et al.*, 2019). Observed values of the longitudinal dispersivity range over several orders of magnitude, from 10^{-1} to 10^2 m, for both porous media and fractured media. This effect is mainly due to the influence of heterogeneities. The effect of heterogeneities becomes more pronounced as a function of scale and therefore may be expected to lead to higher dispersivity values at larger length scales (Gelhar *et al.*, 1992). However, a recent critical review of available field data by Zech *et al.* (2019) found no scale dependence and no fixed relation between the longitudinal and transverse dispersivities.

The mass conservation equation, eq. (11.1), can be rewritten to incorporate a source or sink s [kg/m^3 s] of the solute:

$$\phi(\mathbf{x}) \frac{\partial C}{\partial t} = \nabla \cdot [\mathbf{D}(\mathbf{x}) \nabla C(\mathbf{x}, t) - \mathbf{u}(\mathbf{x}) C(\mathbf{x}, t)] + s(\mathbf{x}, t). \tag{11.4}$$

Equation (11.4) can describe the advection and dispersion of a nonreactive solute or tracer through a continuum or through a fracture network that is described geometrically by a number of discrete fractures. A continuum or macroscopic approach assumes that the fractured matrix is a continuum and also assumes that there is a representative elementary volume that captures the properties of the small scale, such as porosity and permeability, providing a representative average of the medium. However, it is quite difficult to identify these representative elementary volumes for fractured media due to the multi-scale nature of fracture networks. Therefore, methods are often used that incorporate discrete representations of fractures with consistent, local properties, which can then capture the variations of apertures explicitly.

Depending on the scale at which ϕ and $\mathbf{D}$ are defined, this equation can be used to describe the transport within a homogeneous or heterogeneous medium, by assuming either an average or a distributed value of fluid flow velocities. For homogeneous media, this equation can be solved directly to obtain estimates of the solute transport yielding Gaussian distributions of solutes.

When the transport equation is solved analytically or semi-analytically, solutions must often assume that rock properties are uniform. For a heterogeneous medium, this equation is best solved numerically to account for the variations in rock properties. In some cases, these heterogeneities can be captured by upscaling both ϕ and $\mathbf{D}$. In the case of fractured media, discontinuities are strongly anisotropic multi-scale heterogeneities that cannot be represented by simply modifying the porosity and dispersivity. For example, using the finite element method, the dispersion tensor $\mathbf{D}$ captures dispersion at the scale of a single finite element, and fractures and matrix are captured distinctly by different domains, which have different properties that affect porosity and dispersion, such as aperture and permeability. Solving the solute transport equation numerically requires a solution for the flow field and requires the matrix domain, including the fractures, to be spatially discretized. This introduces small errors in the form of numerical dispersion, which leads to nonphysical dispersion being introduced as an artifact of the numerical model. In systems containing multiple solutes, the nonphysical mixing of agents in the numerical model may also introduce dispersion, which is an artifact of the simulation process. These effects can be reduced by employing numerical techniques such as sub-dimensional representation of fractures and the use of hybrid numerical techniques such as those described in Section 11.3.

11.3 Numerical Solution of the Advection–Dispersion and Solute Transport Equations

Fluid velocity variation is a direct result of the spatial variation of the permeability. In a continuum, this is in general a consequence of variations in the properties of the matrix. In a fractured rock, discontinuities introduce local permeability variations that strongly affect transport. In some subsurface systems, fractures may add regions of significantly increased permeability, whereas in other systems in which fractures are filled with impermeable material, the fractures may actually act as barriers to flow, forcing the fluid to travel in a specific direction guided by their pattern. In a fractured rock mass, fracture networks will have a variation of fracture orientations, with variations of permeability that depend on their orientation, their relative positioning within the system, and the *in situ* stresses. Therefore, it is difficult to model such a system in terms of average equations (Nick *et al.*, 2011). Numerically, the problem is usually subdivided into first computing the flow, whether dependent or independent of stresses or mechanical deformation, and subsequently computing the solute transport.

A number of assumptions usually underlie these simulations. Some methods only model flow and transport through the fracture network and disregard the effect of the matrix, effectively assuming an impermeable matrix. Some approaches assume that *in situ* stresses can be projected directly onto the fractures, using the remote stresses to compute aperture distributions, while other approaches model local stress field interactions and aperture

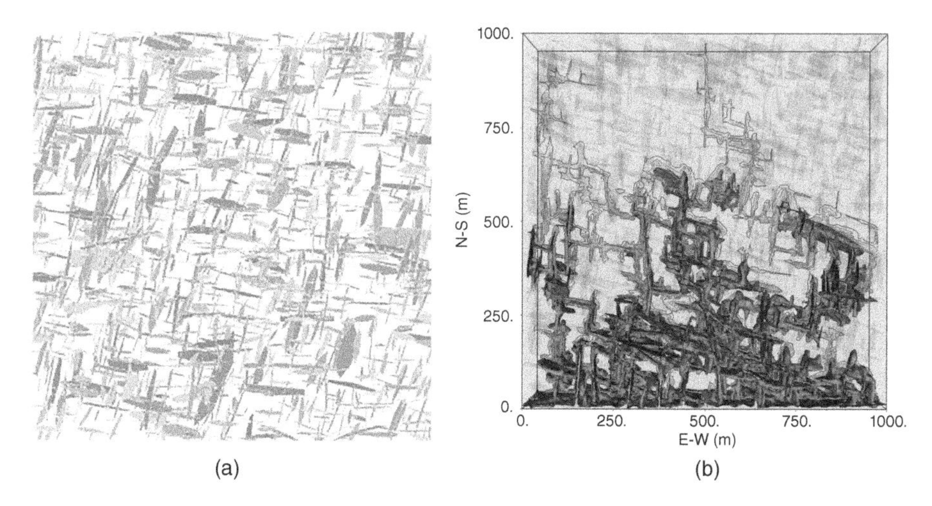

Figure 11.1 Top view of the numerical simulation of transport of a solute through a network of 2000 fractures, from the bottom boundary to the top, driven by a fluid pressure differential (Paluszny *et al.*, 2007). (a) Fracture network. (b) Penetration of solute into the rock mass after 3.5 months. See text for details. Source: Adapted from Paluszny *et al.* (2007).

variability due to *in situ* conditions and fluid flow. At fracture intersections, solutes may be assumed to fully mix with each other, whereas more complex approaches attempt to compute the precise manner in which solute is redistributed among the fractures connected at a given intersection. Fractures may be assumed to be stochastically generated, or may have been geomechanically generated to reflect mechanical interaction and self-organization. Finally, due to the complexity of the modeling of these processes, the geometry of the modeled system may be investigated in two dimensions, or more rarely, in three dimensions.

Figure 11.1 shows an example of a simulation conducted in 3D of solute transport through a rock mass containing a stochastically generated network of two thousand disk-shaped fractures. The fractures, whose statistics are based on observations of a fracture network in the San Andreas Formation in California, have a mean radius of 28 m. The simulated region was a box of dimensions 1000 m × 1000 m × 300 m (vertically), and the rock matrix was assumed to be impermeable. The image shows a numerical solution of the transport and advection equations using the hybrid finite element-finite volume method. The solute, marked by the dark contour surface, took 3.5 months to penetrate the entire fractured rock mass.

A variety of combinations of numerical methods are available to model transport within fractured media. Flow and mechanical deformation can be resolved using the finite element method, discrete element method, or extended finite element method, usually applied to resolve the flow field within a discrete fracture network, with or without contributions from the matrix. These methods are usually based on the fluid pressure equation based on Darcy's law and conservation of mass, as described in Chapter 9, which is solved implicitly using the standard Galerkin finite element method. Finite element and discrete element methods that model the mechanical deformation of the medium can consider poroelastic effects, fracture closure, and shear dilation, and can also account for dynamic changes to the fracture apertures of individual fractures within the network. This is important because, within fracture networks and in heterogeneous media, *in situ* stresses may vary within the network, resulting in a system where remote *in situ* stresses cannot always be regarded as being projected directly onto the individual fractures of the system. Instead, *in situ* stresses can locally become rotated as a function of large features such as faults, bedding planes, or interacting fractures, resulting in reorientation of the remote stresses. These effects can be modeled numerically but are computationally expensive.

As the pressure equation is elliptic and the transport equation is hyperbolic, an "operator splitting" technique can be applied to solve for the pressure using the finite element method and for transport using the finite volume method (Huyakorn and Pinder, 1983). This technique is widely used for modeling transport in porous fractured rocks (Huber and Helmig, 2000) and can be extended to incorporate detailed modeling of phase separation of highly compressible gases to simulate multi-phase flow, such as steam, in the context of thermohaline convection simulations (Geiger *et al.*, 2006).

In the case where the finite element method and the finite volume method are combined, the finite element solution can be relied upon to enforce a mass conservation property on the finite volume method (Baliga and Patankar, 1980). This operator-splitting, decoupling strategy allows the systems to be solved efficiently, with the capability of considering large variations in permeability and porosity within the rock mass. The solution strategy is flexible enough to allow both fluid pressure and transport to be computed, with or without time dependency, with a common strategy being to solve the fluid pressures implicitly, or quasi-statically, followed by an explicit transport solution that takes into account time dependency. The time step is controlled by the size of the smallest finite volume – and very small elements require very small time steps, leading to longer computing times. This difficulty can be avoided by applying time-domain methods that can apply different time steps to different regions of a geometric domain within the same solution space. Depending on the application, one approach may be preferred over the other. The Courant–Friedrichs–Lewy time-step criterion (Courant *et al.*, 1928) limits the size of the time steps that are allowable when using an explicit solution procedure, but implicit methods will not be subject to these time-step limitations. In general, if the modeled process is expected to be slow, such as a fluid slowly diffusing through a matrix or a solute slowly moving through a tight fracture network, the implicit method will offer a solution that is stable and efficient. For a process that occurs more rapidly, such as transport through a fracture network with large variations in flow velocity, an explicit method should be preferred.

Equation (11.4) can be discretized spatially using the finite volume framework and temporally using a fully implicit Backward-Euler finite difference scheme. Finite volumes can be solved together with finite elements by sharing the discretization of the geometry (Paluszny *et al.*, 2007). For each finite-volume cell, the formulation can be expressed in the form of the following volume and surface integrals (Matthäi *et al.*, 2010):

$$\phi_i \int_{V_m} \mathbf{M}_i \frac{\partial C_i}{\partial t} dV + \int_{\Gamma_{out}} (\mathbf{n} \cdot \mathbf{u}) C_c dS + \int_{\Gamma_{in}} (\mathbf{n} \cdot \mathbf{u}) C_u dS - \int_{V_m} \mathbf{M}_i s_i \, dV = 0, \tag{11.5}$$

in which ϕ_i is the porosity of finite volume i, $\mathbf{M}_i$ is the vector of finite-volume interpolation functions, C_i is the solute concentration in finite volume i, $\mathbf{n}$ is the outward unit normal vector to the finite volume facet (Fig. 11.2), $\mathbf{u}$ is the Darcy velocity, C_c is the current solute concentration, C_u is the upstream solute concentration, and s_i is the solute source term

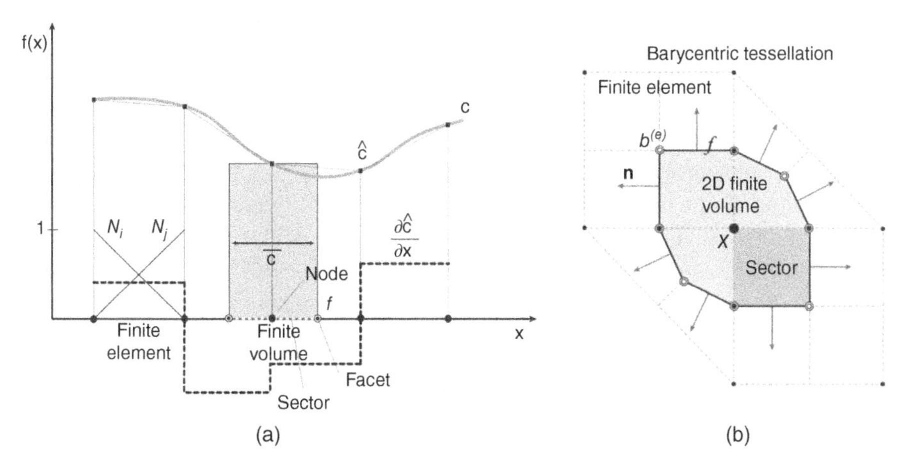

Figure 11.2 Combined finite element/finite volume method for solving the equations of flow and transport. The two images illustrate the node-centered approach, in 2D. (a) Finite volumes (in gray) are built around the nodes of the individual finite elements. (b) Multiple types of finite elements are subdivided into sectors, which form finite volumes around each node. The methodology can be extended to 3D. Source: Paluszny *et al.* (2007)/John Wiley & Sons.

in finite volume i. Γ_{in} denotes the surfaces over which the fluid is flowing into the finite volume, *i.e.*, for which $\mathbf{n}^f \cdot \mathbf{u} < 0$, and Γ_{out} denotes the surfaces over which the fluid is flowing out of the finite volume, *i.e.*, for which $\mathbf{n}^f \cdot \mathbf{u} > 0$, for a given facet with outward normal vector $\mathbf{n}^f$. Diffusive fluxes can also be added to the mass balance, as outlined by Paluszny *et al.* (2007).

Using this scheme, the dispersion of the tracer is modeled simultaneously with advection by projecting concentration gradients onto the facet normal vectors. To solve the system of equations, the contributions of each finite volume are accumulated into a sparse set of linear algebraic equations, with a matrix $\mathbf{A}$ that has coefficients $A_{j,i}$, and a right-side vector, $\mathbf{b}$, with components b_j:

$$A_{j,j} = \frac{\phi_j V_j}{\Delta t} + \sum_{k=1}^{N_f} (1 - H_{j,k}) A_{j,k} (\mathbf{n}_{j,k} \cdot \mathbf{u}), \tag{11.6}$$

$$A_{j,i} = \sum_{k=1}^{N_f} H_{j,k} A_{j,k} (\mathbf{n}_{j,k} \cdot \mathbf{u}), \tag{11.7}$$

$$b_j = \frac{\phi_j V_j}{\Delta t} c_j^t + s_j^{t+1}, \tag{11.8}$$

where the summation in eq. (11.6) is taken over each of the N_f facets of finite volume j, the summation in eq. (11.7) is taken only over those facets that are shared by cells i and j, and the function $H_{j,k}$ is defined such that $H_{j,k} = 1$ if $\mathbf{n} \cdot \mathbf{u} < 0$, and $H_{j,k} = 0$ otherwise. In eq. (11.8), superscript t is used as an index for the discrete time steps. When using this technique, known as "upstream weighting," only incoming fluxes create couplings between adjacent finite volumes, *i.e.*, create off-diagonal terms in the solution matrix. This results in an unconditionally stable first-order scheme (Baliga and Patankar, 1980, 1983). Further details of this scheme are discussed by Geiger *et al.* (2006) and Paluszny *et al.* (2007).

11.4 Non-Fickian Transport

Local differences in the hydraulic conductivity of the system, such as the permeability variations introduced by fracture networks, have a substantial effect on the bulk fluid flow in a rock. Instead of moving homogeneously through the rock, in the presence of fractures, distinct flow paths can be observed, through which most of the fluid travels, as discussed in Section 8.9. Preferential flow paths were first inferred from column experiments (Scheidegger, 1974), then measured with magnetic resonance imaging in artificial porous media (Hoffman *et al.*, 1996), and have subsequently been directly observed using micro-CT scanning of fractured rock samples (Bijeljic *et al.*, 2013). These experimental observations systematically characterize, in three dimensions, the dominant flow paths that form during the migration of solutes. These effects are observed at the pore scale as well as at the meter scale (Levy and Berkowitz, 2003) and kilometer scale (Kosakowski *et al.*, 2001), in which fractures of many sizes interact to modify flow and solute migration. Simulations on discrete fracture networks have shown the same behavior by tracking particles within computational models (Tsang and Neretnieks, 1998) that capture geometric and aperture distribution effects while studying the effect of stresses on solute transport.

As a consequence of these effects, although the ADE may well hold at the local scale, on a larger scale, solute transport cannot generally be modeled by the ADE, using some "effective" value of the dispersion coefficients. The Gaussian concentration profiles that are predicted by the classical ADE, and the profiles that are often observed in the field, were originally referred to as "normal" and "anomalous," respectively. However, due to the ubiquitous occurrence of this type of "anomalous" behavior, the term "non-Fickian" is perhaps more appropriate.

Migration curves of fractured media tend to display early breakthroughs, as connections between fractures accelerate the speed at which the fluid travels through the interconnected fracture network, while also partially trapping solutes, thereby leading to a long concentration tail at later times. The behavior can be observed schematically in Fig. 11.3a, which contrasts the one-dimensional migration of Fickian and non-Fickian concentration profiles. In the Fickian case, a finite amount of injected solute advects at a constant velocity, spreading out due to dispersion, while maintaining a Gaussian shape. The mean location of the solute plume scales with t, whereas the width of the plume in the direction of mean flow grows as $t^{1/2}$ (Berkowitz *et al.*, 2000). In the non-Fickian case, some of the solute moves downstream at a faster rate, leading to early breakthrough. But other solute particles have a longer residence time in the fracture network, causing the concentration front to have a long tail, due to variability in the hydraulic conductivity of the rock (Berkowitz and Scher, 1997). This behavior leads to rapid dispersion of solutes, as compared to classical Fickian predictions, and an uneven distribution of the solute concentration within the fractures and rock matrix.

Di Donato *et al.* (2003) used fine-grid streamline simulations to show that non-Fickian transport can indeed arise from purely advective movement through a heterogeneous medium, even in the absence of molecular diffusion. They simulated solute transport through a region of 366 m × 670 m × 52 m extent, containing 1,122,000 computational cells, using properties based on a North Sea oil field containing high permeability sand channels surrounded by low permeability shale. Figure 11.3b shows the breakthrough

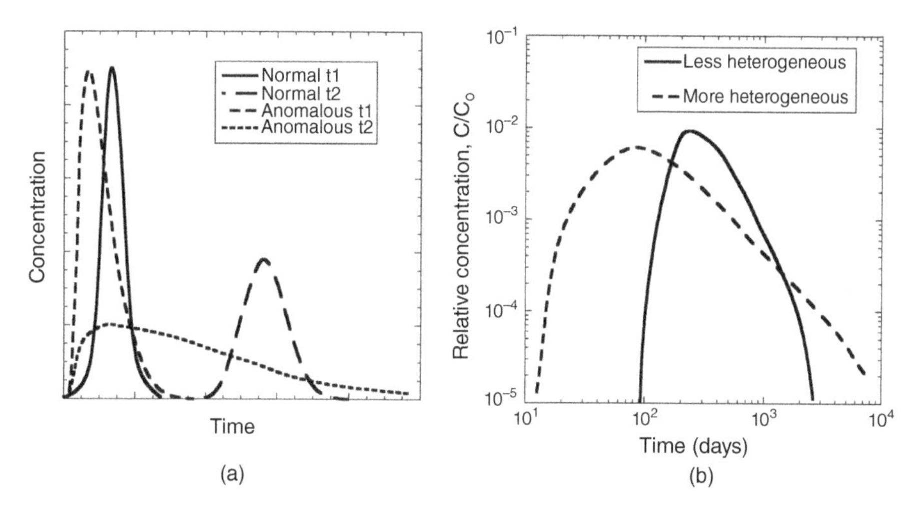

Figure 11.3 (a) Schematic concentration profiles for two values of time, $t_2 > t_1$, showing that non-Fickian "anomalous" profiles disperse more severely, leading to both earlier breakthrough, and a longer tail; after Scher *et al.* (1991). Source: Adapted from Scher *et al.* (1991). (b) Breakthrough curves computed by Di Donato *et al.* (2003) for heterogeneous media, showing the characteristic shape of anomalous transport, and the effects of material heterogeneity. Source: Adapted from Di Donato *et al.* (2003).

curves for two cases, with differing degrees of heterogeneity in the local permeability field. In the case of greater heterogeneity, the earliest time of solute arrival at the extraction well occurs sooner, but some solute remains in the system much longer, which is to say that the breakthrough curve becomes more spread out in time.

Several approaches to modeling solute transport in fractured media have been developed that do not rely on the advection-dispersion equation (11.1). A few of these approaches are reviewed in Sections 11.5–11.7.

11.5 Channel Models

As opposed to modeling flow and transport through the entire fracture network, some approaches assume that there exists a network of channels within these fracture networks, within which the actual movement of the solute primarily takes place. These channels may differ in size and connectivity, and may also have varying flow rates. As discussed in Chapter 8, these channels can exist at multiple scales, and emerge as a result of variations in the preferential paths of fluids within the fractured rock mass. A discussion of the effect of fractures on the formation of channels during fluid flow can be found in Section 8.9. Figure 11.4 shows an example of channeling at the scale of the fracture network.

Once flow and mechanical effects have been simulated, transport can be modeled numerically using a stochastic or deterministic method. Stochastic methods to model transport include the *particle tracking method* and the CTRW approach (Bijeljic and Blunt, 2006), whereas deterministic methods focus on the solution of the transport

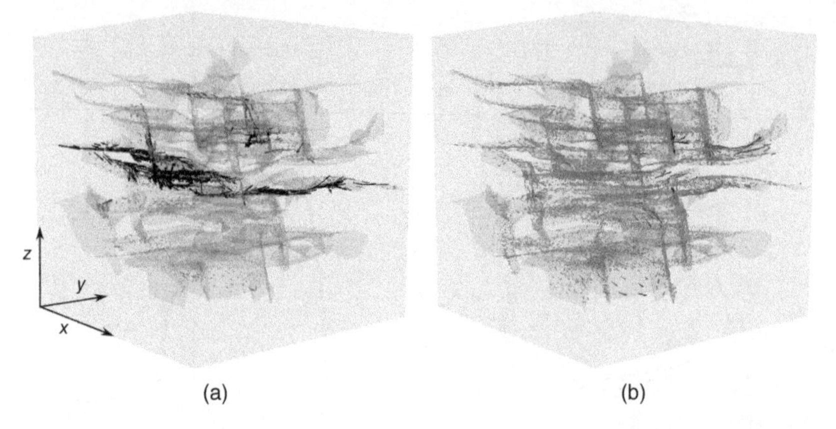

Figure 11.4 Examples of flow channeling in a fracture network. Darker gray indicates higher velocity; scale is only relevant in a relative sense. (a) In this case, the fracture apertures were computed as a consequence of geomechanical fracture growth modeling, leading to more pronounced channeling, whereas in (b) the apertures are uniform. After Thomas *et al.* (2020). Source: Adapted from Thomas *et al.* (2020).

equation using a numerical method such as the mass-conservative finite volume method (Geiger *et al.*, 2006).

Channeling is observed in the subsurface at multiple scales and appears as regions of substantially larger flow. The formation of channels at the small scale is closely related to surface roughness and the distribution of voids within fracture surfaces, which change as a function of stresses, as discussed in Chapters 3 and 8, but also due to thermo-chemical processes, as discussed in Chapter 5. Emerging channels serve as conduits that transport solutes through the rocks. Channeling can occur at a very small scale as a result of variations in the apertures of the fractures, effectively forming preferential paths within the surfaces of individual fractures. When considering a fracture network, preferential flow paths connect a specific subset of fractures of the network, essentially forming a network of a few channels through which most of the fluid travels. If it is assumed that fluid flow and advective solute transport mainly take place through these channels, the modeling of flow can be simplified so as to only consider flow through these channels. Simplifications to the modeling of flow in fracture networks can consider modeling the flow through these channels, taking into account mixing between regions and detailed transport of single or multiple, reactive or nonreactive solutes through the network.

The channel network model of Moreno and Neretnieks (1993) is defined in three dimensions and assumes that flow paths are distinct entities within the overall flow path of the fluid. Thus, when looking at a system, one can identify multiple distinct flow paths that converge and diverge as they traverse the matrix. This type of model, based on the analysis of channels, relies on particle tracking to capture the transit of particles within the modeled channel flow paths.

The channel network model can also incorporate the effects of diffusion of the solute into the rock matrix and the diffusion of particles from the flow channels into "stagnant" flow regions, which are often assumed to exist peripherally to the modeled flow channels. The channel network model assumes that the contact between multiple channels does not lead to immediate mixing, but rather that mixing is a function of the contact time and the length of the channels, among other factors. After a sufficiently long contact time between two streams, the solute will have diffused from the channel with the greatest concentration to the other channels, and concentration equilibrium will be reached. At this stage, the channels can be considered to be indistinguishable from each other. Therefore, the concept of channels goes beyond the mere geometric representation of the dominant flow paths and captures the distribution of solute in the medium in the context of a set of preferential flow paths and their interaction with the medium, which almost always contains other interacting and competing flow paths. Channel models generally assume that multiple migrating solutes are present, all dissolved into a single fluid, and migration is taking place through a rock matrix with multiple interconnected fractures. The analysis becomes even more complex when there are multiple phases, multiple fluids, or temperature effects on the mixing.

In the real system, there possibly will be several solutes that may influence each other and may even produce new solutes as a function of chemical reactions. However, simulations have shown that channels will remain differentiated over long distances, as diffusional mixing of channels over long times and distances is not pronounced (Neretnieks, 2002). The 3D channel model usually assumes that the hydraulic conductivities of the channels are log-normally distributed and are not necessarily correlated in space except for following material property distributions. Channels are difficult to observe in the field, and their width and length are also difficult to approximate unless directly observed in drifts and tunnels. These are, however, considered to be properties of less significance, as compared to the "flow-wetted" surface, which denotes the area of the fracture surface occupied by the channel, which can be approximated by characterizing the local fracture network and by using hydraulic tests. The latter can also be used to measure the overall channel conductivity distribution, which informs the stochastic distribution of the modeled channels (Gylling, 1997).

Channels are routinely used to model fluid flow through unfractured rock matrix units, mixed with channels that represent single or multiple fractures. In the original formulation, the channel network model is solved on a 3D Cartesian grid, in which each segment of the grid corresponds to a channel, and each channel intersection brings together up to six converging channels. The flow Q_{ij} [m^3/s] from channel i to channel j is proportional to an applied pressure gradient, *i.e.*,

$$Q_{ij} = C_{ij}(p_i - p_j), \tag{11.9}$$

in which C_{ij} [m^3/s Pa] is the conductance from channel i to channel j, and $\{p_i, p_j\}$ are the respective fluid pressures at those two nodes. It follows that the pressure field is found by enforcing mass balance at each node of the grid, which corresponds to an intersection point of different channels (Moreno and Neretnieks, 1993):

$$\text{for each node } i, \ \sum_j Q_{ij} = 0. \tag{11.10}$$

This condition ensures that the total fluid flux into each node i equals zero.

Consider the continuous injection of fluid into a fracture embedded in a porous rock mass of infinite extent, with an injected solute concentration of C_o [kg/m^3]. The solute concentration in the fracture will be given by (Neretnieks, 1980; Yamashita and Kimura, 1990;

Moreno and Neretnieks, 1993)

$$\frac{C_f(t)}{C_o} = \text{erfc}\left(\frac{LW}{Q}\sqrt{\frac{K_d\rho_m D_e}{t - t_w}}\right), \tag{11.11}$$

where K_d [m^3/kg] is the linear sorption coefficient, D_e [m^2/s] is the effective diffusivity, ρ_m [kg/m^3] is the density of the rock matrix, LW [m^2] is the flow-wetted area (*i.e.*, length × width of one face of the fracture), Q [m^3/s] is the volumetric fluid flow rate in the channel, t_w [s] is the residence time for a particle in the absence of matrix diffusion, t [s] is the residence time for a particle in the presence of matrix diffusion, and erfc is the complementary error function (Ghez, 1988). This relationship can also be expressed in terms of the fracture aperture h [m], as follows (Tsang and Tsang, 2001):

$$\frac{C_f(t)}{C_o} = \text{erfc}\left(\frac{t_w}{h}\sqrt{\frac{K_d\rho_m D_e}{t - t_w}}\right). \tag{11.12}$$

This model is well suited to investigate the effect of diffusion on the transport process, as simplified solutions of the ADE assume a single effective diffusion coefficient or single "dispersion length" per model. The dispersivity of the solute can be interpreted as a "dispersion length" parameter, which combines the effects of molecular diffusion, fluid flow velocity variations, and the mobilization of fluids between mobile and stagnant regions. Dispersion length is another way of examining diffusion, and when considering a distribution of channels, dispersion is inherently proportional to distance. Dispersion can be found to be a function of the Peclet number when mechanical dispersion dominates.

Finite element solutions of the ADE, with discrete representations of fractures and heterogeneity distributions, assume a distribution of dispersion properties, inherently leading to scale-dependent dispersion lengths, as observed in the field. Despite differences in these models, both advection–dispersion models and channel network models yield similar residence time distributions, in particular for Peclet numbers larger than about 3 (Neretnieks, 2002). In the case where the matrix has a higher permeability, and therefore plays a more significant role in the transport, the channel model provides good approximations for the effluent concentrations and residence times.

11.6 Particle Tracking Methods

In particle tracking methods, the fluid flow equations are first solved, to obtain a distribution of flow velocities. Once the velocity field has been thusly approximated, travel path lines and travel times of particles moving through the system are predicted. This procedure primarily accounts for advective movement within the rock mass and does not usually take into account dispersion, diffusion, or decay. However, it is particularly well suited to calibrate the results of numerical and field experiments in order to compute residence times and estimate transit time distributions for the movement of contaminants within the groundwater in the subsurface. Once the flow paths are established, particles are made to travel through the system from one or more inlets to one or more outlets, following random paths. Any node in the system can serve as either an inlet or outlet. This method models the movement of particles that are injected at fixed or varying concentrations. In the particle tracking method, the simulator usually tracks movement due to advection only, assuming that molecular diffusion is negligible and that the solute concentration is the same in each segment of an individual fracture.

The water residence time of a particle in the fracture segment i is denoted by (Zhao *et al.*, 2011)

$$t_w^i = \frac{V_i}{Q_i}, \tag{11.13}$$

where V_i [m^3] is the volume of the fracture segment (cross-sectional area × length in 3D; aperture × length in 2D), and Q_i [m^3/s] is the volumetric flowrate through the fracture segment. The total water residence time of a particle traveling through fracture network, t_w, scales with $1/Q$, where Q is the total flowrate through the entire system, and is equal to the sum of the residence times through all of fractures that the particle has traversed:

$$t_w = \sum_i t_w^i = \sum_i \frac{V_i}{Q_i}. \tag{11.14}$$

In some codes, each fracture will comprise a single segment, whereas in other implementations, each fracture will comprise multiple connected segments or regions, represented as 2D or 3D objects in space.

The particle tracking method, as well as many of the other methods that track solute transport, keeps track of the mass continuity within the system, as it solves for the flow that results from applying the boundary conditions to the system. The movement of particles is tracked within a given mesh, which could be the underlying structured mesh of a finite difference scheme, or an unstructured mesh corresponding to a finite element solution of flow.

The effects of matrix diffusion and sorption can be modeled by defining a residence time distribution, which relates the total residence time to the initial concentration of the particle. It follows that the relation between the current concentration and the original concentration can be defined as a function of the relation between the material properties of the matrix, surface wetting properties, and the residence time. The migration of solute from the fracture into the matrix is strongly influenced by the size of the contact surface between the fluid within the fracture and the rock matrix, noting that the fluid not always invades the complete surface of the fracture and may instead form channels within the fracture surface. It follows that the concentration, normalized with respect to that of the injected fluid, is given by (Moreno *et al.*, 2006)

$$\frac{C_f(t)}{C_o} = \text{erfc}\left(\frac{A_q}{2Q}\sqrt{\frac{K_d \rho_m D_e}{t - t_w}}\right), \tag{11.15}$$

where A_q/Q is the ratio of the flow-wetted surface area to the volumetric flow rate, and is given by

$$\frac{A_q}{Q} = \sum_i \frac{2L_i W_i}{Q_i}, \tag{11.16}$$

where the sum is taken over all fracture segments in the network, L_i [m] is the length of the ith segment, W_i [m] is the width of the ith segment in the direction normal to the flow, and

Q_i [m^3/s] is the volumetric flowrate through the ith segment. The factor of 2 in eq. (11.16) arises because of the two opposing wetted surfaces of the fracture.

The term $(K_d\rho_m D_e)^{0.5}$ appearing in eq. (11.15), often referred to as the *material properties group*, or MPG (Soler *et al.*, 2022), can also be expressed as $(K_v D_e)^{0.5}$, where K_v [–] is the *volumetric sorption coefficient*. The material property group is assumed to be a constant and depends on the pore diffusivity, the sorption or retardation coefficient, and the matrix porosity. Assuming that D_e is the effective diffusion coefficient in the matrix, and ϕ is the matrix porosity, then D_a is the apparent diffusion coefficient in the matrix, given by

$$D_a = D_e/(\phi + K_v). \tag{11.17}$$

It follows that

$$MPG = \frac{D_e}{\sqrt{D_a}} = \sqrt{D_e(\phi + K_v)} \approx \sqrt{D_e K_v}. \tag{11.18}$$

This model effectively replaces the single residence time per particle as expressed in eq. (11.13). In practice, C_f/C_o is chosen to be a random number between 0 and 1, denoted by $[R]_0^1$, and so eq. (11.15) can be expressed as

$$[R]_0^1 = \text{erfc}\left(\frac{A_q}{2Q}\sqrt{\frac{K_d\rho_m D_e}{t - t_w}}\right). \tag{11.19}$$

Consequently, the total residence time can be expressed as

$$t = t_w + \left(\frac{A_q}{2Q}\frac{\sqrt{K_d\rho_m D_e}}{\text{erfc}^{-1}[R]_0^1}\right)^2. \tag{11.20}$$

This particle tracking method can be used to track solute particles traveling through a network of fractures, or other heterogeneous regions, such as a network of channels, or a network of low-porosity regions embedded within a low-porosity matrix.

11.7 Continuous Time Random Walk Approach

The CTRW approach is an alternative numerical modeling strategy to capture transport, that was originally developed by Montroll and Weiss (1965). It is a flexible formulation that can be regarded as a theoretical framework that can be used to interpret both laboratory and field observations of chemical and particle transport in fractured and porous media. The method is a probabilistic approach to modeling transport in a medium with random properties, in which the fluid is modeled as a set of particles moving from site to site within the volume in a process analogous to electron hopping in solids (Scher *et al.*, 1991; Berkowitz *et al.*, 2006). In the implementation of Berkowitz and Scher (1997), transport is modeled within the context of a two-dimensional fracture network with constant apertures embedded within an impermeable matrix. The model progressively accumulates random transitions on a lattice, based on a prescribed probability distribution that governs the migration of a solute particle from one site to another. These transition probabilities are intended to reflect, in some sense, the microstructure of the medium.

The details of the CTRW method have been developed and described by Scher and Lax (1973), Klafter and Silbey (1980), and Shlesinger (1996), among others, in a quite general context, and by Berkowitz *et al.* (2001) and others in the context of fractured porous media. Only a few key aspects of the method will be discussed here. A key ingredient of this method is the waiting time function, $\psi(s, t)$, which gives the probability per time for a solute particle to move a distance s, with a difference in arrival times of t (Berkowitz *et al.*, 2006). If this function is assumed at long times to approach an exponential function of the form (Scher and Montroll, 1975; Berkowitz *et al.*, 2001),

$$\psi(t) \sim e^{-\lambda t}, \tag{11.21}$$

which corresponds to a microstructure that is in some sense "homogeneous," then the model reproduces "normal," or Fickian-type, transport.

To model the behavior of heterogeneous media, the long-time behavior of the waiting time function is often represented by a power-law of the form

$$\psi(t) \sim t^{-1-\beta}. \tag{11.22}$$

The dimensionless parameter β can capture a range of dispersive behavior, which can be classified into three regimes (Berkowitz *et al.*, 2001). For $\beta > 2$, transport is considered to be "regular," and in this regime, the mean location of the concentration plume, $l(t)$, scales with t; the standard deviation of the plume, $\sigma(t)$, scales with $t^{1/2}$; and hence, $\sigma(t)/l(t) \sim t^{-1/2}$. For $1 < \beta < 2$, transport becomes non-Fickian in this case, the mean plume location scales with t, the standard deviation of the plume scales with $t^{(3-\beta)/2}$, and hence $\sigma(t)/l(t) \sim t^{(1-\beta)/2}$ (Berkowitz *et al.*, 2006). Highly heterogeneous media are modeled with $0 < \beta < 1$, and correspond to a regime in which the mean plume location and the standard deviation of the plume both scale with t^{β}, and hence $\sigma(t)/l(t)$ is constant.

Although this power-law form of the function $\psi(t)$ can capture a large range of behavior and has been matched to many sets of experimental and field data, there are transport modalities observed in the field that cannot be captured by this type of parameterization. An alternative approach is to not predefine a functional form for $\psi(t)$, but instead use a nonparametric inversion algorithm to recover a numerical approximation to this function, by fitting experimental data to the CTRW predictions. This approach was developed by Cortis and Birkholzer (2008) and applied to the modeling of solute exchange between a fracture network and the porous rock matrix.

The CTRW method can reproduce solute migration experiments at the laboratory and field scales (Berkowitz and Scher, 1998; Berkowitz *et al.*, 2000). These experiments, as well as work presented by Kosakowski *et al.* (2001), Levy and Berkowitz (2003), and Cortis and Birkholzer (2004), show porous systems with non-Gaussian dispersion, as well as showing that the observed behavior cannot be captured using traditional averaging approaches. In most of these cases, the authors showed that the CTRW method could reproduce the observed results, provided that the appropriate transition probability times were defined.

Figure 11.5 shows concentrations measured at one of the sampling points in a field test conducted in a block of fractured till by Sidle *et al.* (1998), along with the fits obtained by Kosakowski *et al.* (2001) using several different models. In the effective porous medium (EPM) model, the block was represented by a single uniform value of porosity and permeability. In the discrete fracture and matrix model (DFM), the fractures were modeled as a

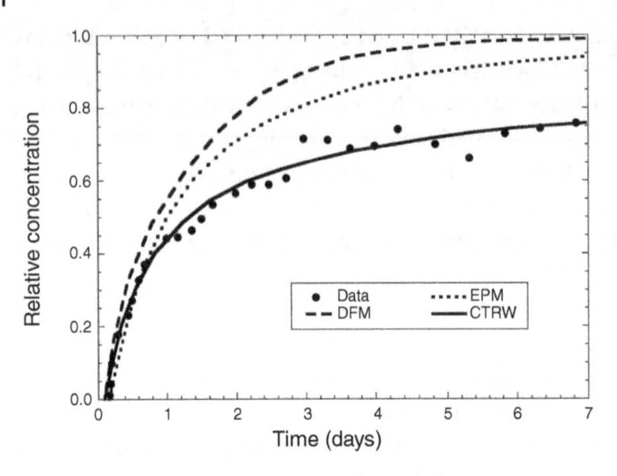

Figure 11.5 Normalized tracer mass *vs.* time, as measured by Sidle *et al.* (1998) during a field test in a block of fractured till, and the fits obtained by Kosakowski *et al.* (2001) using several analytical models: effective porous medium (EPM), discrete fracture and matrix model (DFM), and CTRW.

set of parallel fractures within a hydraulically impermeable matrix, with diffusion allowed into the matrix. Only the CTRW method was able to provide a good match to the solute concentrations at larger values of time.

11.8 Effects of Matrix Permeability

Nick *et al.* (2011) conducted numerical simulations of fluid flow and transport in two-dimensional, geometrically realistic fractured/porous media to study the effect of matrix permeability on solute transport. A fixed pressure gradient was first established across a rectangular region, after which a finite amount of solute was uniformly distributed along a thin vertical slit along the left side of the modeled region. The solute concentration of the fluid emerging at the right edge of the region was then measured. The fracture patterns used in their study were not very well connected, and so flow through the matrix had a major influence on the breakthrough curves (Fig. 11.6). It was observed that the most important parameters controlling solute transport were the matrix permeability, the fracture density, and the fracture-matrix flux ratio. They also found that the apertures of the fracture networks in this scenario could be taken as having a uniform "effective" value while still yielding acceptable predictions of dispersion in these systems.

The role of the matrix can be important in the diffusion of solutes, and the permeability of the matrix controls the rate at which this occurs. The interaction with the matrix serves to extend the time that the solutes travel through the rock mass, as paths become more tortuous once fluid ingresses into the matrix. As fractures appear at multiple scales, once the solute migrates out of a "main" or "large" fracture into the matrix, the solute may travel through a porous rock mass or encounter smaller fractures and void spaces as it travels. This effect is stronger when the fluid pressure gradient is smaller and becomes less prominent when the pressure gradient is larger. This is referred to as a "long tail," which is characteristic of transport in fractured rock masses and describes the long residence of solutes within the rock mass.

Matrix diffusion can also contribute to "delayed breakthrough." When a solute travels from one site to another, for example, from one well to another, the solute is initially absent

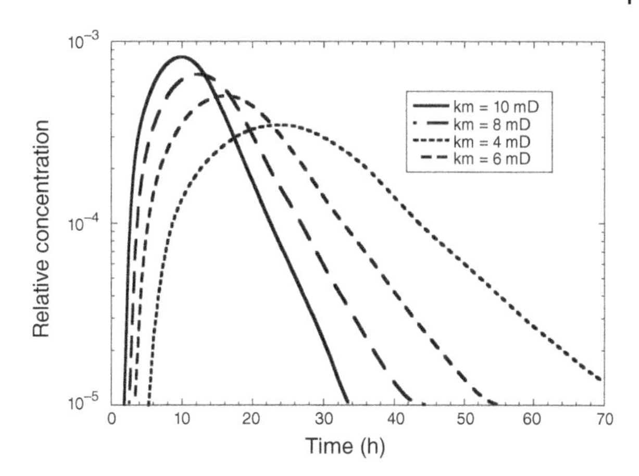

Figure 11.6 Solute concentrations observed at the outlet of the rectangular two-dimensional fractured/porous region in the simulations of Nick *et al.* (2011) for several different matrix permeabilities. A finite amount of solute was injected at one edge of the region, and the concentrations were measured at the opposing face. Note that 1 mD $\approx 10^{-15}$ m^2. Source: Adapted from Nick *et al.* (2011).

from the arrival site, and the moment at which the solute first arrives at the destination, or sampling location, is considered to be the "breakthrough" time. For a rock mass with a permeable matrix, breakthrough is expected to be delayed when the hydraulic gradients are low (Zhao *et al.*, 2013). The interplay between advection and diffusion greatly affects the arrival time, or "breakthrough" time, as well as the residence time in the rock mass, or "tail" of the solute in the medium. Matrix permeability increases dispersion, whereas the fracture network – in general – promotes advection, but is also observed to promote dispersion, depending on the stress state and fluid pressure gradient in the medium. Their relative effects also depend, of course, on the geometry and density of the underlying fracture network. Fractures that are not connected to the main fluid pathways and which are considered to be "inactive," have also been found to contribute to diffusive transfer of solids into the rock mass through micro-fractures and pore-space migration (Anders *et al.*, 2014).

11.9 Effects of *In Situ* Stresses

Closure of fractures has been identified as one of the most important mechanisms controlling permeability changes in response to *in situ* stresses (Zhao *et al.*, 2013). Larger *in situ* stresses have the effect of closing fractures, leading to smaller apertures, and thereby leading to longer residence times of the solute in the medium. For principle stress ratios as large as about three, average residence times also increase as compared to the no-stress scenarios. However, when the stress ratio exceeds three, *i.e.*, for "large" differential stresses, anisotropic compression leads to shear dilation and the formation of preferential flow channels, which locally increase fluid velocity, leading to decreasing residence times (Zhao *et al.*, 2013).

Stresses affect the distribution of apertures. Under isotropic conditions, many fractures close, but under anisotropic conditions, fractures that are preferentially oriented will open, creating permeable pathways. The distribution of these fractures leads to the formation of channelized fluid pathways, which in turn also promotes the channeling of solute migration. This channeling of fluids and solutes is thus regarded as being due to the heterogeneity of the rock mass's permeability, directly influenced by the applied stress and fracture geometry. In addition to these effects, fracture growth and coalescence can also lead to modifications in fracture density and connectivity, which can further influence transport.

Stress anisotropy is an important factor affecting predicted dispersion, as it induces shear dilation in a subset of preferentially oriented fractures. Kang *et al.* (2019) investigated the effects of stress on transport using particle-tracking simulations applied to geologically mapped fracture networks in 2D. They observed that stress anisotropy promotes the formation of preferential flow paths, which then further affect the anomalous behavior of the solute transport.

When fractured rocks are deformed under shear, the apertures and permeability of the fracture will change, which subsequently affects the solute transport in the fractures. Shear stress deforms the walls of the fracture surfaces and has a strong influence on flow properties along the shear direction, as discussed in Chapter 3. This has been observed both experimentally (*e.g.*, Olsson and Barton, 2001) and numerically (Min *et al.*, 2004). Increasing shear displacement of the fracture walls results in a permeability increase, inducing both heterogeneity and anisotropy in the flow. Fractures have a range of orientations and scales and will dilate as a function of both of these factors.

Laboratory studies have shown that when deforming a single fracture under shear, fracture permeability first decreases in the direction of flow before the peak shear stress is reached; see eq. (3.35). During this time, shear deformation causes the fracture to close. This deformation depends on the roughness of the fracture surface, the normal stress, and the stiffness of the fracture. As shear continues to increase, the fracture reopens, effectively changing the distribution of apertures and reorganizing the structure of the void space, subsequently increasing the permeability (Xiong *et al.*, 2011). Therefore, shear dilation ultimately influences permeability, transport, and the travel time distribution of solutes within the medium. Shear can also induce the formation of channels. The redistribution of apertures resulting from shear deformation tends to increase flow rates and travel velocities in the direction perpendicular to the shear.

This effect is also observed at the field scale, for example, by evaluating *in situ* permeability from observed micro-seismic events that occur during shear deformation in a geothermal reservoir (Fang *et al.*, 2018), indicating that the effects of local stress variations can be directly measured in the field. The links between shear deformation, aperture changes, and fracture roughness, have also been observed in numerical experiments (Koyama *et al.*, 2006) that link shear deformation to the changes in permeability of three-dimensional fractured media. These effects can be harnessed to engineer changes in the subsurface permeability, for example, when conducting controlled shear stimulation of geothermal wells.

Problems

11.1 Consider a 100 m × 100 m × 100 m block of granite that contains four parallel disk-like fractures of radius 1 m, that intersect a larger 20 m fracture that is tilted at 45° and positioned at the center of the rock mass. Use one of the methods described in this chapter to approximate the effect of the fracture aperture on the arrival time of a solute traveling from one face of the rock mass to the other.

11.2 What are the effects of chemical processes, such as dissolution and precipitation, likely to be on solute transport through a fractured rock system? Consider the effects of rapid precipitation during injection of a reactive brine, as opposed to long-term precipitation that occurs over geological time.

11.3 How likely are mechanical changes during injection or extraction of fluids likely to influence solute transport through fractured rock masses?

11.4 Write a script that generates random changes of concentration, based on eq. (11.15). What is the relative effect of A_q, Q, and D_e on the resulting distribution of concentrations?

11.5 Provide an explanation of why t_w scales with $1/Q$, based on the relationship between residence time and changes of concentration as a function of flux.

References

Anders, M. H., Laubach, S. E., and Scholz, C. H. 2014. Microfractures: a review. *Journal of Structural Geology*, 69(B), 377–94.

Baliga, B. R., and Patankar, S. V. 1980. A new finite element formulation for convection diffusion problems. *Numerical Heat Transfer*, 3(4), 393–409.

Baliga, B. R., and Patankar, S. V. 1983. A control volume finite-element method for two dimensional fluid flow and heat transfer. *Numerical Heat Transfer*, 6(3), 245–61.

Bear, J. 1988. *Dynamics of Fluids in Porous Media*, Dover Publications, Mineola, N.Y.

Berkowitz, B. 2002. Characterizing flow and transport in fractured geological media – a review. *Advances in Water Resources*, 25(8–12), 861–84.

Berkowitz, B., and Scher, H. 1997. Anomalous transport in random fracture networks. *Physical Review Letters*, 79(20), 4038–41.

Berkowitz, B., and Scher, H. 1998. Theory of anomalous chemical transport in random fracture networks. *Physical Review E*, 57(5), 5858–69.

Berkowitz, B., Scher, H., and Silliman, S. E. 2000. Anomalous transport in laboratory-scale, heterogeneous porous media. *Water Resources Research*, 36(1), 149–58.

Berkowitz, B., Cortis, A., Dentz, M., and Scher, H. 2006. Modelling non-Fickian transport in geological formations as a continuous time random walk. *Reviews of Geophysics*, 44(2), RG2003.

Berkowitz, B., Kosakowski, G., Margolin, G., and Scher, H. 2001. Application of continuous time random walk theory to tracer test measurements in fractured and heterogeneous porous media. *Ground Water*, 39(4), 593–604.

Bijeljic, B., and Blunt, M. J. 2006. Pore-scale modeling and continuous time random walk analysis of dispersion in porous media. *Water Resources Research*, 42(1), W01202.

Bijeljic, B., Mostaghimi, P., and Blunt, M. J. 2013. Insights into non-Fickian solute transport in carbonates. *Water Resources Research*, 49(5), 2714–28.

Cortis, A., and Birkholzer, J. 2004. Anomalous transport in "classical" soil and sand columns. *Soil Science Society of America Journal*, 68, 1539–48.

Cortis, A., and Birkholzer, J. 2008. Continuous time random walk analysis of solute transport in fractured porous media. *Water Resources Research*, 44(6), W06414.

Courant, R., Friedrichs, K., and Lewy, H. 1928. Uber die partiellen Differenzengleichungen der mathematischen Physik [On the partial differential equations of mathematical physics]. *Mathematische Annalen*, 100(1), 32–74.

de Marsily, G. 1986. *Quantitative Hydrogeology*, Academic Press, New York.

Di Donato, G., Obi, E. O., and Blunt, M. J. 2003. Anomalous transport in heterogeneous media demonstrated by streamline-based simulation. *Geophysical Research Letters*, 30(12), 1608.

Fang, Y., Elsworth, D., and Cladahous, T. T. 2018. Reservoir permeability mapping using micro-earthquake data. *Geothermics*, 72(1), 83–100.

Geiger, S., Driesner, T., Heinrich, C. A., and Matthäi, S. K. 2006. Multiphase thermohaline convection in the Earth's crust: I. A new finite element-finite volume solution technique combined with a new equation of state for NaCl-H_2O. *Transport in Porous Media*, 63(3), 399–434.

Gelhar, L. W., Welty, C., and Rehfeldt, K. R. 1992. A critical review of data on field-scale dispersion in aquifers. *Water Resources Research*, 28(7), 1955–74.

Ghez, R. 1988. *A Primer of Diffusion Problems*, Wiley, New York.

Gylling, B. 1997. *Development and Applications of the Channel Network Model for Simulations of Flow and Solute Transport in Fractured Rock*, PhD dissertation, Royal Institute of Technology (KTH), Stockholm.

Hoffman, F., Ronen, D., and Pearl, Z. 1996. Evaluation of flow characteristics of a sand column using magnetic resonance imaging. *Journal of Contaminant Hydrology*, 22(1–2) 95–107.

Huber, R., and Helmig, R. 2000. Node-centered finite volume discretizations for the numerical simulation of multiphase flow in heterogeneous porous media. *Computational Geosciences*, 4, 141–64.

Huyakorn, P. S., and Pinder, G. F. 1983. *Computational Methods in Subsurface Flow*, 3rd ed., San Diego, Academic Press.

Kang, P. K., Lei, Q., Dentz, M., and Juanes, R. 2019. Stress-induced anomalous transport in natural fracture networks. *Water Resources Research*, 55, 4163–85.

Klafter, J., and Silbey, R. 1980. Derivation of continuous-time random-walk equations. *Physical Review Letters*, 44(2), 55–58.

Kosakowski, G., Berkowitz, B., and Scher, H. 2001. Analysis of field observations of tracer transport in a fractured till. *Journal of Contaminant Hydrology*, 47(1), 29–51.

Koyama, T., Fardin, N., Jing, L., and Stephansson, O. 2006. Numerical simulation of shear-induced flow anisotropy and scale-dependent aperture and transmissivity evolution of rock fracture replicas. *International Journal of Rock Mechanics and Mining Sciences*, 43(1), 89–106.

Levy, M., and Berkowitz, B. 2003. Measurement and analysis of non-Fickian dispersion in heterogeneous porous media. *Journal of Contaminant Hydrology*, 64(3, 4), 203–26.

Matthäi, S. K. 2003. Fluid flow and (reactive) transport in fractured and faulted rock. *Journal of Geochemical Exploration*, 78, 79, 179–82.

Matthäi, S. K., Nick, H. M., Pain, C. C., and Neuweiler, I. 2010 Simulation of solute transport through fractured rock: a higher-order accurate finite-element finite-volume method permitting large time steps. *Transport in Porous Media*, 83(2), 289–318.

Min, K.-B., Rutqvist, J., Tsang. C.-F., and Jing, L. 2004. Stress-dependent permeability of fractured rock masses: a numerical study. *International Journal of Rock Mechanics and Mining Sciences*, 41(7), 1191–1210.

Montroll, E. W., and Weiss, G. H. 1965. Random walks on lattices. II. *Journal of Mathematical Physics*, 6(2), 167–81.

Moreno, L., and Neretnieks, I. 1993. Fluid flow and solute transport in a network of channels, *Journal of Contaminant Hydrology*, 14(3, 4), 163–92.

Moreno, L., Crawford, J., and Neretnieks, I. 2006. Modeling radionuclide transport for time varying flow in channel network. *Journal of Contaminant Hydrology*, 86(3, 4), 215–38.

Neretnieks, I. 1980. Diffusion in the rock matrix: an important factor in radionuclide retardation? *Journal of Geophysical Research*, 85(B8), 4379–97.

Neretnieks, I. 2002. A stochastic multi-channel model for solute transport—analysis of tracer tests in fractured rock. *Journal of Contaminant Hydrology*, 55(3, 4), 175–211.

Nick, H. M., Paluszny, A., Blunt, M. J., and Matthäi, S. K. 2011. Role of geomechanically grown fractures on dispersive transport in heterogeneous geological formations. *Physical Review E*, 84, 056301.

Olsson, R., and Barton, N. 2001. An improved model for hydromechanical coupling during shearing of rock joints. *International Journal of Rock Mechanics and Mining Sciences*, 38(3), 317–29.

Paluszny, A., Matthäi, S. K., and Hohmeyer, M. 2007. Hybrid finite element-finite volume discretization of complex geologic structures and a new simulation workflow demonstrated on fractured rocks. *Geofluids*, 7(2), 186–208.

Scheidegger, A. E. 1974. *The Physics of Flow Through Porous Media*, 3rd ed., University of Toronto Press, Toronto.

Scher, H., and Lax, M. 1973. Stochastic transport in a disordered solid, I, Theory, *Physical Review B*, 7(10), 4491–4502.

Scher, H., and Montroll, E. 1975. Anomalous transit-time dispersion in amorphous solids. *Physical Review B*, 12(6), 2455–77.

Scher, H., Shlesinger, M. F., and Bendler, J. T. 1991. Time-scale invariance in transport and relaxation. *Physics Today*, 44(1), 26–34.

Shlesinger, M. F. 1996. Random processes. *Encyclopedia of Applied Physics*, 16, 45–70.

Sidle, C., Nilson, B., Hansen, M., and Fredericia, J. 1998. Spatially varying hydraulic and solute transport characteristics of a fractured till determined by field tracer tests, Funen, Denmark, *Water Resources Research*, 34(10), 2515–27.

Soler, J. M., Neretnieks, I., Moreno, L., Liu, L., *et al.* 2022. Predictive modeling of a simple field matrix diffusion experiment addressing radionuclide transport in fractured rock. *Nuclear Technology*, 208, 1059–73.

Thomas, R. N., Paluszny, A., and Zimmerman, R. W. 2020. Permeability of three-dimensional numerically grown geomechanical discrete fracture networks with evolving geometry and mechanical apertures *Journal of Geophysical Research: Solid Earth*, 125(4), e2019JB018899.

Tsang, C. F., and Neretnieks, I. 1998. Flow channeling in heterogeneous fractured rocks. *Reviews of Geophysics*, 36(2), 275–98.

Tsang, Y. W., and Tsang, C. F. 2001. A particle-tracking method for advective transport in fractures with diffusion into finite matrix blocks. *Water Resources Research*, 37(3), 831–36.

Xiong, X., Li, B., Jiang, Y., Koyama, T., *et al.*, 2011. Experimental and numerical study of the geometrical and hydraulic characteristics of a single rock fracture during shear. *International Journal of Rock Mechanics and Mining Sciences*, 48(8), 1292–1302.

Yamashita, R., and Kimura, H. 1990. Particle-tracking technique for nuclide decay chain transport in fractured porous media. *Journal of Nuclear Science and Technology*, 27(11), 1041–49.

Zech, A., Attinger, S., Bellin, A., Cvetkovic, V., *et al.*, 2019. Critical analysis of transverse dispersivity field data. *Groundwater*, 57(4), 632–39.

Zhao, Z., Jing, L., Neretnieks, I., and Moreno, L. 2011. Numerical modeling of stress effects on solute transport in fractured rocks. *Computers and Geotechnics*, 38(2), 113–26.

Zhao, Z., Rutqvist, J., Leung, C., Hokr, M., *et al.*, 2013. Impact of stress on solute transport in a fracture network: a comparison study. *Journal of Rock Mechanics and Geotechnical Engineering*, 5(2), 110–23.

12

Two-Phase Flow in Fractured Rocks

12.1 Introduction

All previous chapters of this book have focused on situations in which the pore space of the fracture, or the fractured rock mass, is filled with a single-phase fluid. However, in many scenarios of practical importance, the void space contains more than one fluid. In some situations, the two fluids are *miscible*, in which case they will mix together on a molecular scale and essentially form a single-phase fluid with properties that are homogeneous on a scale that is relevant to flow processes occurring in a rock. The most common example of two miscible fluids would be (any) two gases, such as, for example, nitrogen and oxygen. These two gases easily mix together to form "air," which behaves on a macroscopic scale as a single-phase gaseous fluid.

However, many fluid pairs are essentially immiscible, such as water and most liquid hydrocarbons, or water and air. If these fluid pairs are present in a porous or fractured rock, on the scale of a small portion of the fracture or on a pore scale, the two fluids will segregate themselves into separate regions that contain one fluid or the other, but not both, with a definable interface separating the two fluids. On the other hand, macroscopic regions of rock, on the scale of, say, centimeters or greater, will contain some amount of each fluid.

One example of a situation in which two-phase immiscible fluid flow occurs is an oil reservoir, which will always contain a combination of oil, water, and possibly gaseous hydrocarbons. Another example is the *vadose zone* between the water table and the ground surface, where the void space contains a combination of water and air. This zone is also often referred to as the *unsaturated zone*, and fluid flow under these conditions is known as "unsaturated flow" (Hillel, 1980). Vapor-dominated geothermal systems, in which the pore space of the rock is filled with a mixture of liquid water and steam (Bodvarsson and Tsang, 1982), are another example of a subsurface situation in which two-phase flow is important.

Understanding and modeling two-phase flow in fractured rocks requires the consideration of a few additional concepts that do not arise when studying single-phase flow, such as phase saturation, relative permeability, and capillary pressure. These concepts, along with the governing equations for two-phase flow, are discussed in Section 12.2. The conceptual and computational model of the flow of two fluid phases through a single rock fracture that was developed by Pruess and Tsang (1990) is presented in Section 12.3. Additional models and observations of two-phase flow in a single fracture are discussed

Fluid Flow in Fractured Rocks, First Edition. Robert W. Zimmerman and Adriana Paluszny.

Companion website: www.wiley.com/go/zimmerman/fluidflowinfracturedrocks

in Section 12.4. Dual-porosity and dual-permeability models of two-phase flow through fractured rock masses are discussed in Section 12.5. Finally, recent work on numerically modeling two-phase flow through fractured rocks, using discrete-fracture networks, is briefly reviewed in Section 12.6.

12.2 Basic Concepts of Two-Phase Flow

Most of the main concepts governing two-phase flow in porous media were originally developed for non-fractured porous media but have generally been found to also apply to fractures and fractured rocks. Many of the classic monographs on fluid flow in porous media, such as de Marsily (1986) and Bear (1988), devote one or two chapters to two-phase flow. Marle (1981) is devoted entirely to the mathematics of two-phase flow, with a specific focus on oil-water systems. Blunt (2017) presents a modern view of multi-phase flow processes in porous media, with an emphasis on pore-scale phenomena. A useful and insightful review of the topic of two-phase flow in rock fractures can be found in the broader review article on fluid flow in single fractures by Phillips *et al.* (2020). In this section, the basic ideas and equations of two-phase flow will be presented in a general form that applies equally well to oil-water systems, water-air systems, brine-CO_2 systems, or any situation in which the void space contains more than one fluid phase.

Consider a small region of a fracture, such as shown in Fig. 12.1a, portions of which are filled by fluid A and portions of which are filled by fluid B. In most cases, the rock surface will have more affinity for one fluid than for the other fluid; this former fluid is said to "wet" the rock surface and is referred to as the "wetting phase." The two fluids will then tend to configure themselves in such a way that, subject to the constraints imposed by the fact that fractures will generally contain differing amounts of the two fluids, there will be more surface contact between the rock surface and the wetting phase than between the rock surface and the non-wetting phase. In Fig. 12.1a, fluid A is the wetting phase, and fluid B is the non-wetting phase. In unsaturated rocks that contain a mixture of air and water, water is usually the wetting phase. The situation is more complex in oil reservoirs, which may be oil-wet, water-wet, or in an intermediate state of mixed wettability (Blunt, 2017). The following discussion will assume that there is a clearly defined wetting phase and a non-wetting phase.

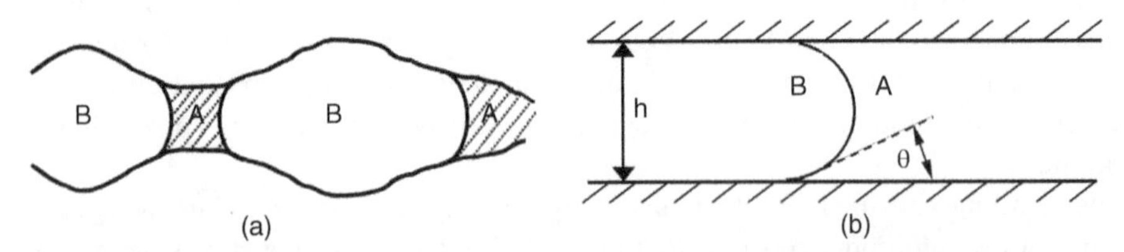

Figure 12.1 Slice, normal to the nominal fracture plane of a rock fracture saturated with a wetting phase liquid, A, and a non-wetting phase liquid, B. (a) Rough-walled fracture, with wetting phase preferentially located in the regions of smaller aperture. (b) Interface between two fluids in a fracture. Assuming that the shape of the interface is the same for all planes parallel to the page, the radius of curvature R will be given by $R = h/2 \cos \theta$, where h is the local aperture and θ is the contact angle.

If two fluids are separated by a flat interface, such as occurs when air sits on top of a pool of water, the pressures in the two phases will necessarily be equal at the interface. But when the interface between the two fluids is curved, the pressures in the two phases will differ by an amount that is proportional to the surface tension (also referred to as the interfacial tension) γ_{wn} [N/m] of the interface between the two phases and inversely proportional to the mean radius of curvature of the interface, R [m]. The difference between these two pressures is known as the *capillary pressure*, p_c, *i.e.*,

$$p_c \equiv p_n - p_w = \frac{2\gamma_{wn}}{R}, \tag{12.1}$$

where the subscripts $\{w, n, c\}$ denote wetting phase, non-wetting phase, and capillary, respectively. According to eq. (12.1), the pressure will be greater in the non-wetting phase than in the wetting phase.

In general, the mean radius of curvature is defined as (Ye *et al.*, 2015)

$$\frac{1}{R} = \frac{1}{2}\left(\frac{1}{R_1} + \frac{1}{R_2}\right), \tag{12.2}$$

where R_1 and R_2 are the two principal radii of curvature of the surface. If one phase exists as a spherical bubble of radius R surrounded by the other phase, the two principal radii of curvature will be $R_1 = R_2 = R$, and so the mean radius of curvature of the interface will be equal to the radius of the bubble, R.

A geometrical configuration that is of more relevance to rock fractures is shown in Fig. 12.1b, in which the shape of the interface is assumed to be the same for all planes parallel to the page. For this configuration, the radius of curvature R_1 in the out-of-the-page direction is infinite, and a simple trigonometric calculation shows that the radius of curvature in the plane of the page is $R_2 = h/2\cos\theta$, where h is the local aperture of the fracture, and the contact angle θ is defined in the figure. Hence, for this configuration, it follows from eqs. (12.1) and (12.2) that (Kueper and McWhorter, 1991; Glass *et al.*, 2003; Phillips *et al.*, 2020)

$$p_c = \frac{2\gamma_{wn}\cos\theta}{h}. \tag{12.3}$$

The contact angle θ is not arbitrary but is in fact controlled by the surface tensions acting between pairs of the three relevant phases: the wetting phase fluid, the non-wetting phase fluid, and the solid rock. It can be shown from thermodynamic considerations (Huinink, 2016, eq. 3.13) that

$$\cos\theta = \frac{\gamma_{ns} - \gamma_{ws}}{\gamma_{wn}}, \tag{12.4}$$

where γ_{wn} is the surface tension of the interface between the two fluid phases, γ_{ws} is the surface tension between the wetting phase fluid and the rock, and γ_{ns} is the surface tension between the non-wetting phase fluid and the rock.

The most salient feature that emerges from the above analysis is that capillary pressures in a rock fracture will be proportional to surface tensions and inversely proportional to aperture. Typical values of the surface tension for an interface between fluid pairs that are commonly present in fractured rocks are about 0.07 N/m for air–water, 0.02 N/m for oil-water systems, and 0.03 N/m for CO_2-water (brine). Hence, capillary pressures in a fracture that has an aperture of 1 mm will be on the order of 100 Pa, whereas capillary pressures in a fracture that has an aperture of 10 μm will be on the order of about 10 kPa, for example.

When averaged over a larger area of the nominal fracture plane, the void space of the fracture will contain some amount of wetting phase fluid and some amount of non-wetting phase fluid. The *saturation* of the wetting phase, denoted by S_w, is defined as the fraction (by volume, not by area) of the fracture void space that is filled with the wetting phase fluid. Similarly, S_n is the fraction of the fracture void space that is filled with the non-wetting phase. For two-phase conditions, it is necessarily the case that $S_w + S_n = 1$.

Although on a "micro-scale" the capillary pressure is related to the mean radius of curvature of the interface between the two fluids, on a "macro-scale" the capillary pressure is a function of the phase saturation, *i.e.*, it is described by some function $p_c(S_w)$. The specific shape of this function depends not only on the three surface tensions $\{\gamma_{ns}, \gamma_{ws}, \gamma_{wn}\}$ but also on the geometry of the void space. The capillary pressure function $p_c(S_w)$ is therefore a property of the fracture + fluid system. Moreover, most porous media display *hysteresis* with respect to the capillary pressure as a function of saturation, in the sense that the $p_c(S_w)$ relation also depends on whether the wetting phase saturation is increasing or decreasing.

Consider the process of *primary drainage*, in which a porous medium is initially fully saturated with the wetting phase, at some pressure p_w, and then the wetting fluid is displaced by the non-wetting phase. In this process, the *wetting* phase is *draining* out of the porous medium; by convention, this process is known as "drainage." When the wetting phase fluid displaces the non-wetting phase fluid, that process is known as "imbibition." Both definitions are therefore based on the motion of the wetting phase fluid. According to eq. (12.1), the non-wetting phase would need to have a pressure greater than p_w in order to begin to displace the wetting phase. If a fracture is idealized as comprising a set of channels of different apertures, eqs. (12.1) and (12.3) imply that the non-wetting phase will not be able to begin to invade the fracture until p_n reaches $p_w + (2\gamma \cos\theta / h_{\max})$, where $h_{\max}$ is the aperture of the largest channel. This minimum capillary pressure that is needed in order for drainage to commence, $2\gamma \cos\theta / h_{\max}$, is called the "entry pressure," p_e. If p_n continues to increase, smaller and smaller channels will be invaded by the non-wetting phase. Hence, since $p_c = p_n - p_w$, the capillary pressure will be an increasing function of the non-wetting phase saturation, which is to say, it will be a decreasing function of the wetting phase saturation.

During primary drainage, some of the wetting phase fluid invariably gets trapped behind, in isolated ganglia that become immobile, and so the wetting phase saturation does not reduce to zero, but rather reaches some finite value known as the irreducible saturation, S_{wi}. "Typical" shapes of the capillary pressure function for a porous medium or a rock fracture during primary drainage are shown in Fig. 12.2a.

If a fracture is partially filled with a non-wetting phase fluid, the presence of this fluid will to some extent hinder the ability of the wetting phase to flow through the fracture, and *vice versa* with regards to the mobility of the non-wetting phase. To account for this phenomenon, Darcy's law is traditionally modified by assuming that the volumetric

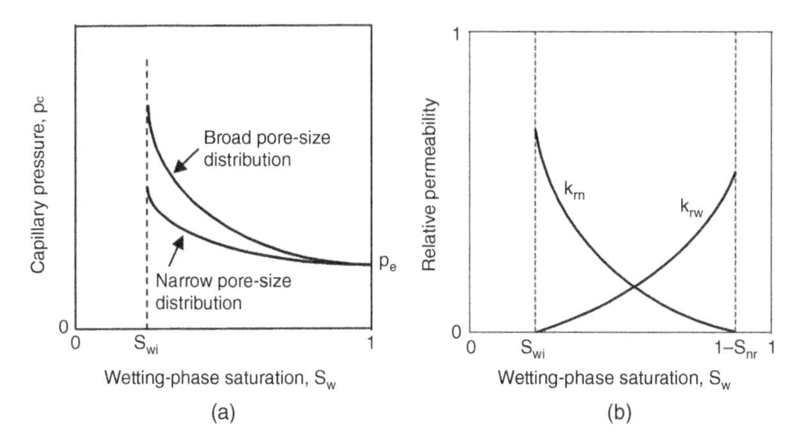

Figure 12.2 (a) Schematic capillary pressure functions during primary drainage, for a porous rock or a rock fracture, showing the finite "capillary entry pressure," p_e. (b) Schematic relative permeability curves, showing the irreducible wetting phase saturation, S_{wi}, and the residual non-wetting phase saturation, S_{nr}.

flux of each phase is proportional to the pressure gradient in that phase and that the permeability of the rock to that phase is reduced by some fractional amount that is represented by a dimensionless *relative permeability function*. Hence, under two-phase flow conditions, Darcy's law takes the following forms in one dimension (Pruess and Tsang, 1990):

$$q_w = -\frac{kk_{rw}(S_w)}{\mu_w}\left(\frac{dp_w}{dx} - \rho_w g_x\right), \quad q_n = -\frac{kk_{rn}(S_w)}{\mu_n}\left(\frac{dp_n}{dx} - \rho_n g_x\right), \tag{12.5}$$

where g_x is the component of the gravitational acceleration in the x-direction, and $k_{rw}(S_w)$ and $k_{rn}(S_w)$ are the relative permeability functions of the wetting phase, and non-wetting phase, respectively.

"Typical" relative permeability curves are shown schematically in Fig. 12.2b. Although $k_{rw}(S_w)$ is an increasing function of S_w, and $k_{rn}(S_w)$ is always a decreasing function of S_w, it is generally not the case that $k_{rw} + k_{rn} = 1$, and there is no general or simple functional relation between $k_{rw}(S_w)$ and $k_{rn}(S_w)$ for a given fracture. Essentially, by definition, when the wetting phase saturation reaches S_{wi}, this phase will become immobile, and so $k_{rw}(S_{wi})$ will become zero. Since the trapped wetting phase will, to some extent, still impede the flow of the non-wetting phase, $k_{rn}(S_{wi})$ will be less than 1. Note that the relative permeability of each phase is reckoned relative to the state in which the saturation of that phase is 1.

The pressures in the two phases will be related to each other through the capillary pressure function,

$$p_n - p_w = p_c(S_w). \tag{12.6}$$

Since $S_n = 1 - S_w$, either phase saturation can be used as the mathematical variable in the capillary pressure and relative permeability functions, although it is common to use S_w.

12.3 Pruess–Tsang Model of Two-Phase Flow in a Single Fracture

For single-phase flow, the parallel-plate model leads to the cubic law and therefore provides a useful first-order model for the fracture transmissivity under single-phase conditions. Unfortunately, the parallel-plate model does not lead to any meaningful or realistic predictions for two-phase flow. A useful model for two-phase flow in a rock fracture must necessarily account for the spatially varying aperture. The first model to attempt to relate the capillary pressure function and the relative permeability functions to the statistics of the aperture distribution was developed by Pruess and Tsang (1990). Their model was based on the aforementioned idea that a capillary pressure at least as large as $2\gamma\cos\theta/h$ is needed in order for the wetting phase to enter a region having aperture h. Therefore, at a given capillary pressure p_c, all regions of the fracture plane that have an aperture less than the "critical aperture," $h_c = 2\gamma\cos\theta/p_c$, are assumed to be filled with the wetting phase fluid, and all regions of the fracture plane that have an aperture greater than h_c are assumed to be filled with the non-wetting phase. This model ignores "accessibility" issues, in which, for example, some regions having smaller aperture may be inaccessible to the wetting phase fluid, by virtue of being completely surrounded by the non-wetting phase (Bertels *et al.*, 2001).

Pruess and Tsang (1990) began by defining an aperture distribution function $f(h)$, such that the fraction of the nominal fracture plane that has an aperture between h and $h + dh$ is equal to $f(h)dh$, subject to the normalization condition that $\int_0^\infty f(h)dh = 1$. The mean aperture can be computed as

$$\langle h\rangle = h_m = \int_0^\infty hf(h)dh. \tag{12.7}$$

As all regions with aperture less than h_c are assumed to be filled with the wetting phase fluid, the wetting phase saturation can be computed as

$$S_w = \frac{1}{h_m}\int_0^{h_c} hf(h)dh. \tag{12.8}$$

Evaluation of this integral gives the saturation as a function of the critical aperture, and hence gives the saturation as a function of the capillary pressure, through the relationship $h_c = 2\gamma\cos\theta/p_c$. Inversion of the $S_w(p_c)$ relation then yields the desired capillary pressure function, $p_c(S_w)$.

Consider a commonly observed aperture distribution function such as the log-normal distribution, which is given by (see Problem 12.1)

$$f(h) = \frac{1}{\sqrt{2\pi}\sigma}\frac{1}{h}e^{-[\ln(h/h_o)]^2/2\sigma^2}, \tag{12.9}$$

in which h_o is the *median* aperture, which has the property that exactly half of the fracture plane has an aperture less than h_o, and half of the fracture plane has an aperture greater than h_o. According to eqs. (12.7)–(12.9), the *mean* aperture is given by

$$h_m = \frac{1}{\sqrt{2\pi}\sigma}\int_0^\infty e^{-[\ln(h/h_o)]^2/2\sigma^2}dh = h_o e^{\sigma^2/2}, \tag{12.10}$$

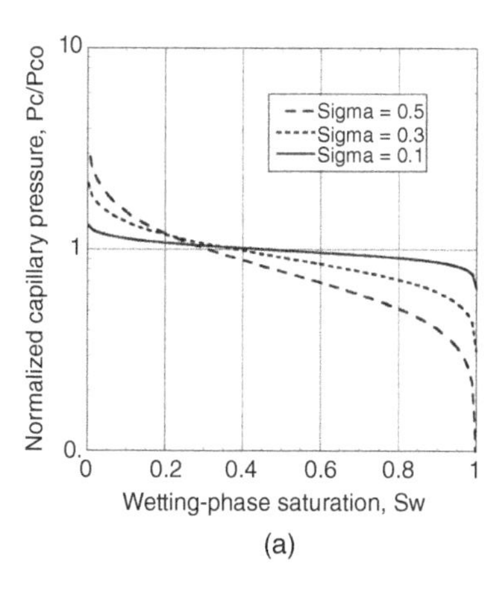

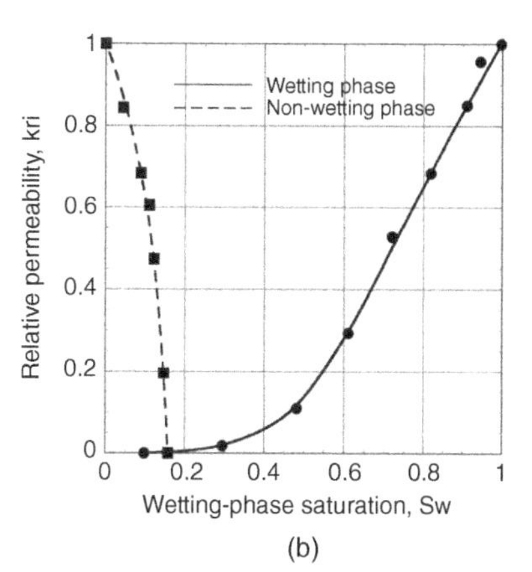

Figure 12.3 (a) Capillary pressure functions predicted by the Pruess–Tsang model, normalized against the entry pressure of the median aperture, for several values of the standard deviation of the aperture distribution. (b) Relative permeability curves predicted by the Pruess–Tsang model for a fracture having a mean aperture of $h_m = 81.8$ μm, an aperture standard deviation of $\sigma = 0.43$, and an isotropic correlation structure.

and the saturation varies with the critical aperture according to

$$S_w = \frac{1}{2}\text{erfc}\left[\frac{\ln(p_c/p_{co}) + \sigma^2}{\sqrt{2}\sigma}\right], \tag{12.11}$$

where "erfc" denotes the complementary error function (Ghez, 1988) and $p_{co} = 2\gamma\cos\theta/h_o$ is the capillary entry pressure that corresponds to the median aperture, h_o. The $S_w(p_c)$ relation given by eq. (12.11) can be explicitly inverted to yield

$$p_c(S_w) = p_{co}\exp[\sqrt{2}\sigma\,\text{erfc}^{-1}(2S_w) - \sigma^2], \tag{12.12}$$

where erfc^{-1} is the inverse complementary error function. The normalized capillary pressure curves that are predicted by this model are shown in Fig. 12.3a for several values of σ.

Pruess and Tsang (1990) computed the flow of each phase of fluid by discretizing the fracture plane into a square grid of computational cells and assigning an aperture to each cell according to the log-normal distribution function, eq. (12.9). They assumed a (possibly anisotropic) exponential spatial covariance with given correlation lengths in the x- and y-directions. The transmissivity of each cell was then assigned according to the local cubic law. For each value of the capillary pressure, cells having an aperture less than the cut-off aperture $h_c = 2\gamma\cos\theta/p_c$ were assumed to be filled with the wetting phase, and the standard boundary conditions of no flow on two opposing edges of the fracture plane and constant pressures on the other two opposing edges were applied. The flux of each fluid was then computed using an integrated finite difference scheme. After dividing the computed flux by the computed single-phase transmissivity, the relative permeability of the non-wetting phase was thereby determined. An analogous calculation was carried out for the wetting phase.

The relative permeability curves obtained for a fracture having a mean aperture of $h_m = 81.8$ μm, a standard deviation of $\sigma = 0.43$, and an isotropic correlation structure are shown in Fig. 12.3b. The calculations of Pruess and Tsang (1990) predicted an extremely high immobile non-wetting phase saturation of about 84% and a very small "window" of saturations for which both phases are mobile. The rapid drop-off of $k_{rn}(S_w)$ with S_w is due to the large number of small aperture cells that result from a log-normal distribution, which makes it relatively easy for a "percolating" connected pathway to exist for the wetting phase but difficult for such a connected pathway to exist for the non-wetting phase, except at very high values of S_n, which is to say, at very low values of S_w.

However, in two dimensions, it is not possible for both phases to simultaneously percolate through the fracture plane from "east to west" and from "north to south." This can easily be seen by noting, for example, that if any continuous path exists for phase A to percolate across the fracture from east to west, this path will necessarily prevent phase B from percolating across the fracture plane from north to south. Similar conclusions apply if a percolating path exists for phase A from north to south. Hence, the existence of a "window" of saturations for which both k_{rn} and k_{rw} are nonzero, such as was predicted by Pruess and Tsang (1990), may have been a numerical artifact arising from the integrated-finite-difference flow calculations conducted on a coarse computational grid. The above-mentioned topological considerations do not, however, rule out the possibility that both phases may percolate across the fracture plane in one direction but not in the other direction.

12.4 Other Models and Observations of Two-Phase Flow in a Single Fracture

Reitsma and Kueper (1994) measured the capillary pressure during primary drainage and subsequent imbibition in a natural fracture in a massive dolomitic limestone obtained from an outcrop in Ontario. Water was used as the wetting fluid, and electrical insulating oil was used as the non-wetting fluid. A large degree of hysteresis was observed, with the capillary pressure curve during imbibition lying below the capillary pressure curve during primary drainage (Fig. 12.4a). A noticeable capillary entry pressure was observed, with no non-wetting fluid entering the fluid until the capillary pressure reached about 300 Pa.

Reitsma and Kueper (1994) fit the primary drainage capillary pressure data using the two capillary pressure functions that are the most widely used for three-dimensional porous media: the Brooks–Corey function and the van Genuchten function. The model of Brooks and Corey (1966) assumes a power-law function of the form

$$p_c(S_w) = p_e\left(\frac{S_w - S_r}{1 - S_r}\right)^{-1/\lambda}, \tag{12.13}$$

in which p_e [Pa] is the capillary entry pressure, S_w is the wetting phase saturation, S_r is the residual wetting phase saturation, and λ [−] is a fitting parameter, generally satisfying $\lambda > 1$, that is intended to reflect the pore size (or aperture) distribution. The model of van Genuchten (1980) takes the form

$$p_c(S_w) = p_o\left[\left(\frac{S_w - S_r}{S_s - S_r}\right)^{-1/m} - 1\right]^{1-m}, \tag{12.14}$$

in which p_o [Pa] is some characteristic capillary pressure, S_s (usually very close to 1) is the wetting phase saturation at which the capillary pressure reaches zero, and m [−] is a

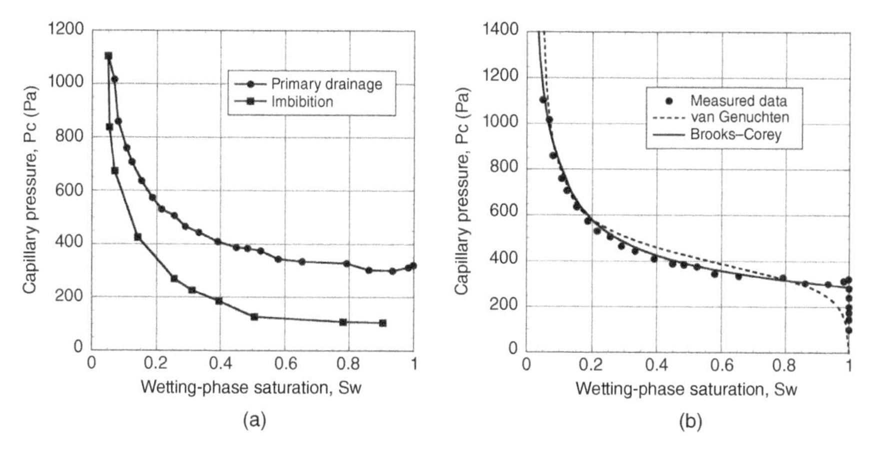

Figure 12.4 (a) Capillary pressures measured by Reitsma and Kueper (1994) in a natural fracture in a massive dolomitic limestone, during primary drainage and imbibition. Source: Adapted from Reitsma and Kueper (1994). (b) Capillary pressures measured during primary drainage, fit by a Brooks–Corey function and a van Genuchten function; see text for details.

fitting parameter that is also expected to reflect in some sense the pore size (or aperture) distribution, but whose value, which must lie in the range $0 < m < 1$, is usually chosen so as to allow eq. (12.14) to provide the best fit to measured data.

In their experiments, Reitsma and Kueper (1994) monitored the amount of water entering or leaving the fracture but were not able to measure the total pore volume of the fracture. Consequently, they reported their data in terms of "change in water volume," rather than "water saturation." These data can be converted to saturations by treating the total pore volume, and the residual wetting phase saturation, as fitting parameters. This conversion has essentially no influence on the conclusions regarding hysteresis, the value of the entry pressure, or the ability to fit the data with either of the theoretical model curves since an error in estimating the total volume or the residual saturation S_r would merely stretch or compress the saturation scale.

The best fits of these two models to the data of "test 1" of Reitsma and Kueper (1994) are shown in Fig. 12.4b. The best-fitting parameters for the Brooks–Corey model were found to be $p_e = 285$ Pa, $S_r = 0.013$, and $\lambda = 2.431$. The best-fitting parameters for the van Genuchten model were found to be $p_o = 393$ Pa, $S_r = 0.039$, $S_s = 1.0$, and $m = 0.816$. Both models provide a reasonably accurate fit to the measured data, although the Brooks–Corey model is somewhat more accurate and captures the finite "entry pressure" effect, which is not accounted for in the van Genuchten model.

Chen and Horne (2006) conducted two-phase flow experiments through an artificial fracture bounded by a flat aluminum plate and a rough-walled glass plate, using water and nitrogen as the wetting and non-wetting phase fluids, respectively. The mean aperture was 240 μm, and the aperture standard deviation was 50 μm. The fracture was initially fully saturated with water, after which the water and gas injection rates were controlled so as to slowly decrease the water saturation. The time-varying saturations were determined by imaging the flow process using a digital camcorder and automatic image processing methods. The measured relative permeabilities are shown in Fig. 12.5a.

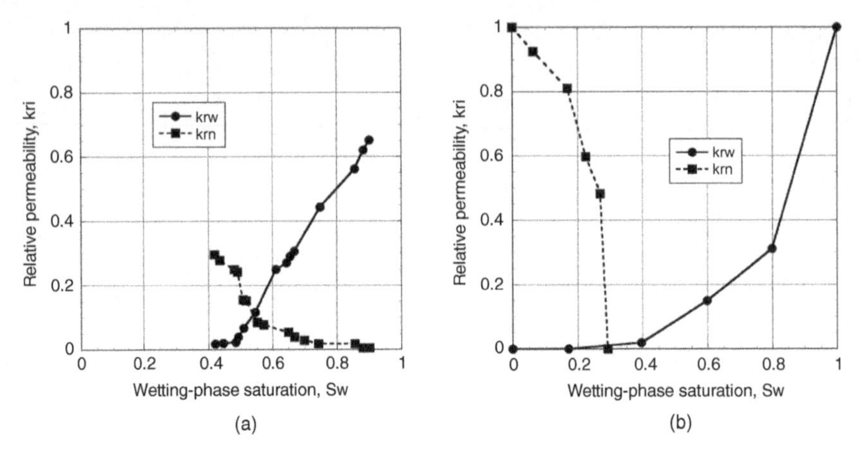

Figure 12.5 (a) Relative permeability functions measured by Chen and Horne (2006) on an artificial "randomly rough" fracture, with water as the wetting phase, and air as the non-wetting phase. Source: Adapted from Chen and Horne (2006). (b) Relative permeability functions computed numerically by Watanabe *et al.* (2015) for a granite fracture, with water as the wetting phase and decane as the non-wetting phase. Source: Adapted from Watanabe *et al.* (2015).

Watanabe *et al.* (2015) computed the relative permeabilities for a granite fracture, based on an aperture field that was measured with a laser profilometer that had positioning accuracy of ±20 μm in the fracture plane and ±1 μm with regard to asperity height. Water was the wetting phase, and decane was the non-wetting phase. The computed relative permeabilities are shown in Fig. 12.5b, along with curve fits of the form

$$k_{rw} = (S_w)^4, \tag{12.15}$$

$$k_{rn} = 1 - (S_w/0.3)^3 \text{ for } S_w < 0.3, \quad k_{rn} = 0 \text{ for } S_w > 0.3, \tag{12.16}$$

Huo and Benson (2016) studied the effect of normal stress on the relative permeability curves of a saw-cut fracture in an immature feldspathic greywacke sandstone. Ranjbaran *et al.* (2017) developed a model in which the wetting phase fluid adheres to the rough fracture walls, with the non-wetting phase fluid traveling through the central region of the fracture aperture, and used this model to investigate the saturation end-points of the relative permeability curves. Yang *et al.* (2019) used a modified invasion percolation model to numerically investigate the effect of aperture field anisotropy on the relative permeabilities and found that aperture anisotropy leads to increased relative permeabilities in both phases in the direction parallel to the fluid displacement and decreased relative permeabilities in the direction perpendicular to the displacement. Guiltinan *et al.* (2020) used lattice-Boltzmann simulations to investigate the effect of a heterogeneous distribution of the wettability of the fracture surface on the displacement of water by supercritical CO_2.

Two-phase flow of a liquid and a gas through a fracture may exhibit other types of behavior that cannot be fit into the paradigm that is described solely by capillary pressure functions and relative permeability curves. Tokunaga and Wan (1997) showed that, in air–water systems at low capillary pressures, gravity-driven flow of water may occur in the form of thin films along the rough-walled fracture surfaces. Further investigations by Tokunaga and Wan (2001) indicated that this type of film flow may be important in fractures having apertures greater than 30 μm, if the host-rock matrix permeability is less than 10^{-14} m^2. Su *et al.* (1999) conducted water flow experiments in partially saturated transparent epoxy replicas of a natural fracture in granite and found that even if care was taken to maintain constant boundary conditions, the system exhibited "intermittent" flow, in which finger-like flow channels would snap off, drain, and reform. Nevertheless, despite the small-scale spatial and temporal randomness of the flow field, the average flux showed a relatively smooth variation in time, when averaged over the entire fracture outflow surface. Jones *et al.* (2018) found that, particularly for fractures having large apertures, in which case capillary forces are negligible, flow may occur as film flow, "slug" flow of gas pockets dispersed in the liquid, small gas bubbles dispersed in the liquid, or liquid bubbles dispersed in the gas. AlQuaimi and Rossen (2017) investigated and modeled the process by which trapped ganglia of the non-wetting phase liquid can be dislodged and mobilized by a sufficiently large pressure gradient, even if the relative permeability of the non-wetting phase has been reduced to zero due to a lack of phase connectivity.

12.5 Dual-Porosity and Dual-Permeability Models for Two-Phase Flow

Two-phase flow in fractured rock masses has often been modeled using some version of the dual-porosity model, which was discussed in detail in Chapter 9 in the context of single-phase flow. A dual-porosity model for single-phase flow contains two flow equations based on mass conservation, one operating within the fracture system and one within the matrix blocks. In a dual-porosity model for two-phase flow, there will be four such equations, one for each fluid phase, in each of the two porosity systems. The presence of relative permeability curves and capillary pressure functions, in both the fracture network and the matrix blocks causes the resultant model to be intrinsically nonlinear.

In the case of air–water systems in the vadose zone above the water table, the situation can be simplified by appealing to the classical Richards approximation (Hillel, 1980), in which the air phase is assumed to be "infinitely" mobile due to its very low viscosity. This assumption allows the pressure in the air phase to be taken as being controlled by hydrostatic equilibrium, and consequently, a flow equation for the air phase is not needed. This model therefore reduces to an essentially single-phase flow model, with a pressure-dependent (or saturation-dependent) permeability. Although the resultant nonlinear model is not amenable to analytical solutions of boundary value problems such as flow to a well, dual-porosity models based on the Richards equation can be implemented numerically.

Gerke and van Genuchten (1993a) developed a dual-permeability model for vertical infiltration of water under unsaturated conditions. A dual-permeability model is conceptually slightly different than a dual-porosity model, in that both the fracture system *and* the matrix system are assumed to contain fully percolating pore networks. Hence, although fluid can be exchanged between the fractures and matrix blocks, as can occur in a dual-porosity model, in a dual-permeability model, fluid can also flow macroscopically through the matrix porosity system itself. Modifying their notation to be more consistent with that used in the present chapter, and ignoring the generally small contribution to the storativity due to the fluid compressibility, the equations of Gerke and van Genuchten can be written as

$$\frac{\partial S_{wf}}{\partial t} = \frac{\partial}{\partial z}\left[\frac{(kk_{rw})_f}{\mu}\frac{\partial}{\partial z}(p_f - \rho g z)\right] - \frac{\Gamma_w}{\phi_f}, \tag{12.17}$$

$$\phi_m \frac{\partial S_{wm}}{\partial t} = \frac{\partial}{\partial z}\left[\frac{(kk_{rw})_m}{\mu}\frac{\partial}{\partial z}(p_m - \rho g z)\right] + \frac{\Gamma_w}{1-\phi_f}, \tag{12.18}$$

in which S_w [–] is the saturation of the water phase, p [Pa] is the pressure in the water phase, μ [Pa s] is the water viscosity, ρ [kg/m^3] is the water density, k [m^2] is the permeability, k_{rw} [–] is the relative permeability of the water phase, z [m] is the vertical depth below the surface, Γ_w [1/s] is the fracture-matrix water flux term, and the subscripts f and m refer to the fracture and matrix pore systems, respectively. The saturations are related to the fluid pressures through the appropriate (fracture or matrix) capillary pressure functions, along with the assumption that the air phase is essentially at atmospheric pressure. The fracture-matrix flux, Γ_w, is assumed to be governed by a Warren–Root-type equation, similar to eq. (10.1) used in single-phase dual-porosity models:

$$\Gamma_w = \frac{\alpha k_{eff}}{\mu}(p_f - p_m), \tag{12.19}$$

where μ [Pa s] is the viscosity of the water, α [1/m^2] is the matrix block shape factor, and k_{eff} [m^2] is an "effective" permeability of the matrix block in the vicinity of the fracture-matrix interface, which implicitly varies with saturation through the relative permeability function. Gerke and van Genuchten (1993a) also included sink terms on the right-hand sides of eqs. (12.17) and (12.18), which represent water taken out of the fracture or matrix pore systems by plant roots.

In a subsequent paper, Gerke and van Genuchten (1993b) examined different methods of computing the effective permeability k_{eff} that appears in the fracture-matrix transfer function. Five different options for computing k_{eff} were investigated: (a) evaluating $(kk_{rw})_m$ at pressure p_m, (b) evaluating $(kk_{rw})_m$ at pressure p_f, (c) using the arithmetic mean of the two previous values, (d) using the geometric mean of the two previous values, and (e) taking an average value of $(kk_{rw})_m$ over the pressure range between p_m and p_f. They also considered the possibility of incorporating an exponent different than unity for the term $(p_f - p_m)$ in the fracture-matrix expression, eq. (12.19), *i.e.*, replacing $(p_f - p_m)$ with $(p_f - p_m)^\beta$. The performance of the fracture-matrix transfer model, for values of β between 1 and 2, was investigated in detail. These various models were tested for a geometry consisting of equally spaced vertical fractures, in which case the fracture-matrix flux can actually be computed "exactly" by solving a horizontal version of eq. (12.18), with gravity and the term Γ_w set to zero. In this fine-scale approach, the fracture serves as a boundary condition for the matrix block, and flux term Γ_w is computed as part of the solution, by evaluating the local water flux at the fracture-matrix interface. Unfortunately, none of the approaches discussed above for choosing the parameters that appear in eq. (12.19) were able to predict fluxes that matched well with the "exact" values. This finding was consistent with the point made in Chapter 9 for the single-phase dual-porosity model, which is that a Warren–Root-type model is inherently unable to match the fracture-matrix flux over all time regimes.

In the single-phase case, Dykhuizen (1990) showed that this difficulty could be circumvented by using a Warren–Root-type flux term at late times, but at early times replacing the right-hand side of eq. (12.19) with a term that depended nonlinearly on the two pressures. Zimmerman *et al.* (1993) modified this idea by finding a single nonlinear expression on the right-hand side of eq. (12.19) that could be used for all times, and which led to fracture-matrix fluxes that had the correct behavior in both the early time and late-time regimes. Zimmerman *et al.* (1996) extended this approach to unsaturated flow, with a fracture-matrix flux term that can be written, in the present notation, as

$$q_{fm} = \frac{\alpha(1-\phi_f)(kk_{rw})_m}{\mu}\left[\frac{(p_f-p_i)^2-(\overline{p}_m-p_i)^2}{2(\overline{p}_m-p_i)}\right], \tag{12.20}$$

in which p_i is the initial pressure of the water in the fractured rock mass, and the other parameters and variables have their usual meanings. Using rock properties that were representative of the Topopah Spring member of the Paintbrush tuff at Yucca Mountain, Nevada, the predictions of this model were validated against fine-grid simulations of vertical and horizontal infiltration, and excellent agreement was found. Computational times for the nonlinear dual-porosity simulations, in all cases for two-dimensional geometries, were 96% less than for the fully discretized simulation for the horizontal flow problem, and 83% less for the vertical flow problem.

Dual-porosity models are frequently used to model the two-phase flow of oil and water. Kazemi *et al.* (1976) generalized the Warren–Root transfer function by first writing separate flow equations for the oil and water phases in the fracture network and incorporating the appropriate relative permeabilities for each phase. Flow between the fracture network and matrix blocks was modeled using a Warren–Root-type equation for each of the two fluid phases. Gilman and Kazemi (1983) incorporated gravity effects, by adding a gravity-related component of the driving force to the fracture/matrix pressure difference. Quandalle and Sabathier (1989) improved this model by treating vertical flow and horizontal flow differently, with the fluxes computed individually over each of the six faces of the assumed brick-like matrix blocks. Lu *et al.* (2008) developed a "general transfer function" that accounts for fluid expansion, capillary imbibition, gravity drainage, and diffusion, with the fluxes due to each of the four mechanisms computed separately, and then summed to yield the total flux. They used the transfer function proposed by Zimmerman *et al.* (1993) to compute the flux due to fluid expansion. Abushaikha and Gosselin (2008) provided a detailed critical review of these models and showed comparisons of the behavior of the various methods under different conditions, such as different matrix block shapes, and different relative effects of gravity and capillary forces. Van Heel *et al.* (2008) presented a fully transient dual-permeability model that accounted for gravity and thermal effects and used it to study gas/oil gravity drainage during steam injection. Their model used different shape factors for "early" and "late" times, with the boundary between the early and late times, for a given matrix block, taken to be the time at which the pressure change in the matrix block has reached 50% of the eventual change between the initial matrix block pressure and the final equilibrium pressure.

12.6 Discrete-Fracture Network Models for Two-Phase Flow in Fractured Rock Masses

The previous section discussed dual-porosity and dual-permeability approaches for modeling two-phase flow processes in fractured rock masses. In those approaches, individual fractures are not modeled explicitly, but the interconnected fractures are "homogenized" and represented as a continuum, having an absolute permeability, relative permeability functions, and a capillary pressure function. Dual-porosity and dual-permeability models are amenable to both analytical solutions for some idealized geometries, as well as to numerical solutions using methods such as finite elements or finite differences. Another approach to numerically modeling two-phase flow processes in fractured rock masses is to represent each fracture as a discrete entity in the computational mesh, such as was discussed in detail in Chapter 8 in the context of single-phase flow. This type of discrete fracture modeling is more computationally demanding, and poses some additional numerical difficulties with regards to meshing, but potentially allows for a more detailed representation of the fluxes and saturation profiles.

Karimi-Fard *et al.* (2004) presented a "finite volume" approach to modeling multi-phase flow through 2D or 3D fractured porous media. In their 3D version, the rock matrix is discretized by convex polyhedra, and the fractures are discretized with two-dimensional planar polygons. For 2D problems, the matrix rock is represented by polygons, and the fractures are represented using line segments (Fig. 12.6a). In either dimension, the fractures as represented in the computational mesh have no thickness/aperture, but their aperture is taken into account when performing the mass balance calculations, and in computing transmissivities between fracture cells and matrix cells.

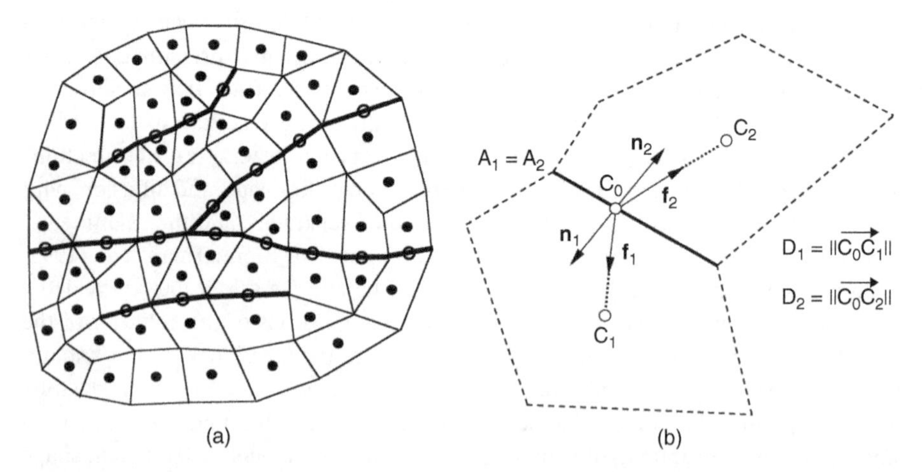

Figure 12.6 (a) Computational mesh used for a two-dimensional fractured-porous medium, using line segments for the fractures, and polygons for the matrix. Thick lines denote fracture elements, and thin lines demarcate matrix cells. Circles indicate the centroid of the element. (b) Two adjacent matrix cells, with C_i being the centroid of cell i, and C_0 the centroid of the interface, as used for computing the inter-cell transmissivity. See text, and Karimi-Fard *et al.* (2004), for details. Source: Adapted from Karimi-Fard *et al.* (2004).

Each mesh object, *i.e.*, line segment or polygon in 2D, polygon or polyhedron in 3D, constitutes a "control volume," also called a "cell." A node is placed at the centroid (*i.e.*, center of mass) of each control volume, and variables such as pressure and saturation are associated with each node. When computing the fluid fluxes, the pressure is assumed to be associated with the nodal point, whereas when performing a mass balance on a cell, the saturation is assumed to represent the mean saturation within that cell. The flowrate Q_{12} [m^3/s] of a given fluid phase, from cell 1 to cell 2, is computed as

$$Q_{12} = T_{12}\frac{k_r}{\mu}(p_1 - p_2), \tag{12.21}$$

where the flowrate, pressure, relative permeability, and viscosity are those of the phase in question, and T_{12} [m^3] is the transmissibility between cell 1 and cell 2, which is discussed in more detail below. Upstream weighting is used, so that the relative permeability is taken to be that of the cell in which the pressure is higher.

The transmissibility factor between cell 1 and cell 2, denoted by T_{12}, is the same for each phase, and is computed as (see Fig. 12.6b)

$$T_{12} = \frac{\alpha_1\alpha_2}{\alpha_1 + \alpha_2}, \quad \text{where} \quad \alpha_i = \frac{A_i k_i}{D_i}\mathbf{n}_i \cdot \mathbf{f}_i, \tag{12.22}$$

in which A_i is the area of the interface between the two cells, k_i is the absolute permeability of cell i, D_i is the distance between the centroid of the interface (C_o) and the centroid of cell i, $\mathbf{n}_i$ is the unit normal vector pointing from the interface into cell i, and $\mathbf{f}_i$ is the unit vector along the direction from the centroid of the interface to the centroid of cell i. The transmissibility factor plays a role that is very similar to the role played by the shape factor that appears in dual-porosity models, although these two parameters are not directly equivalent, and in fact have different units. Further details of the computation of the transmissivity factors for intersecting fracture segments, or for adjacent fracture and matrix cells, as well as methods for handling anisotropic matrix permeabilities, are discussed by Karimi-Fard *et al.* (2004). See also Reichenberger *et al.* (2006) and Paluszny *et al.* (2007) for further developments on this topic.

Hoteit and Firoozabadi (2008) developed a two-phase flow simulator for fractured rock masses that was based on the mixed finite element (MFE) method. The discontinuous Galerkin method was used to approximate the saturation equation, which reduces the numerical dispersion. A first-order implicit time scheme is used to approximate the time derivative in the mass balance equation in the fractures, whereas an explicit method is used for the matrix cells, thereby allowing different time steps in the matrix and the fractures. By imposing continuity of pressure only at the matrix-fracture interface, rather than between the fracture cell and the adjacent matrix cell, their approach allows the use of relatively large matrix cells near the fractures.

Aghili *et al.* (2019) discussed a nonlinear transmissivity model that is based on imposing continuity of the normal flux at each fracture-matrix interface. This approach was reported to provide better accuracy for problems involving large capillary pressure in the fractures, albeit with an increased computational burden. Gläser *et al.* (2017) proposed a cell-centered finite volume scheme in which the two-point flux approximation was replaced with a more accurate multi-point flux approximation. Several researchers have presented methods for solving the coupled problem of two-phase fluid flow plus poromechanical deformation of a fractured rock mass; these methods have been reviewed by Hawez *et al.* (2021).

Problems

12.1 With reference to the configuration shown in Fig. 12.1b, derive expression (12.3) for the capillary pressure. Note that, since the interface must have constant mean curvature, the trace of the interface in the plane of the figure must be an arc of a circle.

12.2 With the aid of a standard table of integrals, verify that, for the Pruess–Tsang model: (a) h_o is indeed the median of the aperture distribution given by eq. (12.9); (b) the mean aperture is related to the median aperture and the standard deviation as in eq. (12.10); and (c) the saturation is related to the capillary pressure as given by eq. (12.11).

12.3 Draw graphs of the Corey and van Genuchten capillary pressure functions, as given by eqs. (12.13) and (12.14), respectively, as functions of the effective saturation, $S_e = (S_w - S_r)/(1 - S_r)$, for a few different values of λ and m. Take $S_s = 1$ in the van Genuchten equation. In order to normalize and non-dimensionalize the graphs, use p_c/p_e and p_c/p_o as the vertical axes.

12.4 Consider the behavior of the Brooks–Corey and van Genuchten capillary pressure functions at low values of saturation. Find a relationship between the parameters that appear in each equation, so that the two curves become asymptotically equivalent at low saturations. How does this calculation compare with the parameters found by Reitsma and Kueper (1994) for the data from their "test 1"?

References

Abushaikha, A. S. A. and Gosselin, O. R. 2008. Matrix-fracture transfer function in dual-porosity medium flow simulation: Review, comparison, and validation. *SPE Europec Annual Conference and Exhibition*, Rome, 9–12 June 2008. Paper SPE 113890.

Aghili, J., Brenner, K., Hennicker, J., Masson, R., *et al.* 2019. Two-phase Discrete Fracture Matrix models with linear and nonlinear transmission conditions. *International Journal on Geomathematics*, 10, article 1.

AlQuaimi, B. I. and Rossen, W. R. 2017. New capillary number definition for displacement of residual nonwetting phase in natural fractures. *Geophysical Research Letters*, 44, 5368–73.

Bear, J. 1988. *Dynamics of Fluids in Porous Media*, Dover Publications, Mineola, N.Y.

Bertels, S. P., DiCarlo, D. A., and Blunt, M. J. 2001. Measurement of aperture distribution, capillary pressure, relative permeability, and in situ saturation in a rock fracture using computed tomography scanning. *Water Resources Research*, 37(3), 649–62.

Blunt, M. J. 2017. *Multiphase Flow in Permeable Media: A Pore-Scale Perspective*, Cambridge University Press, Cambridge.

Bodvarsson, G. S. and Tsang, C. F. 1982. Injection and thermal breakthrough in fractured geothermal reservoirs. *Journal of Geophysical Earth Solid Earth*, 87(B2), 1031–48.

Brooks, R. H. and Corey, A. T. 1966. Properties of porous media effecting fluid flow. *Journal of the Irrigation and Drainage Division ASCE*, 92(IR2), 61–88.

Chen, C. Y. and Horne, R. N. 2006. Two-phase flow in rough-walled fractures: experiments and a flow structure model. *Water Resources Research*, 42, W03430.

de Marsily, G. 1986. *Quantitative Hydrogeology*, Academic Press, San Diego.

Dykhuizen, R. C. 1990. A new coupling term for dual-porosity models. *Water Resources Research*, 26(2), 351–56.

Gerke, H. H. and van Genuchten, M. T. 1993a. A dual-porosity model for simulating the preferential movement of water and solutes in structured porous media. *Water Resources Research*, 29(2), 305–19.

Gerke, H. H. and van Genuchten, M. T. 1993b. Evaluation of a first-order water transfer term for variably saturated dual-porosity flow models. *Water Resources Research*, 29(4), 1225–38.

Ghez, R. 1988. *A Primer of Diffusion Problems*, Wiley, New York.

Gilman J. R. and Kazemi, H. 1983. Improvement in simulation of naturally fractured reservoirs. *Society of Petroleum Engineers Journal*, 23(4), 695–707.

Gläser, D., Helmig, R., Flemisch, B., and Class, H. 2017. A discrete fracture model for two-phase flow in fractured porous media. *Advances in Water Resources*, 110, 335–348.

Glass, R. J., Rajaram, H., and Detwiler, R. L. 2003. Immiscible displacements in rough-walled fractures: competition between roughening by random aperture variations and smoothing by in-plane curvature. *Physical Review E*, 68(6), 061110.

Guiltinan, E. J., Santos, J. E., Cardenas, M. B., Espinoza, D. N., and Kang, Q. 2020. Two-phase fluid flow properties of rough fractures with heterogeneous wettability: analysis with lattice Boltzmann simulations. *Water Resources Research*, 56, e2020WR027943.

Hawez, H. K., Sanaee, R., and Faisal, N. H. 2021. A critical review on coupled geomechanics and fluid flow in naturally fractured reservoirs. *Journal of Natural Gas Science and Engineering*, 95, 104150.

Hillel, D. 1980. *Fundamentals of Soil Physics*, Academic Press, New York.

Hoteit, H. and Firoozabadi, A. 2008. An efficient numerical model for incompressible two-phase flow in fractured media. *Advances in Water Resources*, 31(6), 891–905.

Huinink, H. 2016. *Fluids in Porous Media: Transport and Phase Changes*, Morgan & Claypool Publishers, San Rafael, Calif.

Huo, D. and Benson, S. M. 2016. Experimental investigation of stress-dependency of relative permeability in rock fractures. *Transport in Porous Media*, 113, 567–90.

Jones, B. R., Brouwers, L. B., and Dippenaar, M. A. 2018. Partially to fully saturated flow through smooth, clean, open fractures: qualitative experimental studies. *Hydrogeology Journal*, 26, 945–61.

Karimi-Fard, M., Durlofsky, L. J., and Aziz, K. 2004. An efficient discrete-fracture model applicable for general-purpose reservoir simulators. *Society of Petroleum Engineers Journal*, 9(2), 227–36.

Kazemi, H., Merrill, L. S., Porterfield, K. L., and Zeman, P. R. 1976. Numerical simulation of water-oil flow in naturally fractured reservoirs. *Society of Petroleum Engineers Journal*, 16(4), 317–26.

Kueper, B. H. and McWhorter, D. B. 1991. The behaviour of dense, non-aqueous phase liquids in fractured clay and rock. *Groundwater*, 29(5), 716–28.

Lu, H. Y., Di Donato, G., and Blunt, M. J. 2008. General transfer functions for multiphase flow in fractured reservoirs. *Society of Petroleum Engineers Journal*, 13(3), 289–97.

Marle, C. 1981. *Multiphase Flow in Porous Media*, Editions Technip, Paris.

Paluszny, A., Matthäi, S. K., and Hohmeyer, M. 2007. Hybrid finite element–finite volume discretization of complex geologic structures and a new simulation workflow demonstrated on fractured rocks. *Geofluids*, 7(2), 186–208.

Phillips, T., Kampman, N., Bisdom, K., Forbes Inskip, N. D., *et al.* 2020. Controls on the intrinsic flow properties of mudrock fractures: a review of their importance in subsurface storage. *Earth-Science Reviews*, 211, 103390.

Pruess, K. and Tsang, Y. W. 1990. On two-phase relative permeability and capillary pressure of rough-walled rock fractures. *Water Resources Research*, 26(9), 1915–26.

Quandalle, P. and Sabathier, J. C. 1989. Typical features of a multipurpose reservoir simulator. *SPE Reservoir Engineering*, 4(4), 75–80.

Ranjbaran, M., Shad, S., Taghikhani, V., and Ayatollahi, S. 2017. A heuristic insight on end-point calculation and a new phase interference parameter in two-phase relative permeability curves for horizontal fracture flow. *Transport in Porous Media*, 119, 499–519.

Reichenberger, V., Jakobs, H., Bastian, P., and Helmig, R. 2006. A mixed-dimensional finite volume method for two-phase flow in fractured porous media. *Advances in Water Resources*, 29(7), 1020–36.

Reitsma, S. and Kueper, B. H. 1994. Laboratory measurement of capillary pressure – saturation relationships in a rock fracture. *Water Resources Research*, 30(4), 865–78.

Su, G. W., Geller, J. T., Pruess, K., and Wen, F. 1999. Experimental studies of water seepage and intermittent flow in unsaturated, rough-walled fractures. *Water Resources Research*, 35(4), 1019–37.

Tokunaga, T. K. and Wan, J. 1997. Water film flow along fracture surfaces of porous rock. *Water Resources Research*, 33(6), 1287–95.

Tokunaga, T. K. and Wan, J. 2001. Approximate boundaries between different flow regimes in fractured rocks. *Water Resources Research*, 37(8), 2103–11.

van Genuchten M. T. 1980. A closed form equation for predicting the hydraulic conductivity of unsaturated soils. *Soil Science Society of America Journal*, 44(5), 892–98.

van Heel, A. P. G., Boerrigter, P. M., and van Dorp, J. J. 2008. Thermal and hydraulic matrix-fracture interaction in dual-permeability simulation. *SPE Reservoir Engineering and Evaluation*, 11(4), 735–49.

Watanabe, N., Sakurai, K., Ishibashi, T., Ohsaki, Y., *et al.* 2015. New ν-type relative permeability curves for two-phase flows through subsurface fractures. *Water Resources Research*, 51(4), 2807–24.

Yang, Z. B., Li, D. Q., Xue, S., Hu, R., *et al.* 2019. Effect of aperture field anisotropy on two-phase flow in rough fractures. *Advances in Water Resources*, 132, 103390.

Ye, Z. Y., Liu, H. H., Jiang, Q. H., and Zhou, C. B. 2015. Two-phase flow properties of a horizontal fracture: the effect of aperture distribution. *Advances in Water Resources*, 76, 43–54.

Zimmerman, R. W., Hadgu, T., and Bodvarsson, G. S. 1996. A new lumped-parameter model for flow in unsaturated dual-porosity media. *Advances in Water Resources*, 19(5), 317–27.

Zimmerman, R. W., Chen, G., Hadgu, T., and Bodvarsson, G. S. 1993. A numerical dual-porosity model with semi-analytical treatment of fracture/matrix flow. *Water Resources Research*, 29(7), 2127–37.

List of Symbols

Many of the symbols listed in the table below are used throughout the book. If a symbol is used only within a specific chapter(s), these chapters are generally indicated in the table. In some cases, the defining equation or figure is also listed. Symbols that are used or mentioned only once in the book are generally not listed. The “dimension” of the parameter is listed in brackets in terms of SI units; [–] indicates a dimensionless parameter.

Lowercase Roman Letters

a	half-length of void in row-of-voids model, Fig. 3.2 [m]; Chapter 3
a	weak inertia flow parameter, eq. (4.7) [kg s/m^{11}]; Chapter 4
a	radius of circular contact region, eq. (5.1) [m]; Chapter 5
a	radius of spherical (or cylindrical) matrix block [m]; Chapter 9
b	pre-factor in fracture length distribution, eq. (1.2) [units vary with exponent]; Chapter 1
b	Forchheimer coefficient, eq. (4.10) [kg/m^8]; Chapter 4
c	contact area fraction of fracture plane [–]; Chapters 2, 5
c	total compressibility, $c = c_F + c_\phi$ [1/Pa]; Chapter 9
c_F	fluid compressibility [1/Pa]; Chapter 9
c_f	total compressibility of fracture network [1/Pa]; Chapter 9
c_m	total compressibility of matrix rock [1/Pa]; Chapter 9
c_ϕ	pore compressibility [1/Pa]; Chapter 9
d	pore diameter [m]; Chapter 2
d_Q	flow channeling density indicator, eq. (8.19) [1/m]; Chapter 8
erfc	complementary error function, eq. (6.11) [–]; Chapters 6, 11
f	molar Helmholtz free energy [J/mol]; Chapter 5
f	roughness factor [–]; Chapter 5
g	gravitational acceleration [m/s^2]
h	fracture aperture [m]
h	height of fracture surface above some nominal plane [m]; Chapter 1
h_c	critical fracture aperture in the Pruess–Tsang model [m]; Chapter 12
h_G	geometric mean of aperture [m]; Chapter 2
h_H	hydraulic aperture [m]

Fluid Flow in Fractured Rocks, First Edition. Robert W. Zimmerman and Adriana Paluszny.

Companion website: www.wiley.com/go/zimmerman/fluidflowinfracturedrocks

h_m mean aperture [m]
h_o median fracture aperture in the Pruess–Tsang model [m]; Chapter 12
h_x aperture of orthogonal fractures lying in y–z plane [m]; Chapter 7
j total solute flux, eq. (6.4) [kg/m^2 s]; Chapter 6
$\mathbf{j}_d$ diffusive solute flux vector, eq. (6.1) [kg/m^2 s]; Chapter 6
j_d diffusive solute flux, eq. (6.1) [kg/m^2 s]; Chapter 6
$\mathbf{j}_a$ advective solute flux vector [kg/m^2 s]; Chapter 6
k permeability [m^2]
$\mathbf{k}$ permeability tensor [m^2]; Chapter 8
k_{eff} effective permeability of fractured rock mass [m^2]; Chapter 7
k_f permeability of individual parallel-plate fracture, eq. (7.15) [m^2]; Chapters 7, 8
k_{fn} permeability of fracture network [m^2]; Chapter 7
k_m permeability of matrix rock [m^2]; Chapters 7–9
k_{rn} relative permeability of non-wetting phase fluid [–]; Chapter 12
k_{rw} relative permeability of wetting-phase fluid [–]; Chapter 12
k_S permeability of fractured rock mass, according to Snow's model [m^2]; Chapter 7
k_{xy} xy component of permeability tensor [m^2]; Chapter 8
k_+ dissolution rate constant [mol/m^{2}s]; Chapter 5
k_- precipitation rate constant [mol/m^{2}s]; Chapter 5
k_+^0 pre-exponential factor for the dissolution reaction [mol/m^{2}s]; Chapter 5
k_-^0 pre-exponential factor for the precipitation reaction [mol/m^{2}s]; Chapter 5
k_i ith eigenvalue of permeability tensor, ordered as $k_1 \geq k_2 \geq k_3$ [m^2]; Chapter 7
m mass of solute inside a region of porous rock [kg]; Chapter 6
m stress ratio, eq. (8.18) [–]; Chapter 8
m mass of fluid inside a region of porous rock [kg]; Chapter 9
m power-law exponent in the van Genuchten model, eq. (12.14) [Pa]; Chapter 12
$\mathbf{n}$ outward unit normal vector to surface [–]; Chapters 2, 11
n exponent in fracture radius distribution, eq. (8.1) [–]; Chapter 8
p pressure [Pa]
$\widehat{p}$ reduced pressure, eq. (2.2) [Pa]; Chapter 2
p_c capillary pressure, eq. (12.1) [Pa]; Chapter 12
p_{Df} dimensionless pressure in fractures, eq. (9.30) [–]; Chapter 9
p_{Dm} dimensionless pressure in matrix blocks, eq. (9.31) [–]; Chapter 9
p_e entry pressure of non-wetting phase [Pa]; Chapter 12
p_f pressure in fractures [Pa]; Chapters 8, 9, 10, 12
p_i pressure at fracture inlet [Pa]; Chapter 2
p_i initial pressure in matrix block [Pa]; Chapter 9
p_m local pressure in matrix block [Pa]; Chapters 9, 12
$\overline{p}_m$ mean pressure in matrix block [Pa]; Chapter 9
p_n pressure in non-wetting phase fluid [Pa]; Chapter 12
p_o pressure at fracture outlet [Pa]; Chapter 2
p_o characteristic pressure in the van Genuchten model, eq. (12.14) [Pa]; Chapter 12
p_w pressure at well [Pa]; Chapter 9
p_w pressure in wetting phase fluid [Pa]; Chapter 12
q spatial frequency or wavenumber of fracture [1/m]; Chapter 1

$\mathbf{q}$ fluid flux vector [m/s]; Chapter 7
q_{mf} fluid flux from matrix blocks to fractures, per unit volume [1/s]; Chapter 9
q_n volumetric flux of non-wetting phase [m/s]; Chapter 12
q_w volumetric flux of wetting phase [m/s]; Chapter 12
q_x volumetric flux per unit length in transverse direction [m^2/s]; Chapter 2
r radius of disk-like fracture [m]; Chapter 1
r ratio of fracture permeability to matrix permeability [-]; Chapter 7
r radial coordinate [m]; Chapter 9
r_D dimensionless radial coordinate, eq. (9.29) [-]; Chapter 9
r_w radius of well [m]; Chapter 9
s fracture spacing [m]; Chapter 1
s standard deviation of the asperity height distribution [m]; Chapter 3
s fluid source–sink term, per unit volume [kg/m^3s]; Chapter 9
s Laplace transform variable [1/s]; Chapter 10
s solute source–sink term, per unit volume [kg/m^3s]; Chapter 11
t time [s]
t_D dimensionless time, eq. (9.28) [-]
t_w residence time of solute particle [s]; Chapter 11
$\mathbf{u}$ fluid velocity vector [m/s]
u_x x-component of velocity vector [m/s]
$\overline{u}_x$ depth-averaged value of x-component of velocity vector [m/s]; Chapter 2
v_m volume of matrix block [m^3]; Chapter 9
w width of fracture in fracture plane, transverse to mean flow [m]; Chapters 2, 6
w thickness of pore fluid film at the asperity interface [m]; Chapter 5

Uppercase Roman Letters

A cross-sectional area normal to flow [m^2]
A cross-sectional area of 2D matrix block [m^2]; Chapter 10
A_{ex} excluded area around 2D fractures [m^2]; Chapter 1
A_f surface area of fracture [m^2]; Chapter 1
B rate of increase of pressure in fracture [Pa/s]; Chapter 10
C solute concentration [kg/m^3] or [-]; Chapters 5, 6, 11
C pre-factor in fracture radius distribution, eq. (8.1) [units vary with exponent]
C dimensionless shape factor [-]; Chapter 10
C_n Fourier coefficients for pressure diffusion problem in matrix block [-]; Chapter 9
C_{ij} channel conductance, eq. (11.9) [m^3/s Pa]; Chapter 11
C_o solute concentration in injected fluid [kg/m^3]; Chapters 6, 11
D diffusion (or dispersion) coefficient [m^2/s]; Chapters 5, 6, 11
D_e effective diffusion coefficient [m^2/s]; Chapter 11
$\mathbf{D}$ diffusion (or dispersion) tensor [m^2/s]; Chapters 6, 11
D_L longitudinal dispersion coefficient [m^2/s]; Chapters 6, 11
D_{Th} transverse horizontal dispersion coefficient [m^2/s]; Chapter 11

D_{Tv}	transverse vertical dispersion coefficient [m^2/s]; Chapter 11
D_f	fractal dimension [–]; Chapter 1
E	Young's modulus [Pa]
E_D	activation energy for diffusion [J/mol]; Chapter 5
Ei	Exponential integral function, eq. (9.39) [–]; Chapter 9
E_{k_+}	activation energy coefficient of the dissolution reaction [J/mol]; Chapter 5
E_{k_-}	activation energy coefficient of the precipitation reaction [J/mol]; Chapter 5
E_m	latent heat of fusion of mineral [J/mol]; Chapter 5
F	normal force acting on an asperity [N]; Chapter 5
F_n	eigenfunctions of pressure diffusion equation [–]; Chapter 9
G	shear modulus [Pa]
H	Hurst exponent [–]
J	diffusive mineral flux, eq. (5.21) [kg/s]; Chapter 5
J_0	Bessel function of the first kind, order zero [–]; Chapter 10
K_d	distribution coefficient [m]; Chapter 6
K_d	linear sorption coefficient [m^3/kg]; Chapter 11
K_v	volumetric sorption coefficient [–]; Chapter 11
L	length of fracture in direction of flow [m]; Chapters 2, 4, 6
L	length of cubic matrix block [m]; Chapter 10
L_f	length of fracture [m]; Chapters 1, 7
L_Ω	size of the domain over which permeability is evaluated [m]; Chapter 7
M_{diss}	dissolved mass [kg]; Chapter 5
P	perimeter of 2D matrix block [m]; Chapter 10
P_e	Peclet number [–]; Chapter 6
P_{ij}	fracture persistence/density/intensity measures [m^{j-i}]; defined in Chapter 1
Q	volumetric flowrate [m^3/s]
Q_{mf}	fluid flux from matrix blocks to fractures [m^3/s]; Chapter 9
Q_i	flowrate through fracture i [m^3/s]; Chapter 8
$\mathscr{R}$	universal gas constant [Pa m^3/kg K]; Chapters 4, 5
R	radius of curvature of asperity [m]; Chapters 3, 5
R	retardation coefficient, eq. (6.25) [1]; Chapter 6
R	radius of oblate spheroidal fracture, Fig. 7.2 [m]; Chapter 7
R	radius of curvature of the fluid interface [m]; Chapter 12
Re	Reynolds number, eq. (2.21) [–]
S	source–sink term for a region of porous rock [kg/s]; Chapter 9
S	surface area of 3D matrix block [m^3]; Chapter 10
S_i	surface area of fracture i [m^2]; Chapter 8
S_n	non-wetting phase saturation [–]; Chapter 12
S_{nr}	residual non-wetting phase saturation [–]; Chapter 12
S_r	residual wetting phase saturation in the Brooks–Corey model [–]; Chapter 12
S_x	spacing of orthogonal fractures in x-direction, Fig. 7.1 [m]; Chapter 7
S_w	wetting phase saturation [–]; Chapter 12
S_{wi}	irreducible wetting phase saturation [–]; Chapter 12
T	fracture transmissivity [m^4]
T	absolute temperature [K]; Chapter 5

T_m melting point of mineral [K]; Chapter 5
T_o fracture transmissivity at low Reynolds numbers [m^4]; Chapter 4
T_{ij} transmissibility between cells i and j [m^3]; Chapter 12
U mean fluid velocity [m/s]; Chapter 11
U_x characteristic fluid velocity in x-direction [m/s]
V volume of 3D matrix block [m^3]; Chapter 10
V_{ex} excluded volume around 3D fractures [m^3]; Chapter 1
V_f macroscopic volume of a region of rock [m^3]; Chapter 9
V_m molar volume of mineral [m^3/mol]; Chapter 5
V_p pore volume of a region of rock [m^3]; Chapter 9

Lowercase Greek Letters

α fracture aspect ratio, eq. (1.4) [–]
α dimensionless weak inertia flow parameter, eq. (4.9) [–]; Chapter 4
α matrix block shape factor [1/m^2]; Chapters 9, 10
α_L longitudinal dispersivity [m]; Chapters 6, 11
α_{Th} transverse horizontal dispersivity [m]; Chapter 11
α_{Tv} transverse vertical dispersivity [m]; Chapter 11
β geometrical parameter for elliptical contact region, eq. (2.53) [–]; Chapter 2
β dimensionless Forchheimer coefficient, eq. (4.12) [–]; Chapter 4
β geometrical parameter for ellipsoidal fractures, eq. (7.19) [–]; Chapter 7
β power-law exponent in CTRW waiting time function, eq. (11.22); Chapter 11
β_n normal compliance of fracture, eq. (3.8) [m/Pa]; Chapter 3
γ Euler's number, 1.781 [–] NB: some books/papers define $\gamma = \ln(1.781) = 0.5772$
γ interfacial energy between solid and fluid [N/m]; Chapter 5
γ_i fitting parameters for relation between aperture and contact fraction; Chapter 5
γ_{ns} surface tension between non-wetting phase fluid and solid [N/m]; Chapter 12
γ_{wn} surface tension between two fluid phases [N/m]; Chapter 12
γ_{ws} surface tension between wetting phase fluid and solid [N/m]; Chapter 12
δ aperture variation of sinusoidal fracture, eq. (2.22) [–]; Chapter 2
δ fracture closure, Fig. 3.2 [m]; Chapter 3
δ normal displacement of asperity [m]; Chapter 5
ε dimensionless fracture density, eq. (7.13) [–]; Chapter 7
ε_c critical fracture density at percolation limit [–]; Chapter 7
θ absolute temperature [K]; Chapter 4
θ contact angle, Fig. 12.1b [–]; Chapter 12
κ parameter in the Forchheimer equation for gas flow, eq. (4.23) [m]; Chapter 4
κ mean curvature of the solid–fluid interface [1/m]; Chapter 5
κ ratio of matrix permeability to fracture permeability, eq. (7.16) [–]; Chapter 7
κ_n normal stiffness of fracture, eq. (3.4) [Pa/m]; Chapter 3
λ fracture wavelength [m]; Chapters 2, 4
λ spacing of voids in row-of-voids model, Fig. 3.2 [m]; Chapter 3
λ radioactive decay constant [1/s]; Chapter 6

λ transmissivity ratio, eq. (9.35) [–]; Chapter 9
λ power-law exponent in the Brooks–Corey model, eq. (12.13) [–]; Chapter 12
λ_n eigenvalues for pressure diffusion problem [m^{-2}]; Chapter 9
μ fluid viscosity [Pa s]
μ chemical potential [J/m]; Chapter 5
μ friction coefficient [–]; Chapter 5
μ_n viscosity of non-wetting phase fluid [Pa s]; Chapter 12
μ_w viscosity of wetting phase fluid [Pa s]; Chapter 12
ν Poisson's ratio [–]
ρ fluid mass density [kg/m^3]
ρ_m rock mass density [kg/m^3]; Chapter 11
ρ_s mineral density [kg/m^3]; Chapter 5
ρ_{2D} two-dimensional fracture density [$1/m^2$]
ρ'_{2D} dimensionless two-dimensional fracture density, eq. (1.22) [–]
ρ_{3D} three-dimensional fracture density [$1/m^3$]
ρ'_{3D} dimensionless three-dimensional fracture density, eq. (1.23) [–]
σ normal stress acting on fracture plane [Pa]
σ_a disjoining pressure [Pa]; Chapter 5
σ_h standard deviation of fracture aperture [m]
σ_i ith eigenvalue of the stress tensor, ordered such that $\sigma_1 \geq \sigma_2 \geq \sigma_3$ [Pa]; Chapter 8
τ dummy time-like variable of integration [s]; Chapter 9
ϕ porosity [–]
ϕ_f porosity of fracture network [–]; Chapters 7, 9
ϕ_m porosity of matrix rock [–]; Chapter 9
ψ waiting time function in CTRW method; Chapter 11
ω storativity ratio, eq. (9.34) [–]; Chapter 9

Uppercase Greek Letters

Γ boundary of contact region, Fig. 2.5; Chapter 2
Γ_w fracture-matrix water flux term [1/s]; Chapter 12

Mathematical Operators

$\langle x \rangle$ mean value of variable x
$\bar{x}$ mean value of variable x
∇ gradient operator
$\nabla\cdot$ divergence operator
∇^2 Laplacian operator
$\bullet$ dot product operator

Index

Fluid Flow in Fractured Rocks, First Edition. Robert W. Zimmerman and Adriana Paluszny.

Companion website: www.wiley.com/go/zimmerman/fluidflowinfracturedrocks